MICROSURGERY

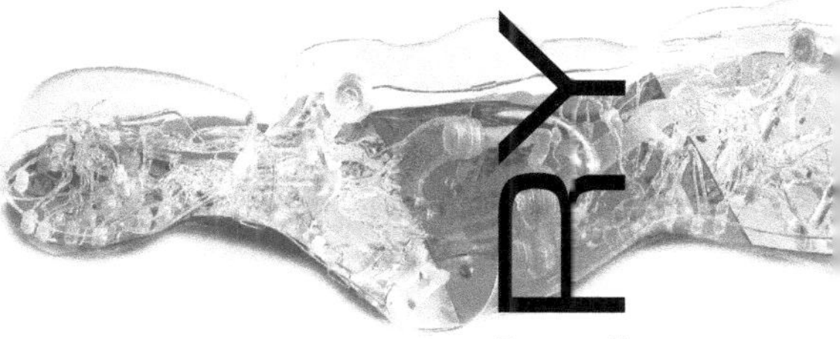

MICROSURGERY

ADVANCES, SIMULATIONS, AND APPLICATIONS

edited by
Toshio Fukuda
Carlos Tercero

PAN STANFORD PUBLISHING

Published by

Pan Stanford Publishing Pte. Ltd.
Penthouse Level, Suntec Tower 3
8 Temasek Boulevard
Singapore 038988

Email: editorial@panstanford.com
Web: www.panstanford.com

British Library Cataloguing-in-Publication Data
A catalogue record for this book is available from the British Library.

Microsurgery: Advances, Simulations, and Applications

ISBN 978-981-4364-69-0 (Hardcover)
ISBN 978-981-4364-70-6 (eBook)

Printed in the USA

Preface

The development of the endovascular evaluator (EVE) was motivated by the lack of an *in vitro* simulation environment reproducing patient-specific vasculature morphology with high accuracy for the research and development of new technology for endovascular intervention. This book was developed to show the scientific context where the EVE project was born and also to present the applications of this vasculature modeling technique for the quantitative representation of tissue respect in endovascular surgery simulation, flow control human blood pressure simulation, computer fluid dynamics on vasculature models, catheter insertion robot control, and tailor-made scaffolds for blood vessel regeneration. The development of patient-specific silicone vasculature models and the EVE is a breakthrough that is improving medical training and facilitating research and development in industry and academia. These contributions will have impact in other domains of medicine through the research on materials for human tissue modeling, and will enable the development of cyber-physical systems for medical training.

Contributors

Chapter 1

Carlos Tercero, Seiichi Ikeda, Tomoyuki Uchida, Toshio Fukuda
Department of Micro-Nano Systems Engineering, Nagoya University, Japan

Fumihito Arai
Department of Mechanical Science and Engineering, Nagoya University, Japan

Makoto Negoro
Department of Neurosurgery, Fujita Health University, Japan

Chapter 2

Motoharu Hayakawa, Takeya Watabe, Jumpei Oda, Yuichi Hirose
Department of Neurosurgery, Fujita Health University, Japan

Chapter 3

Seiichi Ikeda, Carlos Tercero, Toshio Fukuda
Department of Micro-Nano Systems Engineering, Nagoya University, Japan

Fumihito Arai
Department of Mechanical Science and Engineering, Nagoya University, Japan

Makoto Negoro
Department of Neurosurgery, Fujita Health University, Japan

Ikuo Takahashi
Department of Neurosurgery, Anjo Kosei Hospital, Japan

Chapter 4

Carlos Tercero, Motoki Matsushima, Seiichi Ikeda, Toshio Fukuda
Department of Micro-Nano Systems Engineering, Nagoya University, Japan

Fumihito Arai
Department of Mechanical Science and Engineering, Nagoya University, Japan

Makoto Negoro
Department of Neurosurgery, Fujita Health University, Japan

Ikuo Takahashi
Department of Neurosurgery, Anjo Kosei Hospital, Japan

Chapter 5

Masahiro Kojima
Department of Micro-Nano Systems Engineering, Nagoya University, Japan

Yasuhiko Sakai, Kouji Nagata
Department of Mechanical Engineering, Nagoya University, Japan

Haruo Isoda
Department of Radiology, Hamamatsu University School of Medicine, Japan

Chapter 6

Carlos Tercero*, Motoki Matsushima, Seiichi Ikeda, Toshio Fukuda
Department of Micro-Nano Systems Engineering, Nagoya University, Japan

Erick Tijerino*
Department of Mechanical, Materials and Aerospace Engineering, University of Central Florida, USA
**Department of Electronics Engineering, Del Valle de Guatemala University, Guatemala*

Makoto Negoro
Department of Neurosurgery, Fujita Health University, Japan

Ikuo Takahashi
Department of Neurosurgery, Anjo Kosei Hospital, Japan

Chapter 7

Carlos Tercero, Seiichi Ikeda, Shi Chaoyang, Toshio Fukuda
Department of Micro-Nano Systems Engineering, Nagoya University, Japan

Fumihito Arai
Department of Mechanical Science and Engineering, Nagoya University, Japan

Makoto Negoro
Department of Neurosurgery, Fujita Health University, Japan

Ikuo Takahashi
Department of Neurosurgery, Anjo Kosei Hospital, Japan

Chapter 8

Tomoyuki Uchida, Hiroyuki Oura, Seiichi Ikeda, Chengzhi Hu, Carlos Tercero, Toshio Fukuda
Department of Micro-Nano Systems Engineering of Nagoya University, Japan

Fumihito Arai
Department of Mechanical Science and Engineering, Nagoya University, Japan

Makoto Negoro
Department of Neurosurgery, Fujita Health University, Japan

Toshio Fukuda and Carlos Tercero

Editors

Contents

Chapter 1

Introduction

As far as research is done in fields of medicine and engineering, it is needed to develop joint projects between both fields to provide physicians of new tools and methods for training [1–3], for disease diagnosis [4–5], for treatment [6–7], therapy and rehabilitation [8–9]. This book was written with the motivation to make a contribution to develop such kind of tools and methods, specifically for endovascular intervention. Following that idea, a modeling technique for *in vitro* representation of human vasculature is presented with some of its applications. To set the context of this simulation tool and its applications, we will make an overview of medical robotics and regenerative medicine techniques related to the contents of the following chapters.

Medical robotics is divided in three domains that are defined according to the time were the application of medical robotics enters in contact with the patient:

- Pre-surgery: diagnostic, path planning, simulation and training
- Intra-surgery: robotic surgery, drug delivery systems, telesurgery, organ regeneration and implants
- Post-surgery: rehabilitation, prosthesis and assistance

1.1 Minimally Invasive Surgery

Minimally invasive surgery (MIS) was created following physicians' desire to reduce recovery time of patients, by reducing the damage

Microsurgery: Advances, Simulations, and Applications
Edited by Toshio Fukuda and Carlos Tercero

ISBN 978-981-4364-69-0 (Hardcover), 978-981-4364-70-6 (eBook)
www.panstanford.com

of healthy tissue surrounding the area to be treated. Minimally invasive procedures exist in fields of gastroenterology [10–11], pneumology [12], neurology [1–3], cardiology [13], ophthalmology [14], and urology [15], among others. MIS is usually more complex than conventional surgery techniques because the physicians do not have direct contact to the area to be treated, as in natural orifice trans-luminal endoscopic surgery (NOTES), guided needle insertion, laparoscopic surgery and intravascular intervention.

1.1.1 Endovascular Surgery

Endovascular surgery appeared as the solution for the treatment of vascular diseases such as stenosis and aneurysms, providing an alternative treatment method in cardiology and neurology. In those surgeries, a catheter of few millimeter of diameter is introduced from a blood vessel of the leg or arm to the diseased region of the vasculature. As seen in Figs. 1.1–1.4, during an endovascular surgery the catheter is seen using a fluoroscope, exposing physicians and patients to X-rays [16]. With the two-dimensional image, the surgeon has to drive the catheter through the complex structure of vasculature to reach the target, avoiding puncture in the blood vessel wall.

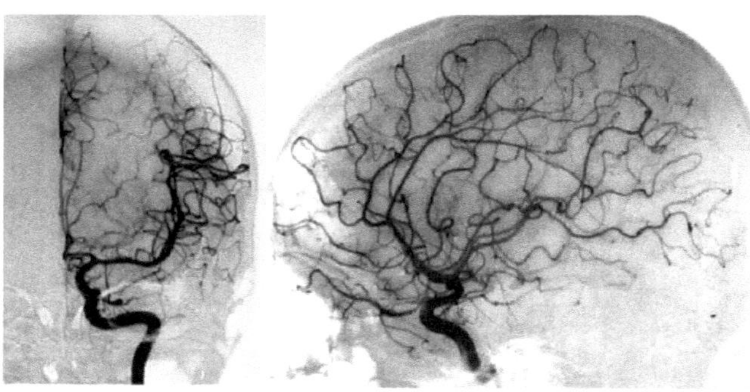

Figure 1.1 Fluoroscopic images of human cerebral arteries structure obtained with digital subtraction angiography.

1.1.1.1 Neuro-endovascular treatments

In this section, we will provide a summarized explanation of Neuro-endovascular treatments. These treatments will be explained in detail in Chapter 2. Catheter treatments are categorized into embolism and vascular reconstruction. Embolism is a treatment for aneurysm and cerebral arteriovenous malformation, where diseases are treated by blocking the blood flow in a specific vasculature area. For treating brain cancer, the blood vessel that irrigates the cancer cells is blocked during the treatment. In case of arterial aneurysms, a platinum coil is deployed inside the aneurysm to prevent the entrance of the blood flow and hemorrhage.

Angioplasty is a method of dilating the blood vessel by inserting a balloon catheter and swelling the narrowed area with the balloon, and improving the blood stream. In this case, a self-expandable metallic cylinder called stent is deployed to prevent the shrinking of the lumen to the original state after angioplasty.

When blood vessels are obstructed by thrombus, the blood flow can be reestablished by dissolving or removing the thrombus using endovascular techniques. Dissolution is achieved delivering thrombolytic drugs directly to the area to be treated in combination with anticoagulants. During revascularization, blood flow reaching the thrombus is stopped using protection tools. Then the clots are destroyed by specialized tools deployed using the guide catheter, and the debris are then retrieved by suction through the guide catheter.

Figures 1.2–1.4, respectively, show the comparison before and after the catheter treatment of cerebral aneurysm, stenosis, and embolism.

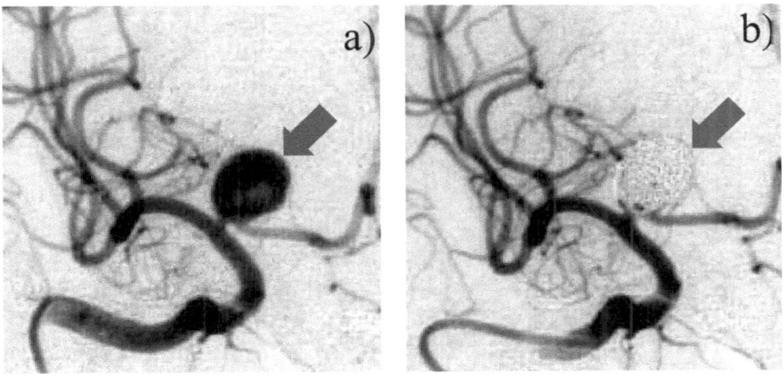

Figure 1.2 (a) Fluoroscopic image of cerebral aneurysm before surgery. (b) Fluoroscopic image of cerebral aneurysm after treatment, contrast media circulating in the blood stream do not enter to the aneurysm. Arrows point at the aneurysm.

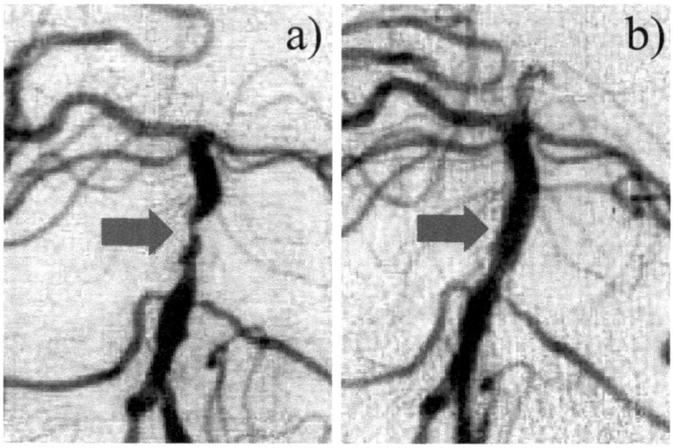

Figure 1.3 (a) Fluoroscopic image of cerebral stenosis before surgery. (b) Fluoroscopic image of cerebral aneurysm after treatment, contrast media circulating in the blood stream shows the enlargement of the blood vessel cross-section. Arrows point at the stenosis.

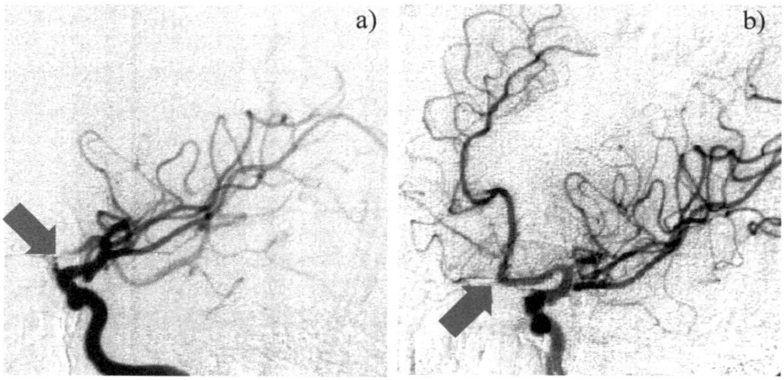

Figure 1.4 (a) Fluoroscopic image of cerebral embolism before surgery. (b) Fluoroscopic image of cerebral aneurysm after treatment. Arrows point at the embolism.

1.1.1.2 Percutaneous trans-luminal coronary angioplasty

The accumulation of lipid in blood vessel wall is called arteriosclerosis; when it occurs, the blood flow through the vessel reduces producing ischemia. Myocardial ischemia produces a chest pain called angina. Percutaneous trans-luminal coronary angioplasty aims improving the state of the ischemia of the myocardium. Moreover, when arteriosclerosis produces complete occlusion of coronary artery by thrombus growth or detachment, it is necessary to restart the blood stream promptly for avoiding brain damage due to the cardiac infarction condition. The lack of blood flow in the myocardium results in its necrosis when this state remains for a long time; therefore it is necessary to expand the narrowed area as soon as possible. Usually this dilation is achieved by the expansion of a balloon catheter. The balloon located in the balloon-catheter tip has originally the same diameter as the catheter body; after the balloon is led into the affected part, it is dilated by injecting physiological saline. However, the dilated vessel might return to the original size again, therefore a cylindrical metallic self-expandable mesh called stent is placed to keep the blood vessel dilated.

1.1.1.3 Trans-catheter aortic valve implantation

Recent artificial heart valves can be deployed into the heart to replace functions of stenotic valves using an endovascular approach. The artificial valve membranes are sewed to a stainless steel stent, and as for angioplasty treatments the stent is brought to its final expansion state with a balloon catheter. The artificial valve is brought into place using a steerable and deflectable catheter [17].

1.1.2 Laparoscopic Surgery and Single-Port Surgery

Laparoscopic surgery appeared as replacement for open-chest or -abdomen surgeries. For this surgery, the procedure is done usually through 3 to 5 incisions of a maximum of 12 mm diameter on the chest or abdomen of the patient. Through 2 to 4 of them, a variety of surgical tools are inserted to manipulate tissue and fluids inside patient. Vision of the surgical procedure is achieved using a laparoscope inserted through the remaining port in patient's chest or abdomen. The laparoscope is used to illuminate and capture intra-surgical images (Fig. 1.5).

With the development of medical robotics, the multiport configuration is changing to single-port surgery (SPS). In this intervention, a specialized port enables to insert multiple tools through the same orifice. The space restrictions presented by the single-port configuration required the development of new manipulation tools for this purpose.

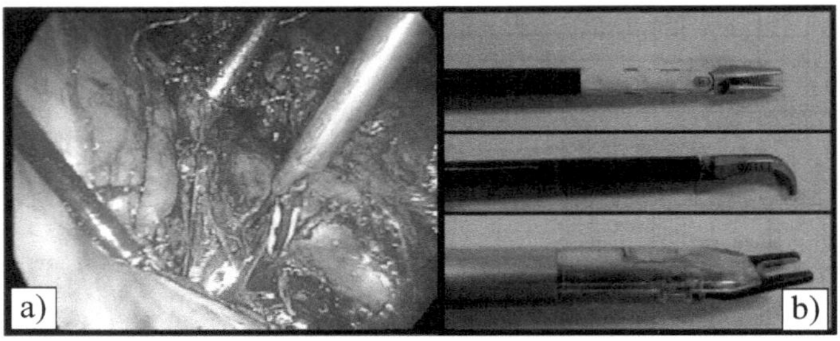

Figure 1.5 (a) Intra-surgical image captured using a laparoscope. (b) Examples of Laparoscopic surgery tools for tissue manipulation.

In most cases, the target of the surgery is surrounded by healthy tissue and fat and is located near major vasculature. This condition and the limited vision present the biggest challenge for the surgeon. For example, during the kidney ablation surgery, the surgeon is required to locate and suture renal artery and vein. Removing the surrounding tissue of those vascular structures is the most time consuming task during this laparoscopic surgery.

1.1.3 Stereotactic and Functional Neurosurgery

This kind of surgery is used to provide treatment to cerebral arteries or tumors, which cannot be treated by endovascular intervention. In this treatment, the skull is fixed to a metal frame, and the treatment is made using accurately positioned endoscopes for reducing as much as possible the damage to healthy tissue. A camera and a therapeutic instrument (forceps and suck tube) are built inside the brain endoscope, and its tip is movable. The position of surgery target is accurately measured by X-ray CT (X-ray Computed Tomography) and magnetic resonance imaging, and the position for opening the skull is decided accurately for avoiding damage to any brain function.

1.1.4 Natural Orifice Trans-Luminal Endoscopic Surgery

1.1.4.1 Digestive tract endoscopies

The endoscope used for digestive organs is required to be flexible to correspond to the twisting of digestive organs. A typical example of the digestive organ endoscope is the Olympus gastro-camera, developed around 1950 by Hayashida and presented in 1965 by G. D. Hadley [18]. This endoscope's tip had a flash light for illumination and a monochrome film camera for registering the images of the digestive tract. The development of glass fiber around 1960 enabled later on the real-time observation on the stomach and the videoscopes that capture these real-time images with a CCD camera and in 2002 the integration of high-fidelity displays. The most novel endoscopic technologies are endoscopic capsules that travel along digestive tract while transmitting wireless video signals [19].

The videoscope's tip motion is achieved by variable tension of wires that link the tip with the control section. The endoscope for a digestive organ is about 10 mm in the diameter, additionally of the optical system for observation of the procedure, it enables the insertion of therapeutic instrument through one of its lumens (forceps, laser irradiation probe, and high-frequency snare wire), and also it includes water pipe rods. Moreover, the tip part bends dynamically to treat effectively in narrow digestive organs. Some of the targeted diseases to be treated are polypus, which may appear in the stomach, rectum, and large intestines and early-stage cancer.

1.1.4.2 Respiratory tract endoscopies

Bronchoscopes are flexible endoscopes used for the inspection of respiratory tracts. The use of these flexible endoscopes for removing pulmonary foreign bodies from children was reported since 1984 [20], and these endoscopes have been in use for pulmonary disease treatments since the end of the decade of 1960 [21]. Bronchoscopes are used principally for diagnostic by the observation of the respiratory tract and for biopsies.

1.1.4.3 Transurethral resection of prostate

This surgery is intended for treating the prostate dilation syndrome. In this surgery, the dilated part of the prostate is excised by inserting a resectoscope through the patient's urethra enabling the observation of the target and manipulation of excision apparatus. The burden to the patient is less with this treatment than with laparotomy.

1.1.5 Extracorporeal Shock Wave Lithotripsy

The target of this procedure is to crush urinary tract stones by generating a shock with an acoustic wave in water and focusing it with an acoustic lens. The stone fragments then move outside the body through the urinary tract. Since the supersonic wave does not attenuate easily in water and it penetrates human body and acts only on the stone, it is possible carry out this treatment with no invasion. Larger stone evacuation through the urinary tract may be assisted using a stent.

1.1.6 Stereotactic Gamma Radiosurgery

This is an operation that uses high-precision irradiation of gamma rays to treat cerebral artery deformity and a metastatic brain tumor among others. Gamma rays are emitted from multiple sources located in a hemisphere configuration, and focused to the same point; the cells in the focus are brought to morbid state. The energy of gamma rays emitted by the individual sources is generally low avoiding surrounding tissue damage. In case of brain cancer, the treatment is achieved by blockading the blood flow to the tumor. Arteriovenous malformations are treated by destroying the vasculature with abnormal morphology.

1.2 Medical Robotics

1.2.1 Control Systems

Control theory is used in robotics and automated systems to adapt a dynamic system to obtain a desired output. The principal signals in a control system are the *reference* and the *output*: The first one represents the expected behavior of the dynamic system and the *output* is the resulting behavior. Conventional *reference* signals used for control analysis in linear systems are the step function, ramp, and parabolic. During the analysis, the response of the dynamic system to these functions is analyzed. The dynamic system to be controlled is often called the *process* and in linear system has transfer function with fixed order and coefficients. The *controller* is another component of the control system and has a transfer function that is adaptable depending on the *process* transfer function and the reference signal to obtain the desired output. The *error* signal is defined as the difference between the *output* and *reference* signals [22] (Fig. 1.6).

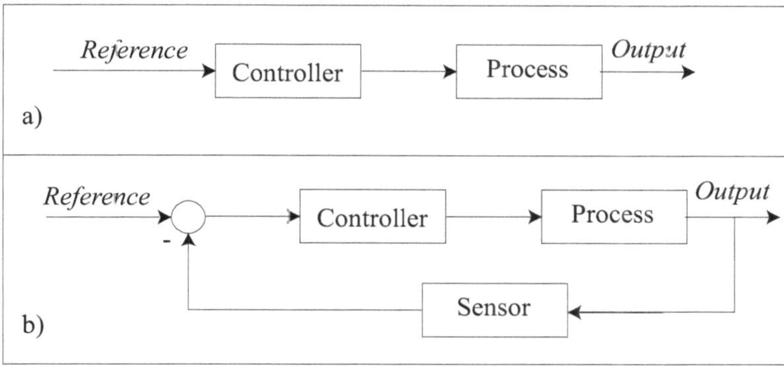

Figure 1.6 (a) Open-loop control system. (b) Feedback Control System.

In an open-loop system, the *output* signal is obtained by calculating the response of the multiplication of the transfer functions of the *controller* and *process* to the *reference* signal. In feedback control, the *output* of the *process* is measured using a *sensor* enabling to calculate the *error* signal, in these systems the *output* of the systems is calculated from the response of the multiplication of the transfer functions of the *controller* and *process* to the *error* signal. For linear or linearized systems the *controller* can be designed to improve the system stability, reduce the response time, to limit the system response overshoot or minimizing the steady error [22]. From these basic control theory fundaments, different control technical areas are being developed. Among them are Adaptive and Learning Systems, Discrete Event and Hybrid Systems, Stochastic Systems, Networked Systems, Control Design, Linear Control Systems, Non-linear Control Systems, Optimal Control and Robust Control [23].

1.2.2 Diagnostic and Training

This domain of medical robotics comprises systems that are used in a pre-surgical stage for disease diagnosis or for the surgeon skills improvement. Recently different organ modeling, visualization and examination systems have been proposed. An endoscope composed of a sheath with controlled stiffness and distal end bending and a fiberscope was used for inspection of a swine colon [30]. Augmented reality systems using sensor fusion of magnetic trackers and ultrasound probes has been proposed for organ modeling in virtual reality for endoscopies [31]. Training systems have been proposed for laparoscopic surgery relying on virtual reality environment and motion capture with inertia sensors in [32].

Global Rating Index for Technical Skills for medical training were introduced for laparoscopic surgery in [24] and extended later on in [25], both of them consider first evaluation criteria the respect for tissue. The technology for simulator-based training in medicine is being developed to correspond to the technological advances in minimally invasive surgery. The benefits of simulator-based training has been shown for endovascular intervention by [26] and for endoscopic intervention by [27]. They found that the use of simulators during training improves the resident's performance, decreases the risk to the patient's safety during supervised practice, and reduces the instruction time. However, it still needed to develop a quantitative evaluation method based on GRITS for endovascular surgery simulation.

Simulators for endovascular intervention may be classified into two groups corresponding to the approach used for human vasculature modeling: hardware or software. Simulators based on virtual reality environments for blood vessel modeling reproduce also in software the motion of the catheters and guide wires as well as other intravascular tools deployment. The intravascular devices and blood vessel interaction is transmitted to the operator using force feedback at the insertion port. These ports accept a limited number of standard catheter and guide wires without lubrication; contrast media injection is simulated by air injection in a secondary port. However, inside the virtual reality, a large number of intravascular devices and blood vessel morphologies and diseases are selectable [26]. In early stages, vasculature modeling in hardware was done using glass [28]. Then merging to softer and transparent materials, models of human vasculature were built from corrosion cast of vascular lumen of diseased human specimen [29]. Nowadays most advanced modeling techniques are based on silicone elastomer and enable DICOM data fusion, reproduce the human vasculature, and will be presented in detail in the following chapters [1]. These simulators offer the advantage that enables to insert any standard intravascular devices into the silicone vasculature and offers a realistic representation of the interaction between these devices, human blood vessel morphology, and flow circulation (Fig. 1.7).

Modeling human vasculature with transparent photoelastic materials enables stress visualization and analysis during the endovascular surgery simulation. As the stress produced by intravascular tools on the blood vessel model wall is measurable, there is a way for numerical representation of respect for tissue during endovascular surgery simulation. In following chapters, we present the theory and applications of photoelastic stress analysis in blood vessels models.

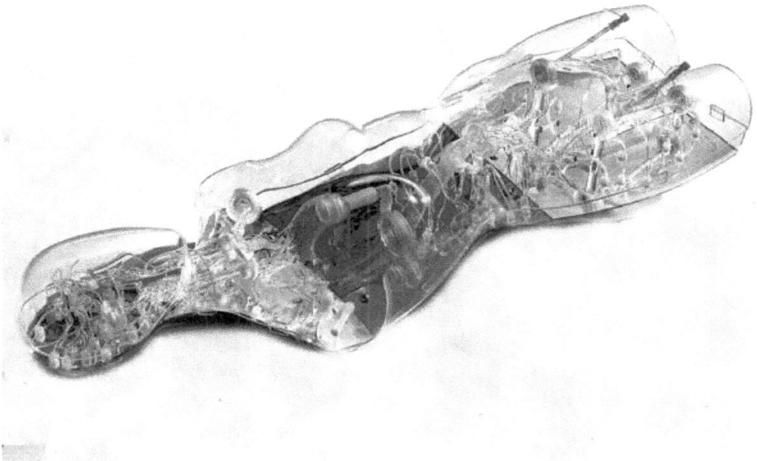

Figure 1.7 Endovascular Evaluator EVE developed at Nagoya University in 2005 and used now for training for endovascular intervention and medical robotics research worldwide.

1.2.3 Catheters and Guide Wires

To ease and increase the number of treatments in endovascular surgery, a large number of specialized catheters have been investigated. The shape of the catheter tip is bent preliminary to pass through blood vessel, and the insertion path is selected by setting the direction of catheter tip in desired direction. However, when the shape of the vessel is not suitable for the shape of the catheter, the insertion becomes difficult. To deal with this problem, active catheters were proposed. An active catheter can be bent arbitrary to fit the shape of catheter to the blood vessel morphology. The addition of a controlled motion of the catheter tip has been studied from different approaches such as the use of Ionic Conductive Polymer Film (ICPF) [33],

Shape Memory Alloy (SMA) [34], continuum active catheters [35], multi-active link catheters [36], string actuated mechanisms [37], and hydraulic systems built using stereo lithography [38] (Fig. 1.8).

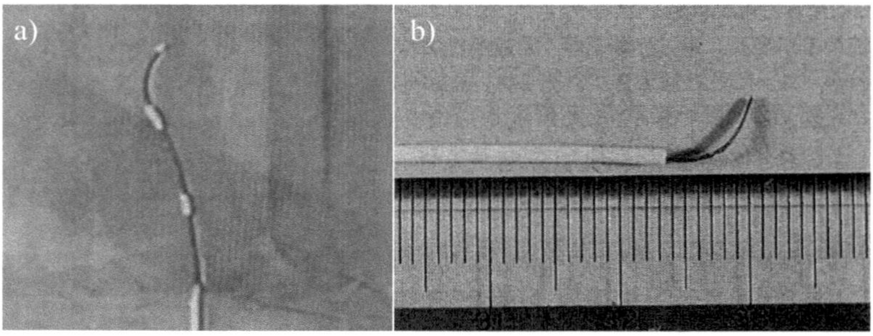

Figure 1.8 (a) Active catheter using a SMA actuator [34]. (b) Active guide wire using ICPF [33].

There is a possibility of excessive stimulation of the vessel wall when the catheter is driven through the blood vessel. The surgery may be interrupted if the excessive stimulation reduces the blood vessel diameter causing convulsions. Therefore, it is desirable to measure the contact force between the catheter and the blood vessel wall. For that purpose, sensors were added on the catheter tip, such as force sensing using piezo-resistive strain gauges [39] or optical fibers [40]. Others sensors were also added, such as microcoils for motion capture [41], and ultrasounds for vasculature imaging and diagnostic. Also intravascular controlled ablation is achieved by applying an electrical current to the target tissue with a catheter that has controlled tip motion and lubrication (Fig. 1.9).

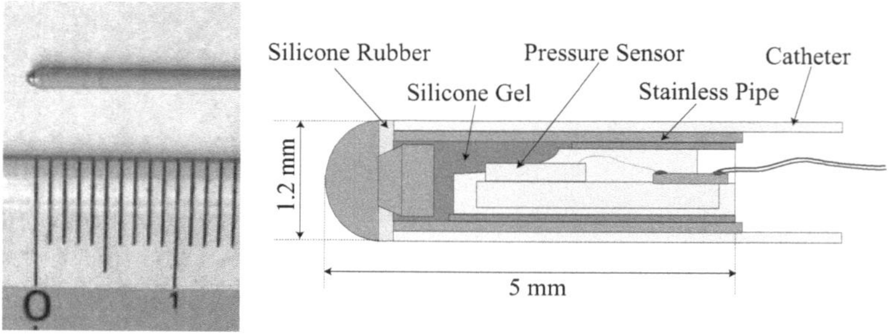

Figure 1.9 Force sensor on catheter tip [39].

1.2.4 Navigation Systems

As the treatment targets of minimally of MIS may be reached through a natural hole with a trans-luminal approach, inserting a needle to a precise location, or through small incisions in the body, it is desirable to use preoperative data or intra-surgery imaging systems information to guide the physicians and reach the target faster, easier and safer. An augmented reality environment allowing navigation is obtained using a computer generated three-dimensional image of the target obtained from medical imaging systems. After registration of this image with landmarks in the patient body; this image is superimposed to intra-surgical images to provide guidance. As the intra-surgical images are obtained from a variable location, the superimposed image must change according to the relative motion of camera that provides the intra-surgical image and the landmarks on the patient's body.

In laparoscopic surgery, the motion capture of the laparoscope is done using optical trackers [42], magnetic trackers, or a hybrid system combining them [12, 43]. The target image generation is done conventionally using preoperative data of the target such as a tomogram or magnetic resonance scan and kinematics calculations. Optical trackers have the property to be the most accurate motion capture sensors, but they have the problem that visual contact is needed between the object to be tracked and the sensor. During the surgery, occlusion may occur due to the constant activity of surgeons around the patient, which makes it difficult to have a continuous motion capture of all the landmarks during the surgery. Those trackers need at least three landmarks to be attached to the object to be tracked.

Magnetic trackers, however, offer a constant tracking even when there are non-ferrous objects between the transmitter and the receiver. But those trackers have a distorted output, due to ferrous objects in the working space and intrinsic properties of the field generated by their transmitter. Such sensors are also affected by oscillating magnetic fields such as produced by cathode ray tube monitors. This distortion is a barrier for the use of preoperative data with magnetic trackers, because it needs to be corrected in order to have both models in a non-distorted frame of reference [44–45]. Hybrid systems, such as the one discussed in [42], correct the distortion of the magnetic trackers using an optical tracker and numerical methods.

For needle guidance and microcannulation, a long, thin, and stiff tube must be inserted through healthy tissue to treatment target. With purpose of cannulation, a microcannula was inserted through the vitreous body of porcine eye to reach a target vessel using a parallel link manipulator. The injection of dye confirmed the cannulation effectiveness compared with the manual approach [46]. Needle insertion into inferior porcine cava vein was achieved relying on guidance information provided from ultrasound imaging and tissue deformation simulation [47]. Another approach uses a palpation probe in combination with the ultrasound probe and needle insertion mechanism for breast tumor treatment [48]. Also 3D augmented reality system for MRI guided surgery was proposed [49] (Fig. 1.10).

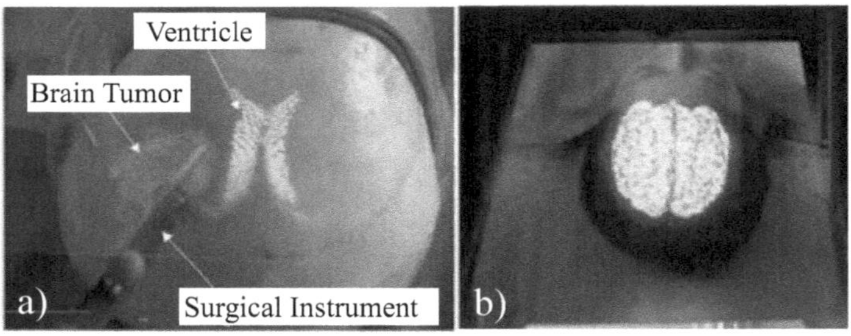

Figure 1.10 3D augmented reality system for MRI guided surgery. (a) Surgical instrument guidance example. (b) Brain model overlaid over the body of a volunteer [49].

1.2.5 Telesurgery

The geographical constraints between specialized surgeons and patients motivated the development of remote treatment or telesurgery; these treatments became possible with the development of robotics and communication systems. It requires exchanging video, audio, and control signals with short delay between the location of the surgeon and the patient. For tele-operation systems an operator moves a human interface that is named the master, the human interface encodes the motion into electric and optical signals that are transmitted to a remote location. At the remote location the signal is received by a slave robot that moves according to it. In unilateral systems, telesurgery is limited to reproduce the motion desired by the operator to manipulate tissue (Fig. 1.11). Bilateral systems not only reproduce the motion desired by the operator, but additionally they reflect to the operator the force produced by the interaction of the slave robot with the manipulated tissue (Fig. 1.12). Other configuration examples for telesurgery are given in Figs. 1.13 and 1.14. In these master–slave robotic systems, the master side must also show to the surgeon the progress of the surgery. For that, the slave side should manipulate the vision system capturing intra-surgical or fluoroscopic images. As the distance increases between the master and the slave units, the time-delay of video, control and audio signals becomes more important.

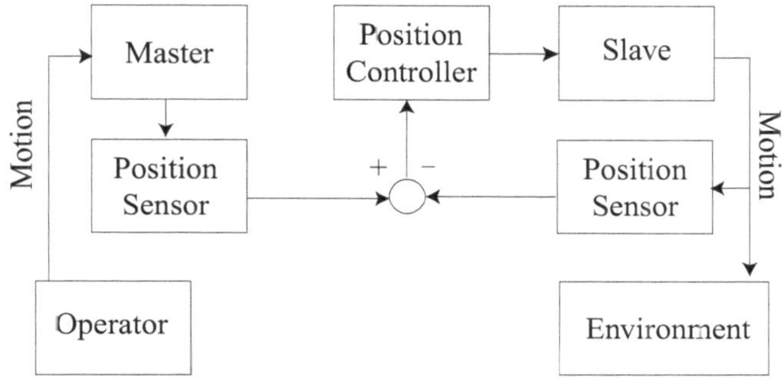

Figure 1.11 Unilateral Control System for Tele-operation, the motion of the operator in the master side is reproduced in the environment by the slave robot.

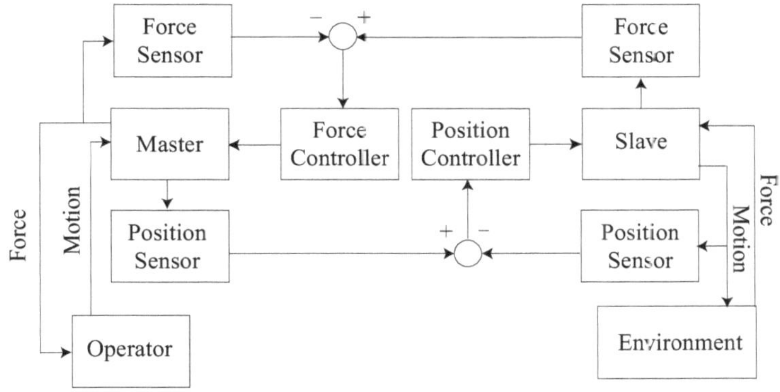

Figure 1.12 Bilateral Control System for Tele-operation, the motion of the operator in the master side is reproduced in the environment by the slave robot. The force produced during the interaction between the slave and the environment is reflected to the operator in the master side.

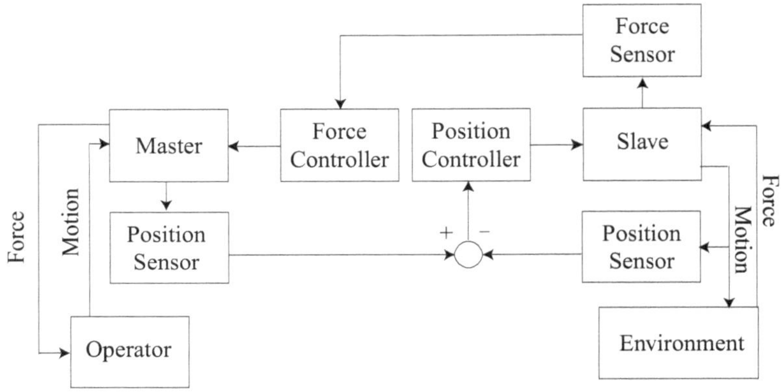

Figure 1.13 Force Reflection Control System for Tele-operation, the motion of the operator in the master side is reproduced in the environment by the slave robot. The force produced during the interaction between the slave and the environment is reflected to the operator in the master side in an open loop. The reflected force is highly dependent to the intrinsic characteristics of the master device.

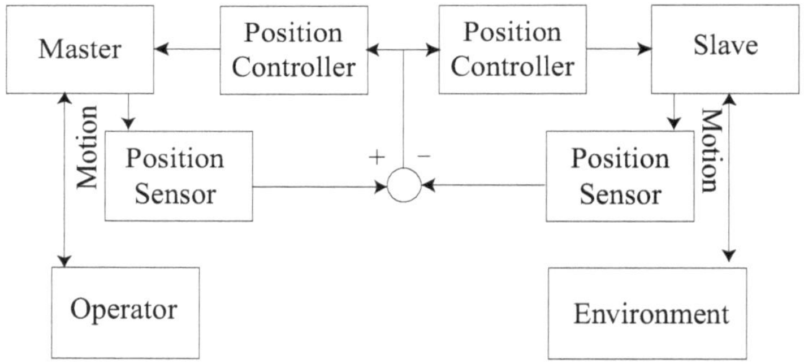

Figure 1.14 Symmetric Position Control System for Tele-operation, this system enables to reproduce in the master side the applied force in the slave side.

The first transatlantic telesurgery was achieved in 2001, where a cholecystectomy was achieved while the surgeon was in New York and the patient in Strasbourg [50], while the 7 DOF master–slave robotic system for minimally invasive surgery DaVinci was presented in 2000 [51] and is actually used in more 1450 hospitals around the world, and its applications extend to a large number of treatments in cardiology, urology, gynecology, colorectal and general surgery.

Telesurgery was proposed for endovascular surgery not only for overcoming the surgical constrains but also as an alternative to reduce X-ray exposure of surgeons during the intervention. In this case, the catheter is driven by the slave mechanism while the physician controls the slave device using a specialized joystick that is placed away from the fluoroscope. Slave catheter insertion devices were developed by Tanimoto *et al.* [52], Arai *et al.* [53], Ikeda *et al.* [54], and Jayender *et al.* [55]. In the first generation proposed in [52], catheter grasping was based on gum rollers and the feedback of catheter position was given using encoders. Jayender *et al.* [55] proposed a device in which the feedback is given by accelerometers placed on the joints of a robot arm. Also magnetically steerable catheters and catheter insertion systems are available in the market [56], but they have not spread widely as DaVinci robot (Figs. 1.15–1.18).

Figure 1.15 Master device for remote operation in neurosurgery.

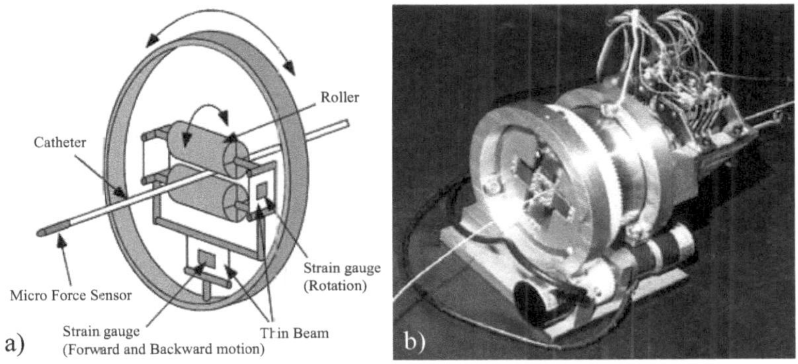

Figure 1.16 Slave device: (a) Structure and (b) mechanism.

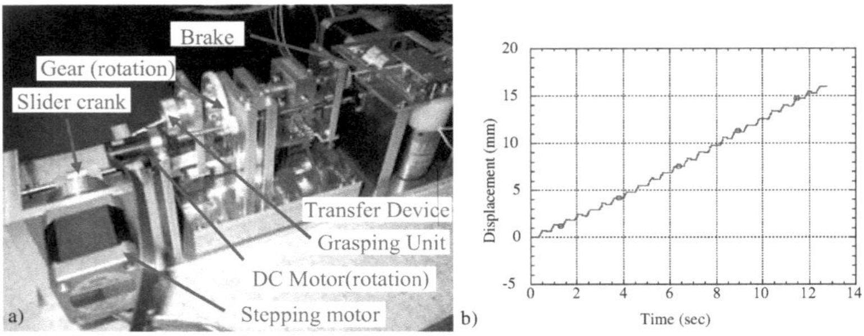

Figure 1.17 (a) LSM slave device by Arai *et al.* (b) Catheter displacement while driven with the LSM [53].

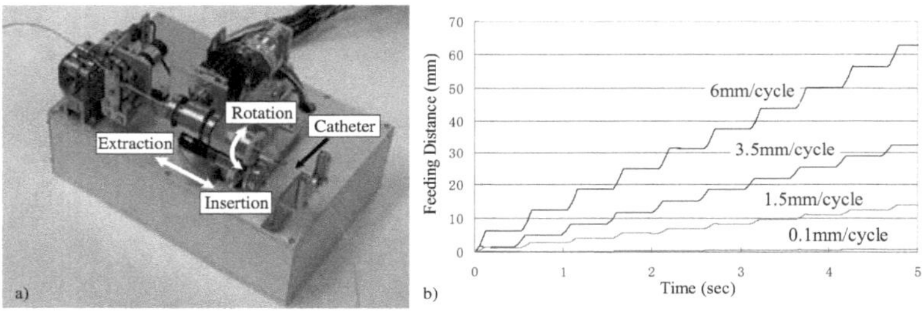

Figure 1.18 (a) LSM2 slave device by Ikeda *et al.* [54]. (b) Catheter displacement while driven with the LSM2 [54].

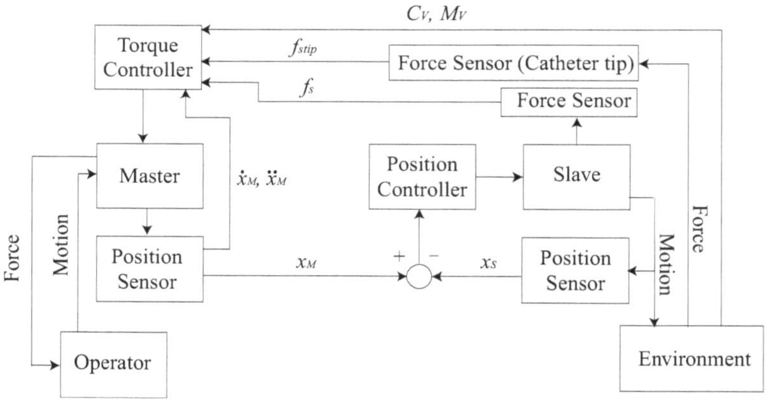

Figure 1.19 Force reflecting system for catheter tele-operation.

When the force sensor presented in Fig. 1.9 is combined with the slave system shown in Fig. 1.16 and the master device shown in 1.15, the tele-operation system for catheter manipulation can be described by the diagram shown in Fig. 1.19 and the following equations [52]:

$$m_M \ddot{x}_M + c_M \dot{x}_M = g_M + f_M \tag{1.1}$$

$$m_S \ddot{x}_S + c_S \dot{x}_S = g_S - f_S \tag{1.2}$$

where the sub-indexes M and S stand for parameters belonging to the master and slave respectively, m_M and m_S the mass of each device, c_M and c_S their viscosity coefficient, x_M and x_S their displacement, g_M and g_S the applied forces from the actuators, f_M the force applied from an operator to the master, and f_S the force applied by the slave to the environment. This system is described by

$$g_S = K_P(x_M - x_S) + K_V(\dot{x}_M - \dot{x}_S) \tag{1.3}$$

where K_P and K_V are the constant coefficients for position and velocity feedback. On the other hand, the master device should have variable impedance that relates with the interaction of the catheter with vasculature. For that g_M is defined as follows:

$$g_M = -(F_{sensor} + F_V) = -(K_{FS} f_s + K_{FStip} f_{stip} + M_V \ddot{x}_M + C_V \dot{x}_M) \tag{1.4}$$

where F_{sensor} is the sum of the force measurements given by the sensors f_s and f_{stip} multiplied by their respective coefficients K_{FS} and K_{FStip}. F_V is a virtual resistant force while mass M_V and viscosity coefficient C_V are variable depending on the diameter of the blood vessel where the tip is located, catheter position (expressed using the axial motion and twisting motion), and blood pressure [52].

For applying telesurgery to endovascular intervention, several additional challenges have to be solved as the end effector is a catheter or guide wire, which are flexible bodies. These intravascular tools are in contact with human blood along the surgery; therefore the grasping mechanism should be compatible with sterilization process and also grasp them firmly without damaging them or without slips. If the catheter slips from the grasping mechanism, the shape of the catheter within the blood vessel may change and the encoders or accelerometers used for motion encoding may provide inaccurate measurements for the amount of catheter inserted and the applied torsion. The interaction between the catheter tip and the vascular wall should be sensed, too, for minimizing the risk of puncturing the vascular wall.

If gum rollers as used as grasping mechanism, they may accumulate blood residues making them unsuitable for sterilization. A linear stepping mechanism (LSM) using a mechanical pencil-like mechanism to grasp the catheter was proposed by Arai *et al.* [53]. During the first half of the cycle, the mechanism closes the chalk to grasp the catheter, and then it pulls the catheter. Then during the rest of the cycle, the chalk is open to release the catheter, and then it is restored to the initial position to start the cycle, and it is compatible with sterilization process. A similar traction system was designed by Ikeda *et al.* [54]. This device uses encoders as unique feedback source. This LSM is able to insert and extract the catheter with a controlled motion from 0.1 mm to 6 mm per stroke and to twist the catheter at different speeds. This slave device gave us the basis for *in vitro* catheter insertion paths reconstruction; this is achieved by placing a microcoil on the catheter tip for motion capture and LSM feedback control. This system is used in Chapter 7 for catheter insertion path reconstruction.

1.2.6 Modular Robots for Endoluminal Surgery

The increment in the space restriction of SPS and natural orifice transluminal surgeries has motivated the application of modular robots to these interventions. In this case the modules are delivered through the surgical port and they integrate inside the patient's body into a complex robotic for providing the treatment. The projects Assembling Reconfigurable Endoluminal Surgical system (ARES) and Array of Robots for Augmenting the Kinematics of Endoluminal Surgery (ARKANES) are being developed by Dario *et al.* [57]. These projects aim to develop a modular robot deployable through the esophagus of the patient into the stomach. Once deployed into the stomach, using specialized modules the robot would provide new diagnosis and treatment options to the stomach and surrounding organs (Fig. 1.20). Magnetic manipulation, levitation, and propulsion [58–59] are also being investigated for modular robots for the digestive tract and may play a key role in the development of concepts as the one proposed in [57] (Fig. 1.21).

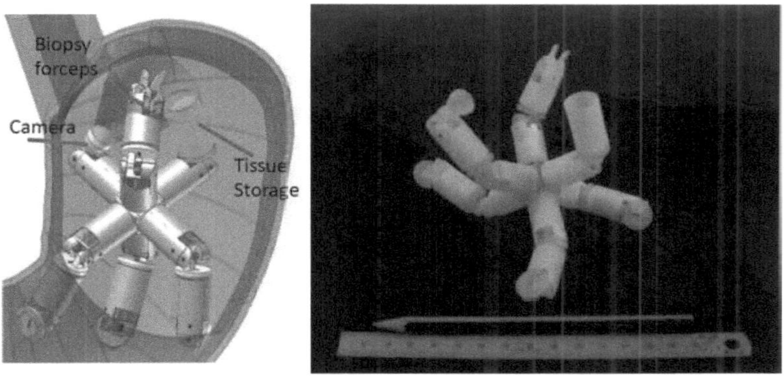

Figure 1.20 Example of a multi-module robot integrating a grasping tool, camera, tissue storage module, and structural modules proposed by projects ARES and ARKANES [57].

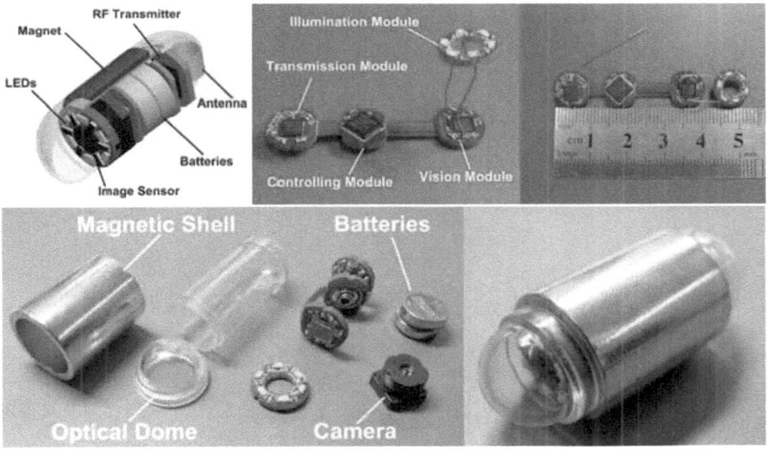

Figure 1.21 Self-propelled endoscopic capsule with 4 motion types proposed by [59].

1.2.7 Drug Delivery Systems

Drug delivery systems are being developed for vascular tissue healing and tumor treatments. One of the approaches for this purpose is the use of intravascular stents for releasing the drug in a specific location. The amounts of drug to be released from the stent and their release time are controlled by alternating coatings of the drug and biodegradable polymers. This reduces the initial drug burst of other treatments and enables the development of programmable chemotherapy for cardiovascular diseases [60–61]. The manipulation of magnetic micro beads traveling in the blood stream has been achieved in *in vitro* and *in vivo* experiments using the magnetic field of magnetic resonance imaging equipment [62–63]. These experiments enabled the development of therapeutic microcarriers encapsulating magnetic nanoparticles and antitumor drugs. The therapeutic microcarriers were delivered to the hepatic artery of a rabbit and controlled chemoembolization was achieved steering the carriers to the target vasculature using the magnetic field of the magnetic resonance imaging equipment [64] (Fig. 1.22). Magnetoliposomes were also proposed for drug delivery systems as they are formed by iron-complex solution and biogenic phospholipids, their size is within the range of 100–400 nm, and they enable the encapsulation of boron compounds for cancer treatments [65].

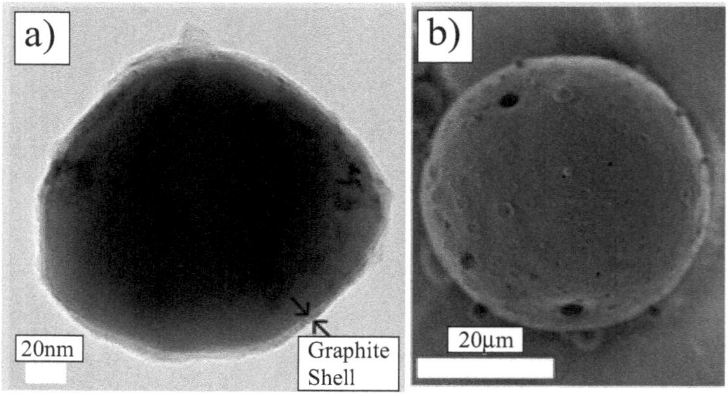

Figure 1.22 (a) Magnetic nanoparticles and (b) therapeutic microcarriers proposed for drug delivery in [64].

1.2.8 Rehabilitation and Recovery

Another domain of medical robotics is composed of the systems developed for rehabilitation and recovery, such as prosthesis, wheelchairs, and robots used for physical and mental therapy. One of the most successful robots for mental therapy is the robot Paro, which is being used in many countries and improves the motivation and socialization of patients and their relation with the caregiver [8]. Simulation systems have been constructed to improve the design of manual wheel chairs [9], and others researchers are developing navigation systems to ease the indoor motion of electric wheelchairs [66]. One of the most recent challenges for prosthetic arm development is the inclusion of controllable impedance in prosthesis to replicate the impedance changes in human arms [67] (Fig. 1.23).

Figure 1.23 Interactive robot for mental therapy used in Japan and through Europe since 2003 and used for dementia prevention in [8].

1.3 Regenerative Medicine and Artificial Organs

The purpose of regenerative medicine is to regenerate lost and damaged tissues by using cells [68]. Tissue is an assembled structure composed of cells, which are the constituent unit of our body. Multiple tissues and cells aggregate to form functional organs. As shown in Fig. 1.24, there are two approaches for regenerative medicine. One of them is the implantation of cells. Especially in cells, the formation of an individual organism starts from a fertile egg (Fig. 1.25). After 6 days' growth, this cell forms a blastocyst that contains inner cell mass (ICM). If we culture ICM, embryonic stem cells (ES cells, multipotent) can be obtained. Figure 1.26 shows a schematic of differentiation from a multipotent ES cell. Therefore, if somatic stem cells destined to form a tissue or further differentiated cells are implanted to damaged part of a tissue, the damaged part will be restored. In this kind of cell implantation-based regenerative medicine, technological development of cell separation, differentiation, and growth are very important.

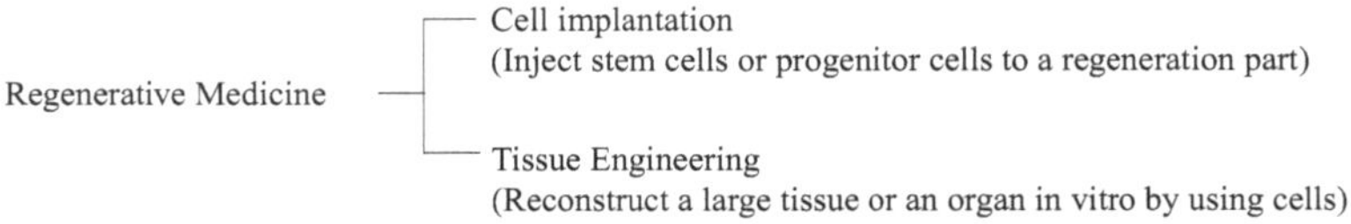

Figure 1.24 Approaches for regenerative medicine.

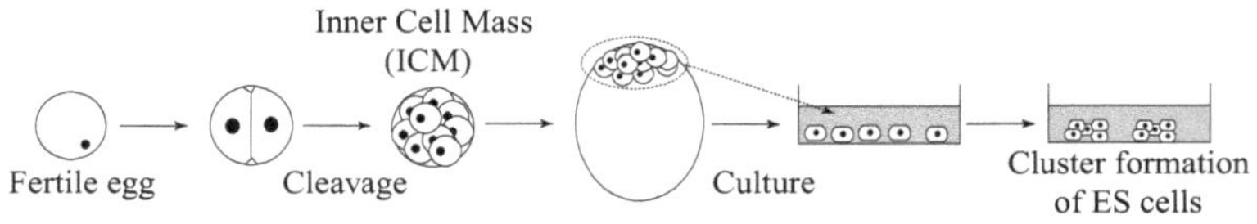

Figure 1.25 Process for obtaining embryonic stem cells from a fertile egg.

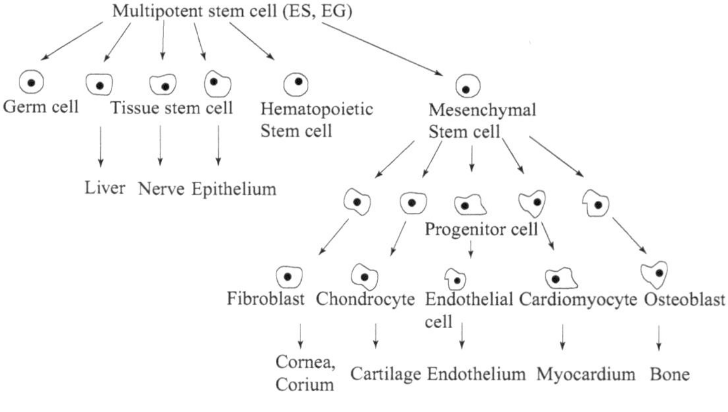

Figure 1.26 Differentiation from a multi-potent stem cell.

The other approach is tissue engineering. Tissue engineering means a technology to regenerate a damaged tissue whose deficit is too heavy to be cured by only cell implantation. The most important issue in tissue engineering is the development of appropriate environment for *in vivo*-like regeneration. In this section, traditional artificial organs, cell implantation-based (*in vivo*) regeneration and tissue engineering-based (*in vitro*) regeneration are introduced by using representative examples. An artificial organ is a man-made structure that is implanted into, or integrated onto, human body to replace in damaged organs a natural function or a group of related functions. This enables the person receiving the artificial organ to stay alive or continue a normal life [69].

Materials for long-term implantable devices and their applications were listed by David Williams [117], as well as some of the generic host response to biomaterials. Titanium alloys are used for dental implants, femoral stems, pacemaker cans, heart valves, fracture plates, and spinal cages. Cobalt–chromium alloys are used for bearing surfaces, heart valves, stents, and pacemaker leads. Platinum group alloys are used for electrodes. Nitinol is used for shape memory alloys applications. Stainless steel is used for stents and orthopedic implants. Alumina and UHMW are used for bearing surfaces. Calcium phosphates are used for bioactive surfaces and bone substitutes. Carbon is used for heart valves. PEEK is used for spinal cages. PMMA is used as bone cement and for intraocular lenses. Silicones are used for tissue augmentation, lead insulation and ophthalmological devices. Polyurethane is used for pacemaker lead insulation. Expanded PTFE and polyester textile are used for heart valves and vascular grafts. Poly(styrene-block-isobutylene-block-styrene) is used for drug eluting stent coating. Some of the responses listed by Williams are: protein absorption and desorption characteristics, generalized cytotoxic effects, neutrophil activation, macrophage activation, fibroblasts behavior, microvascular changes, tissue or organ specific cell response, activation of clotting cascade, platelets adhesion, activation and aggregation, complement activation, antibody production and immune cells responses, acute hypersensitivity and anaphylaxis, mutagenetic response and genotoxicity, reproductive toxicity, and tumor formation.

1.3.1 Artificial Organs

1.3.1.1 Artificial heart

Cardiac diseases are one of the three major causes of death. The main purpose of artificial heart is to replace the function of left ventricle. From 1935, development of artificial heart has been continued to mimic real heart as much as possible. Initially, researchers made exertion mainly on mimicking the shape of heart. However, the point of research transferred gradually to the recreation of the pumping function. Currently main issues in the development of the artificial heart are the hardware longevity and low weight. One of the most popular artificial hearts is *TERUMO HEART*, which is an implantable centrifugal pump that works as an auxiliary artificial heart (Dura Heart). This equipment works as an alternative left ventricle and uses magnetic levitation for controlling centrifugal pump impeller position. This levitation reduces the friction during actuation and thus increases the life time of pump. The controlled magnetic levitation of the impeller prevents thrombus formation.

1.3.1.2 Artificial vascular graft

The number of patients suffering from vascular diseases, such as aneurysm and thrombus, has unfortunately increased due to modern diet. Artificial blood vessels have been developed to replace diseased vessel since 1952. The principal requirements for artificial vascular graft are

1. Biocompatibility
2. Exclusive use of non-toxic materials
3. Robustness and compliance
4. Ease of handling
5. Antithrombogenic
6. Reduced blood leakage

TERUMO also developed artificial blood vessels by using biocompatible polymer. This technology is successful in clinical application for large-diameter vascular graft. The prevention of clot formation in small-diameter artificial arteries (cross-section diameter below 6 mm) is still a challenge.

Matsuda *et al.* have been continuously developing artificial vascular graft of small-diameter arteries by the use of biodegradable polymer [70–72]. In the world's first report of vascular regeneration, Shinoka *et al.* developed regenerated cardiovascular system. Tailor-made blood vessels scaffold were proposed by Uchida *et al.* in 2008 [73] and will be explained in detail in Chapter 7.

A tissue engineered blood vessel is commercialized by Cytograft Inc. It was developed by rolling fibroblasts sheets to produce vascular grafts with an inner layer lined with patient endothelial cells for thrombosis prevention. Tissue engineering was integrated with stent graft technology, by using biological sheets for the stent graft membrane a biological bond is created between device and native tissue.

1.3.1.3 Artificial kidney

Currently in Japan, artificial kidneys are used for 0.2 million of patients with renal failure. Also, 30,000 patients with renal failure are newly starting dialysis therapy every year. Modern artificial dialyzers remove wastes and moisture from blood stream and maintain ion composition and osmotic pressure of biological fluid. There are four kinds of methods of dialysis and are categorized into the following:

1. Hemodialysis
2. Peritoneal dialysis
3. Hemofiltration
4. Hemodialysis filtration

Hemodialysis is the most popular way of treatment. Almost 100% of dialysis membrane is composed of hollow fibers. The world's first artificial dialyzer comprising hollow fibers was developed by Cordis Dow (USA) in 1971. In 1975, *TORAI* developed the world's first artificial dialyzer comprising PMMA hollow fibers. Currently in Japan, *NIPRO* has top share of an artificial dialyzer consisting of hollow fibers.

1.3.1.4 Artificial liver

Liver is called the largest chemical plant in our body. Its function involves the synthesis of proteins, processing of wastes, and detoxification of harmful substances. The recreation of these functions is needed for an artificial liver. However, detoxification is the only achievement in the development of artificial liver. To improve the function of detoxification, bio-artificial liver is mainly investigated now. In the bio-artificial liver, animal hepatocytes are utilized to replace the function of human hepatocytes. In the United States, clinical application of bio-artificial liver to patients with acute kidney failure has already started.

1.3.1.5 Artificial skin

Skin is the largest organ in human body and is the barrier that protects us from external harms. Therefore, the partial loss or the damage of the skin is a possibility of threat to our life. This is cause of death in high level of burn injury cases. However, our skin has a high capacity of regeneration.

Artificial skin contains artificial corium, cultured epidermis, and cultured skin. Cultured skin has been fabricated by culturing skin cells in a clean room. On October 2007, Japan Tissue Engineering Co. Ltd. officially received government approval to culture and to commercialize the autologous cultured epidermis "JACE" for serious burn treatment in Japan. This was the first case of commercial product related to tissue engineering in Japan.

1.3.1.6 Artificial bone

Similarly to skin, bone has high capacity of regeneration. Artificial bone is used to compensate the damaged or lost part of bone. In general, artificial bone is implanted into the patient's body by a surgical procedure that replaces damaged part due to crush fracture or tumor. Before 1990, biocompatible materials such as titanium, ceramics, or coral were used as artificial bone. However, periodic maintenance was necessary due to aged deterioration and user's aging. Since the 1990s, biodegradable materials that are candidates to be alternated gradually by natural and regenerated bone have been investigated and developed. Apatite is a representative material of biodegradable polymer. More recently, 3D printing has been used to achieve mass production of artificial bone.

1.3.2 Cell Implantation-Based Regeneration

Human body consists of 60 trillion of cells. Originally our body comes from only one fertile egg, and it is maintained with continuous growth and differentiation of cells. If we reveal the mechanism of cellular action, there is a possibility of tissue regeneration by injecting cells to a damaged part of tissue. One option for that are somatic stem cells. The function of tissues and organs can be regained by injecting these cells to a patient's body. The regeneration of tissues by means of injecting stem cells inside a patient's body is called "stem cell engineering" [74]. It is difficult to answer to the question of "*When will the cell implantation-based treatment be clinically applied?*" As described below, a part of application has already started and cultured cells are already being implanted as artificial skin. In addition, osteogenesis by means of extraction and implantation of bone marrow fluid has been recognized as a standard method of bone regeneration. On the other hand, it is very difficult to expect the application period of cellular treatment to spinal cord injury, Parkinson's disease, and diabetes owing to scientific and ethical problems. Regarding with regeneration based on embryonic stem cells, several kinds of human embryonic stem cells as a source have just been prepared. Within the cellular treatments that are already clinically applied, there is Angiogenesis by means of implanting mononuclear cells from autologous marrow. This treatment has been applied widely and rapidly as a treatment for patients with peripheral vascular disease (chronic obstructive arteriosclerosis, Buerger's disease). This method consists of following process:

(i) Extract bone marrow fluid from patients with peripheral vascular disease
(ii) Separate mononuclear cells from bone marrow fluid
(iii) Under general anesthesia, medical doctors inject mononuclear cells into skeletal muscle of inferior limb by using syringes. The injection is performed at 40–50 separate sites of skeletal muscle.

As described above, novel type of angiogenesis by use of endothelial progenitor cells or bone marrow cells has already been clinically applied. Cell implantation-based regeneration can be one of supplemental methods to cure ischemic heart disease or chronic obstructive arteriosclerosis.

As a cell source for osteogenesis, bone marrow-derived cells are cultured *in vitro*. These cells are applicable to compensate the damaged bone by combining with an appropriate implant material. Bone marrow-derived cells are expected for not only the treatment of bone deficit but also the treatment of bone fracture, fixation of spinal cord, and regeneration of skull bone. Human chondrocytes are cultured to inject into joints with heavy damage and to help regeneration of joint cartilage.

1.3.3 Tissue Engineering-Based Regeneration

Each year in the United States, millions of people suffer from a variety of diseases that could be aided from therapies such as organ transplantation. However, despite the widespread need for transplantable tissues, many patients die while waiting for donor organs. It is from this need that the field of tissue engineering emerged

[75–76]. Tissue engineering is an interdisciplinary field that involves the use of biological sciences and engineering to develop tissues that restore, maintain and enhance tissue function.

In general, there are three main approaches to tissue engineering:

1. Use isolated cells or cell substitutes as cellular replacement parts.
2. Use acellular biomaterials capable of inducing tissue regeneration.
3. Use a combination of cells and materials (typically in the form of scaffolds).

In this section, we introduce scaffold- and cell-based method of regeneration. In the first step, stem cells are extracted from a human tissue, such as bone marrow or umbilical vein. In the second step, those stem cells are mass cultured *in vitro* to increase the cell number. In the third step, those cultured cells are detached from a dish and seeded on a canned biocompatible/biodegradable scaffold. In the fourth step, cells on the scaffold are cultured three-dimensionally with effective rotation or shaking. After the formation of 3D tissue substrate *in vitro*, the artificial tissue is implanted into the patient's body. As the world's first impressive demonstration of scaffold-based tissue engineering, Vacanti *et al.* accomplished the implantation of chondrocytes-seeded biodegradable scaffold in a nude mouse and the regeneration of cartilage [77]. Owing to the ear-shaped unique scaffold, the report was quite impressive to tissue engineering researchers in the world.

The scaffold-based tissue engineering is composed of three components:

1. Biodegradable scaffold
2. Cell growth factor (CGF)
3. Cells

In general, cells proliferate after adhering on a surface. In our body, cells are enclosed in extracellular matrix (ECM) and maintain homeostasis by communicating with surrounding cells. If a tissue is seriously damaged, both cells and ECM will be lost. Three-dimensional scaffolds help the regeneration of the tissue as an artificial ECM. Therefore, biodegradable scaffold plays an important role for the effective proliferation of cells and tissue regeneration. An original scaffold *in vivo* is ECM. However, if a large portion of a tissue is lost, no ECM will be left around the tissue. This is why ECM is required for *in vitro* creation of tissue substitutes.

Typically scaffolds should have porous structure to maximize the function of cells and to supply enough oxygen and nutrients to cells proliferated all over the scaffold. On the other hand, mechanical strength with appropriate Young's modulus is required to maintain its shape and to fit the surrounding living tissue after implantation. Therefore, the balance between porosity and mechanical strength becomes important. Additionally, biodegradable materials are preferred because scaffolds may disarrange the function of living tissue after the complete regeneration of tissue. Some required properties of biodegradable scaffolds are

1. Cell-adhesive substrate at differentiation and proliferation
2. Securement of regeneration space
3. Configuration of restored tissue
4. Supply route of oxygen and nutrients to cells
5. Storage and slow release of cell growth factor (CGF)
6. Tissue substitute until regeneration (strength, compliance, etc)
7. Bioabsorbability/biodegradability
8. Porous structure to accommodate cells

1.3.4 Production Methods of Biodegradable Scaffolds

The size of cells is around 20 μm. Therefore, cells are surrounded by microenvironments in our body. That is why microscale fabrication technologies are necessary to fabricate biodegradable scaffolds. Biodegradable polymers are fabricated into rope, sheet, tube, and block-like 3D porous scaffolds depending on the shape of a targeted part of tissue. Here, we introduce several representative methods to fabricate biodegradable scaffolds.

1.3.4.1 Electrospinning

Electrospinning has been used as a fiber-forming method since the 1930s. It can produce micro-nano-scale polymer fibers. Figure 1.24 shows a schematic of electrospinning. If high voltage (5–30 kV) is applied between a needle and a collector, the ejection of polymer solution from a syringe yields an electrically charged jet because repulsive force becomes more dominant than surface tension. Jet forms a very thin fiber in an electric field and fibers are piled up on a collector. Some electrospinning characteristics are

1. It can generate nanofibers with 50–500 nm.
2. It can generate fibers with large surface area.
3. The microscopic structure of fiber is controllable.
4. Various types of polymer solution can be used.

Figure 1.27 shows the principle of fiber formation. Highly charged droplets split into jets owing to large repulsive force on the surface. The jet becomes thinner to form fibers, and polymer fiber is generated after the complete evaporation of volatile solvent. Prof. Takehisa Matsuda's group has fabricated electrospun tubular scaffolds. Electrospinning has been widely used in scaffold fabrication. This method is suitable for the fabrication of flat or cylindrical scaffolds. Kidoaki *et al.* used SPU as a biocompatible polymer [78] for electrospinning and proposed a mixing electrospinning technique by using two different polymer solutions [79].

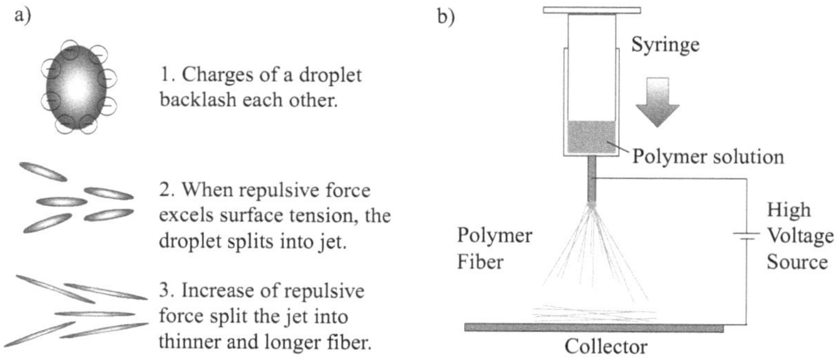

Figure 1.27 (a) Fiber formation principle during electrospinning. (b) Electrospinning system schematic.

Biodegradable polymer PLCL has been investigated to achieve mechano-active artificial vascular graft. Inoguchi *et al.* evaluated the pressure-diameter property of electrospun PLCL fibers [80]. They also cultured HUVECs inside a small-diameter under-fluid shear stress [81].

1.3.4.2 Porogen leaching

Porogen is defined as a soluble particulate to generate pores inside a polymer. Porogen leaching is a method to control the porous structure inside a scaffold by using hydrosoluble particles. In previous studies, salt or sugar particles have been commonly used as porogens. This method has following absolute advantages compared with other techniques:

i. The size of porogens can be controlled easily by using sieves.
ii. Porogens can be packed in complex 3D structure, thereby ensuring 3D porous scaffolds with arbitrary shape.
iii. Porogens such as salt and sugar do not dissolve in general organic solvents and can be removed by water easily.

Salt particles are most commonly used in the fabrication of porous scaffold. Lee *et al.* reported an artificial skin by using salt leaching method in gelatin scaffolds [82]. Murphy *et al.* reported a salt fusion method to enhance the pore connectivity inside a scaffold [83]. They used heat and humidity of CO_2 incubator to fuse adjacent salt particles. Subsequently, Gao *et al.* applied this salt fusion method to tubular scaffolds [84]. Elastic biodegradable polymer of PGS was cast on a fused salt template by using a cylindrical mold. Heijkants *et al.* combined salt leaching and freeze drying methods to achieve hybrid structure of macropores and micropores [85]. Wu *et al.* developed an ear-shaped scaffold by packing salt particles and polymer in an ear-shaped mold [86].

Similarly, sugar particles have been used as porogens. Lee *et al.* packed sucrose particles to a complex mold designed by CAD [87]. After the molding of PLGA and the removal of sucrose particles, patterned indented surface was generated. Wei *et al.* proposed a unique method to enhance pore connectivity by using D-fructose spheres as porogens and achieved ordered and well-connected pores inside PLLA scaffolds [88]. Ice microparticles [89] or paraffin spheres [90] have been used as porogens too. In the case of porogen leaching method, the average diameter of pores and porosity can be truly controlled by the size of porogens and the weight ratio of polymer/porogens, respectively. Therefore, this method has been widely used in scaffold fabrication

1.3.4.3 Three-dimensional printing

Three-dimensional (3D) printing is a mass production method of scaffold fabrication. Therrihault *et al.* developed 3D microvascular networks by direct-write assembly of fugitive inks [91]. After printing wax in the shape of jungle gym, they cast epoxy resin on the wax and removed wax by heat. They also reported improvement of fabrication materials [92]. Vozzi *et al.* reported a microsyringe deposition method controlled by a computer, a stepping motor and a pressure regulator [93]. They printed a PLGA mesh-like ordered structure by controlled positioning of a syringe needle. Sakai *et al.* developed a biodegradable scaffold with ordered and 3D microchannels [94]. They have been developing scaffolds with microvascular networks as a liver substitute

Additionally, some groups are printing cells with scaffold materials. This method is called "bioprinting" or "organ printing." Tan *et al.* developed a strategy for layer-by-layer microfluidic patterning of living cells and biopolymer matrix [95]. Mironov *et al.* proposed organ printing method [96]. Similarly to Tan's group, they developed a cell printer that can print gels, single cells, and cell aggregates. The combination of an engineering approach with the developmental biology concept of embryonic tissue fluidity enables the creation of a new rapid prototyping 3D organ printing technology, which will dramatically accelerate and optimize tissue and organ assembly.

1.3.4.4 Soft lithography

Soft lithography, a set of techniques for microfabrication, is based on printing and molding using elastomeric stamps with the patterns of interest in bas-relief [97]. The technology was established by Whitesides' group in the 1990s. As a technique for fabricating microstructures for biological applications, soft lithography overcomes many of shortcomings of photolithography. In the field of tissue engineering, soft lithography has been used mostly for vascular channel inside scaffolds. Originally PDMS has been used as a soft material for 3D microvascular structure [98]. Biodegradable polymers have been also applied in soft-lithography-based tissue engineering. Borenstein *et al.* and King *et al.* developed biodegradable microchannel by casting PLGA thermoplastic on a PDMS mold with vascular configuration [99–100]. To overcome the brittleness of previous biodegradable polymers, Wang *et al.* reported PGS, as tough and elastic biodegradable polymer [101]. By using PGS in soft lithography, Fidkowski *et al.* developed an endothelialized microvasculature inside PGS [102]. Bettinger *et al.* extended this PGS microvasculature sheet to 3D microfluidic scaffold by piling it up [103]. They cultured hepatocytes in the microchannel to mimic liver structure.

1.3.4.5 Membranous microfluidic device (MeME process)

Ikeuchi *et al.* reported a great breakthrough to achieve membranous microchannel [104]. They developed a multi-layer hybrid scaffold consisting of artificial capillary networks and cell-containing hydrogels for tissue engineering. In earlier techniques to fabricate porous scaffolds, compatible achievement of cell accommodation and microvascular networks was difficult. Especially in soft lithography, cells have been accommodated in microvascular networks. Because the volume of microchannel is very small, 3D mass cultures of cells and tissue reconstruction were very difficult. Also, the space for cell accommodation and nutrient supply had to be the same. According to their membrane micro embossing (MeME) process, the space for cell accommodation and nutrient supply can be separate. Nutrients were supplied through membranous microchannel while cells were cultured with hydrogel in the interspaces between each membrane. In addition, this process contains no concern for cell toxicity because no organic solvent was used in the polymer fabrication. Heat-pressing and sealing methods were used to achieve membranous microfluidic device.

1.3.5 Cell Sheet Engineering

Okano and Yamato *et al.* proposed a novel tissue engineering methodology layering cell sheets to construct 3D functional tissues without any artificial scaffolds [105]. Confluent cells on temperature-responsive culture surfaces can be harvested as a viable contiguous cell sheet only by lowering temperature without any enzymatic digestions. When cells are cultured confluently, they connect to each other via cell-to-cell junction proteins and. With enzymatic digestions, these proteins are disrupted and each cell is released separately. In the case of using PIPAAm-grafted surfaces, cell-to-cell connections are not disrupted and cells are harvested as a contiguous cell sheet by decreasing temperature. Furthermore, adhesive proteins underneath cell sheets are also maintained and they play a desirable role as an adhesive agent in transferring cell sheets onto other culture materials or other cell sheets [106]. These viable cell sheets are composed of cells and biological ECM without any artificial scaffolds. Various types of cell sheets have been successfully lifted up and transferred on other surfaces [107–114].

As cell sheet manipulation, two techniques have been performed according to cell types and objects. One is to manipulate cell sheets directly with forceps or pipetting after the sheets are completely harvested resulting in proportionally shrunk and thicker constructs due to active cytoskeletal reorganization. As indicated by synchronized beating of shrunk cardiomyocyte sheets, cell-to-cell connections are preserved after this procedure [114]. The other is to use support membranes, including a hydrophilically modified poly(vinylidene difluoride) (PVDF) membrane, for preserving cell sheet morphology without any shrinkage. Before cell sheets release, support membranes are placed over the confluent cells. Then the cell sheets physically attached to the support membranes are harvested from PIPA Amgrafted surfaces below 32°C and transferred onto other surfaces. Incubation at 37°C causes reattachment of the cell sheets to new surfaces via remaining adhesive proteins. Finally, only the support membranes are removed. The latter technique has realized the cell sheet manipulation preserving their structure and function.

These cell sheet manipulation techniques without using any biodegradable scaffolds have been applied to tissue engineering in three types of contexts. First is transplanting single-cell sheet for skin and cornea reconstruction. Advantages of skin epithelial cell sheets harvested by using PIPAAm-grafted surfaces have been confirmed in comparison with those harvested by dispase treatments. E-cadherin, which is an essential protein for skin cell-to-cell junctions, and laminin 5, which is a major component of epithelial basement membranes, were retained in skin cell sheets released from PIPAAm-grafted surfaces [109]. It should attenuate the risk of infection after artificial skin transplantation. Second is to layer same cell sheets for reconstructing homogeneous tissues including myocardium. Third is to layer several types of cell sheets for fabricating laminar structures including liver, kidney and vascular. Layered co-culture comprising a hepatocyte sheet and an endothelial cell sheet has revealed the differentiated cell shape and extensive albumin expression of hepatocytes, which have never been seen in hepatocyte mono-culture [114]. Cell sheet engineering technology described above has been applied into regeneration of myocardium [113], cornea [115] and hepatic tissue [116].

References

1. Ikeda S, Arai F, Fukuda T, Negoro M, Irie K, An *in vitro* patient-specific biological model of the cerebral artery reproduced with a membranous configuration for simulating endovascular intervention, *J Robotics Mechatron*, 2005, 17(3): 327–333.
2. Ikeda S, Fukuda T, Arai F, *et al.*, Patient-specific neurovascular simulator for evaluating the performance of medical robots and instruments, in *Proceedings of the IEEE-ICRA*, 2006, 625–630.
3. Ikeda S, Arai F, Fukuda T, Negoro M, Irie K, Patient specific 3D real time vascular stress analysis, in *Proceedings of the 14th Congress of the Japanese Society of Computer Assisted Surgery*, 2005, 77–78.
4. Uchiyama T, Fukuda T, Arai F, Negoro M, Ikeda S, Numerical simulation of blood flow in a cerebral aneurysm, in *Proceedings of the 6th KSME-JSME Thermal and Fluids Engineering Conference*, 2005.
5. Horie T, Morimoto T, Konishi S, Medical MEMS with switcheable sucker for local diagnosis and treatment, in *Proceedings of the 15th Congress of the Japanese Society of Computer Assisted Surgery*, 2006, 131–132.
6. Pitawatchara P, Warisawa S, Mitsuishi M, Gain Scheduling for force feedback control in the minimal invasive telesurgical system, in *Proceedings of the 14th Congress of the Japanese Society of Computer Assisted Surgery*, 2005, 21: 65–66.
7. Mitsuishi M, Warisawa S, Sugita N, *et al.*, Development of a bone cutting robot for minimally invasive arthroplasty, in *Proceedings of the 14th Congress of the Japanese Society of Computer Assisted Surgery*, 2005, 121–122.
8. Wada K, Shibata T, Asada T, Musha T, Robot therapy for prevention of dementia at home, *J Robotics Mechatron*, 2007, 19(6): 691–697.
9. Sasaki M, Kimura T, Matsuo K, *et al.*, Simulator for optimal wheelchair design, *J Robotics Mechatron*, 2008, 20(6): 854–862.
10. Makinishi W, Matunaga T, Esashi M, Haga Y, Active bending endoscope using memory shape alloy, in *Proceedings of the 14th Congress of the Japanese Society of Computer Assisted Surgery*, 2005, 59–60.
11. Suzuki S, Suzuki N, Eto K, Hattori A, Yanaga K, Lapararoscopic surgery simulation for colorectal surgery using soft tissue model with anisotropic characteristics, in *Proceedings of the 14th Congress of the Japanese Society of Computer Assisted Surgery*, 2005, 115–116.
12. Chung A, Edwards P, Deligianni F, Yang G, Freehand co-calibration of optical and electromagnetic trackers for navigated bronchoscopy, in

Proceedings of the Second International Workshop on MIAR, 2004, 320–328.

13. Umeda T, Oura M, Toyoda K, Kawamura K, Kobayashi Y, Okamoto J, Surgical robot system with heartbeat compensation for CABG, in *Proceedings of the 15th Congress of the Japanese Society of Computer Assisted Surgery*, 2006, 149–150.
14. Watanabe Y, Maeda M, Yaji N, *et al.*, Application of pressure driven flexible microactuator for retinal pigment epithelium transplantation, in *Proceedings of the 15th Congress of the Japanese Society of Computer Assisted Surgery*, 2006, 123–124.
15. Won-Ho S, Seong-Yong K, Dong-Soo K, Design of a dexterous and compact laparoscopic assistant robot, in *Proceedings of the SICE-ICASE International Joint Conference*, 2006, 233–237.
16. Ikeda S, Tercero C, Fukuda T, *et al.*, Patient-specific IVR endovascular simulator with augmented reality for medical training and robot evaluation, *J Robotics Mechatron*, 2008, 20(3): 441–448.
17. Lichtenstein S, Cheung A, Ye J, *et al.*, Transapical transcatheter aortic valve implantation in humans, *Circulation*, 2006, 114: 591–596.
18. Hadley, G. D., The gastro-camera, *BMJ*, 1965, 2: 1209–1212.
19. http://www.olympus-global.com/en/corc/history/story/endo/gastro/ Accessed June 2011.
20. Wood RE, Gauderer MWL, Flexible fiberoptic bronchoscopy in the management of tracheobronchial foreign bodies in children: the value of a combined approach with open tube bronchoscopy, *J Pediatr Surg*, 1984, 19(6): 693–698.
21. Pereira W, Jr, Kovnat DM, Snider GL, A prospective cooperative study of complications following flexible fiberoptic bronchoscopy, *CHEST*, 1978 (June), 73(6): 813–816.
22. Dorf R, Bishop R, *Modern Control Systems*, Prentice Hall, 9th Ed., Chapter 1.
23. http://www.ifac-control.org/areas/talist Accessed on June 2011.
24. Moorthy K, Munz Y, Sarker S, Darzi A, Objective assessment of technical skills in surgery, *BMJ*, 2003, 327: 1032–1037.
25. Doyle JD, Webber EM, Sidhu RS, A universal global rating scale for the evaluation of technical skills in the operating room, *Am J Surg*, 2007, 193(5): 551–555.
26. Chaer R, DeRubertis B, Lin S, Bush H, Karwowski J, Birk D, Morrisey N, Faries P, McKinsey J, Craig K, Simulation improves resident performance in catheter-based intervention, *Ann Surg*, , 2006, 244(3): 343–352.

27. Kunkler K, The role of Medical simulation: an overview, *Int J Med Robotics Comput Assist Surg*, 2006, 2: 203–210.

28. Willems MC, Van der Vilet JA, Williams V, Kool LJ, Bergqvist D, Blankensteijn JD, Assessing endovascular skills using the simulator for testing and rating endovascular skills (STRESS) machine, *Eur J Vasc Endovasc Surg*, 2009, 37: 431–436.

29. Gailloud P, Pray JR, Muster M, Piotin M, Fasel JHD, Rufenacht DA, An *in vitro* anatomic model of the human cerebral arteries with saccular arterial aneurysms, *Surg Radiol Anat*, 1997, 19: 119–121.

30. Yamashita H, Zuo S, Masamune K, Liao H, Dohi T, Nonmagnetic rigid and flexible outer sheath with pneumatic interlocking mechanism for minimally invasive surgical approach, MICCAI. 2009, Part I, LNCS 2009, 5761: 418–425.

31. Furukawa Y, Kurazume R, Iwashita Y, *et al.*, Development of 3D superposed display system for endoscopic ultrasound images, in *Proceedings of the JSME Conference on Robotics and Mechatronics*, 2011, 2P2–H01.

32. Lin Z, Uemura M, Zecca M, *et al.*, Waseda bioinstrumentation system WB-3 as a wearable tool for objective laparoscopic skill evaluation, *IEEE-ICRA*, 2011, 5737–5742.

33. Guo S, Fukuda T, Kosuge K, Arai F, Oguro K, Negoro M, Micro catheter system with active guide wire — structure, experimental results and characteristic evaluation of active guide wire catheter using ICPF actuator, *Fifth International Symposium on Micro Machine and Human Science (MHS'94)*, 191–197, Nagoya, Japan, October 2–4, 1994.

34. Guo S, Fukuda T, Kosuge K, Arai F, Negoro M, A study on active catheter system (1st report, Structure, experimental results and characteristic evaluation of multiunits using SMA active catheter with multi D.O.F.), *Trans Japan Soc Mech Eng*, 1994, 60-573C: 1727–1734. In Japanese.

35. Rucker D, Jones B, Webster R, III, A model for concentric tube continuum robots under applied wrenches, in *Proceedings of the IEEE-ICRA*, Anchorage, 2010, 1047–1052.

36. Bailly Y, Amirat Y, Modeling and control of a hybrid continuum active catheter for aortic aneurysm treatment, in *Proceedings of the IEEE-ICRA*, 2005, 924–929.

37. Chen Y, Chang J, Greenlee A, Cheung K, Slocum A, Gupta R, Multi-turn, tension-stiffening catheter navigation system, in *Proceeding of the IEEE-ICRA*, Anchorage, 2010, 5570–5575.

38. Ikuta K, Ichikawa H, Suzuki K, Multi-degree of freedom hydraulic pressure driven safety active catheter, in *Proceedings of the IEEE-ICRA*, Orlando, 2006, 4161–4166.

39. Tanimoto M, Arai F, Fukuda T, *et al.*, Micro force sensor for intravascular neurosurgery, in *Proceedings of the IEEE-ICRA*, 1999, 1561–1566.

40. Polygerinos P, Puangmali P, Schaeffter T, Razavi R, Seneviratne L, Althoefer K. Novel miniature MRI-compatible fiber-optic force sensor for cardiac catheterization procedures, in *Proceedings of the IEEE-ICRA*, Anchorage, 2010, 2598–2603.

41. Tercero C, Okada Y, Ikeda S, Fukuda T, Sekiyama K, Negoro M, Takahashi I, Numerical evaluation method for catheter prototypes using photo-elastic stress analysis on patient-specific vascular model, *Int J Med Robotics Comput Assist Surg*, 2007, 3(4): 349–354 .

42. Nakada K, Nakamoto N, Sato Y, *et al.*, A rapid method for magnetic tracker calibration using a magneto-optic hybrid tracker, in *Proceedings of the MICCAI*, Lecture Notes in Computer Sciences, 2003, 285–293.

43. Hananouchi T, Sugano N, Nishii T, *et al.*, Effect of optical localizer grade on accuracy of registration of HIP arthroplasty with CT based navigation system, in *Proceedings of the 14th Congress of the Japanese Society of Computer Assisted Surgery*, 2005, 147–148.

44. Kindratenko V, Bennett A, A survey of electromagnetic position tracker calibration techniques, in *Proceedings of the Eurographics Workshop*, 2000, 13–22.

45. Livingstone M, State A, Magnetic tracker for improved augmented reality registration, *PRESENCE: Teleoperators Virtual Environ*, 1997, 6(5): 14–20.

46. Ueta T, Nakano T, Ida Y, Sugita N, Mitsuishi M, Tamaki Y, Comparison of robot-assisted and manual retinal vessel microcannulation in an animal model, *Br J Ophthalmol* , 2011, 95: 731–734.

47. Kobayashi Y, Onishi A, Watanabe H, *et al.*, Development of an integrated needle insertion system with image guidance and deformation simulation, *Comput Med Imaging Graph*, 2010, 34: 9–18.

48. Hatano M, Kobayashi Y, Hamano R, *et al.*, *In vitro* and *in vivo* validation of robotic palpation-based needle insertion method for breast tumor treatment, in *Proceedings of the IEEE-ICRA*, 2011, 392–397.

49. Liao H, Inomata T, Sakuma I, Dohi T, 3-D augmented reality for MRI-guided surgery using integral videography autostereoscopic image overlay, *IEEE Trans Biomed Eng*, 2010, 57(6): 1476–1486.

50. Marescaux J, Leroy J, Gagner M, *et al.*, Transatlantic robot-assisted telesurgery, *Nature*, 2001, 413: 379–380.

51. Guthart GS, Salisbury JK, Jr, The Intuitive™ telesurgery system: overview and application, in *Proceedings of the IEEE-ICRA*, 2000, 618.

52. Tanimoto M, Arai F, Fukuda T, *et al.* Augmentation of safety in teleoperation system for intravascular neurosurgery, in *Proceedings of the IEEE-ICRA*, 1998, 2890–2895.

53. Arai F, Fujimura R, Fukuda T, Negoro M, New catheter driving method using linear stepping mechanism for intravascular neurosurgery, in *Proceedings of the IEEE-ICRA*, 2002, 2944–2949.

54. Ikeda S, Fukuda T, *et al.* A new remote control device of catheter using linear stepping mechanism for intravascular neurosurgery, in *Proceedings of the 20th Annual Conference of Robotics Society of Japan.*, 2003, 2F17.

55. Jayender J, Patel R, Nikumb S, Robot-assisted catheter insertion using hybrid impedance control, in *Proceedings of the IEEE-ICRA*, 2006, 607–612.

56. Govindarajan S, Thenkurussi K, Xinyan L, Design and fabrication of a robotic mechanism for remote steering and positioning of interventional devices, *Int J Med Robot Comput Assist Surg*, 2010, 6: 160–170.

57. Menciassi A, Dario P, Single access robots for surgery, Workshop on Mechanisms for Surgical Robotics, IEEE-ICRA 2011.

58. Palagi S, Pensabene V, Beccai L, Mazzolai B, Menciassi A, P. Dario. Design and development of a soft magnetically-propelled swimming microrobot, in *Proceedings of the IEEE-ICRA*, 2011, 5109–5114.

59. Gao M, Hu C, Chen Z, Zhang H, Liu S, Design and fabrication of a magnetic propulsion system for self-propelled capsule endoscope, *IEEE Trans Biomed Eng.*, 2010 , 57(12): 2891–2902.

60. Kothwala D, Raval A, Choubey A, Engineer C, Kotadia H, Paclitaxel drug delivery from cardiovascular stent, *Trends Biomater Artif Organs*, 2006, 19(2): 88–92.

61. Finkelstein A, McClean D, Kar S, *et al.*, Local drug delivery via a coronary stent with programmable release pharmacokinetics, *Circulation*, 2003, 107: 777–784.

62. Martel S, Mathieu JB, Felfoul O, *et al.*, Automatic navigation of an untethered device in the artery of a living animal using a conventional clinical magnetic resonance imaging system, *Appl Phys Lett*, 90(11): 2007, 114105-1.

63. Mathieu JB, Martel S, Steering of aggregating Magnetic microparticles using propulsion gradients coils in an MRI scanner, *Magn Resonan Med.*, 2010, 63: 1336–1345.

64. Pouponneau P, Leroux JC, Soulez G, Gaboury L, Martel S, Co-encapsulation of magnetic nanoparticles and doxorubicin into biodegradable microcarriers for deep tissue targeting by vascular MRI navigation, *Biomaterials*, 2011, 32(13): 3481–3486.

65. Nawroth T, Rusp M, May RP, Magnetic liposomes and entrapping: time-resolved neutron scattering TR-SANS and electron microscopy, *Physica B: Condensed Matter*, 2004, 350(1–3): E635–E638.

66. Peinado G, Urdiales C, Peula JM, Fernandez-Carmona M, Annicchiarico R, Sandoval F, Caltagirone C, Navigation skills based profiling for collaborative wheelchair control, in *Proceedings of the IEEE-ICRA*, 2011, 2229–2234.

67. Blank A, Okamura AM, Whitcomb LL, Task-dependent impedance improves user performance with a virtual prosthetic arm, 2 in *Proceedings of the IEEE-ICRA*, 2011, 2235–2242.

68. Ikada Y. (2001) *Saisei Ikougaku* (Kagakudoujin). In Japanese.

69. Ikada Y. (2002) *Jinkouzouki Monogatari* (Shoukabou). In Japanese.

70. Matsuda T, Biojinkoukekkan no mekanikusu to rimoderingu, *Kekkan Igaku* (Medical Review), 2004, 5: 577–586. In Japanese.

71. Matsuda T. (2005) Tissue Engineering (Nihon Igakukan) pp. 67–75. In Japanese.

72. Matsuda T. (2005) Seitairikigakujou ni kannousuru kokkakukizai no kakou gijutsu to saiseiiryou, *Saisei Iryou* (Medical Review), 4: 15–25. In Japanese.

73. Uchida T, Ikeda S, Oura H, *et al.*, Development of biodegradable scaffolds based on patient-specific arterial configuration, *J Biotechnol*, , 2008, 133: 213–218.

74. Tateishi T, Tanaka J, *Zukai Saisei Iryou Kougaku* (Kougyou Chousakai). In Japanese.

75. Khademhosseini A, Langer R, Borenstein J, Vacanti JP, Microscale technologies for tissue engineering and biology, *Proc Natl Acad Sci USA*, 2006, 103(8): 2480–2487.

76. Langer R, Vacanti JP, Tissue engineering, *Science*, 1993, 260: 920–926.

77. Vacanti CA, Langer R, Schloo B, Vacanti JP, Synthetic polymers seeded with chondrocytes provide a template for new cartilage formation, *Plast Reconst Surg*, 1991, 88: 753.

78. Kidoaki S, Kwon IK, Matsuda TJ, Structural features and mechanical properties of in situ-bonded meshes of segmented polyurethane electrospun from mixed solvents, *Biomed Mater Res B*, 2006, 76B: 219–229.

79. Kidoaki S, Kwon IK, Matsuda T, Mesoscopic spatial designs of nano- and microfiber meshes for tissue-engineering matrix and scaffold based on newly devised multilayering and mixing electrospinning techniques, *Biomaterials*, 2005, 26: 37–46.

80. Inoguchi H, Kwon IK, Inoue E, Takamizawa K, Maehara Y, Matsuda T, Mechanical responses of a compliant electrospun poly(L-lactide-co-epsilon-caprolactone) small-diameter vascular graft, *Biomaterials*, 2006, 27: 1470–1478.

81. Inoguchi H, Tanaka T, Maehara Y, Matsuda T, The effect of gradually graded shear stress on the morphological integrity of a huvec-seeded compliant small-diameter vascular graft, *Biomaterials*, 2007, 28: 486–495.

82. Lee SB, Kim YH, Chong MS, Hong SH, Lee YM, Study of gelatin-containing artificial skin V: fabrication of gelatin scaffolds using a salt-leaching method, *Biomaterials*, 2005, 26: 1961–1968.

83. Murphy WL, Dennis RG, Kileny JL, Mooney DJ, Salt fusion: an approach to improve pore interconnectivity within tissue engineering scaffolds, *Tissue Eng*, 2002, 8: 43–52.

84. Gao J, Crapo PM, Wang YD, Macroporous elastomeric scaffolds with extensive micropores for soft tissue engineering, *Tissue Eng*, 2006, 12: 917–925.

85. Heykants RGJC, Van Tienen TG, De Groot JH, *et al.*, Preparation of a polyurethane scaffold for tissue engineering made by a combination of salt leaching and freeze-drymg of dioxane, *J Mater Sci*, 2006, 41: 2423–2428.

86. Wu LB, Jing DY, Ding JD, A "room-temperature" injection molding/particulate leaching approach for fabrication of biodegradable three-dimensional porous scaffolds, *Biomaterials*, 2006, 27: 185–191.

87. Lee M, Dunn JCY, Wu BM, *et al.*, Study of gelatin-containing artificial skin V: fabrication of gelatin scaffolds using a salt-leaching method *Biomaterials*, 2005, 26: 1961–1968.

88. G.B. Wei, and P.X. Ma, Macroporous and nanofibrous polymer scaffolds and polymer/bone-like apatite composite scaffolds generated by sugar spheres, *J Biomed Mater Res. A*, 2006, 78A: 306–315.

89. Chen GP, Ushida T, Tateishi T, Preparation of poly(L-lactic acid) and poly(DL-lactic-co-glycolic acid) foams by use of ice microparticulates, *Biomaterials*, 2001, 22: 2563–2567.

90. Ma PX, Choi JW, Biodegradable polymer scaffolds with well-defined interconnected spherical pore network, *Tissue Eng*, 2001, 7: 23–33.

91. Therrihault D, White SR, Lewis JA, Chaotic mixing in three-dimensional microvascular networks fabricated by direct-write assembly, *Nat Mater*, 2003, 2: 265–271.

92. Therrihault D, Shepherd RF, White SR, Lewis JA, Fugitive ink for direct-write assembly of three-dimensional microvascular networks, *Adv Mater*, 2005, 17: 395–399.

93. Vozzi G, Flaim C, Ahluwalia A, Bhatia S, Fabrication of PLGA scaffolds using soft lithography and microsyringe deposition, *Biomaterials*, 2003, 24: 2533–2540.

94. Sakai Y, Otsuka M, Hanada S, Nishiyama Y, Konishi Y, Yamashita A, A novel poly-L-lactic acid scaffold that possesses a macroporous structure and a branching/joining three-dimensional flow channel network: its fabrication and application to perfusion culture of human hepatoma Hep G2 cells , *Mater Sci Eng C*, 2004, 24: 379–386.

95. Tan W, Desai TA, Layer-by-layer microfluidics for biomimetic three-dimensional structures, *Biomaterials*, 2004, 25: 1355–1364.

96. Mironov V, Boland T, Trusk T, Forgacs G, Markwald RR, Organ printing: computer-aided jet-based 3D tissue engineering, *Trends Biotechnol*, 2003, 21: 157–161.

97. Whitesides GM, Ostuni E, Takayama S, Jiang XY, Ingber DE, Soft lithography in biology and biochemistry, *Annu Rev Biomed Eng*, 2001, 3: 335–373.

98. Ng JMK, Gitlin I, Stroock AD, Whitesides GM, Components for integrated poly(dimethylsiloxane) microfluidic systems, *Electrophoresis*, 2002, 23: 3461–3473.

99. Borenstein JT, Terai H, King KR, Weinberg EJ, Kaazempur-Mofrad MR, Vacanti JP, Microfabrication technology for vascularized tissue engineering, *Biomed Microdev*, 2002, 4: 167–175.

100. King KR, Wang CCJ, Kaazempur-Mofrad MR, Vacanti JP, Borenstein JT, *Adv Mater*, 2004, 16: 2007–2012.

101. Wang YD, Ameer GA, Sheppard BJ, Langer R, A tough biodegradable elastomer, *Nat Biotechnol*, 2002, 20: 602–606.

102. Fidkowski C, Kaazempur-Mofrad MR, Borenstein J, Vacanti JP, Langer R, and Wang YD, Endothelialized microvasculature based on a biodegradable elastomer, *Tissue Eng*, 2005, 11: 302–309.

103. Bettinger CJ, Weinberg EJ, Kulig KM, *et al.*, *Adv Mater*, 2006, 18: 165–169.

104. Ikeuchi M, Ikuta K, *Seitai Ikougaku*, 2005, 43: 646–652.

105. Shimizu T, Yamato M, Kikuchi A, Okano T, Cell sheet engineering for myocardial tissue reconstruction, *Biomaterials*, 2003, 24: 2309–2316.

106. Kushida A, Yamato M, Konno C, Kikuchi A, Sakurai Y, Okano T, Temperature-responsive culture dishes allow nonenzymatic harvest of differentiated Madin-Darby canine kidney (MDCK) cell sheets, *J Biomed Mater Res*, 2000, 51: 216–223.

107. Kikuchi A, Okuhara M, Karikusa F, Sakurai Y, Okano T, Two-dimensional manipulation of confluently cultured vascular endothelial cells using temperature-responsive poly(N-isopropylacrylamide)-grafted surfaces, *J Biomater Sci Polym Ed*, 1998, 9: 1331–1348.

108. Hirose M, Kwon OH, Yamato M, Kikuchi A, Okano T, Creation of designed shape cell sheets that are noninvasively harvested and moved onto another surface, *Biomacromolecules*, 2000, 1: 377–381.

109. Yamato M, Utsumi M, Kushida A, Konno C, Kikuchi A, Okano T, Thermo-responsive culture dishes allow the intact harvest of multilayered keratinocyte sheets without dispase by reducing temperature, *Tissue Eng*, 2001, 7: 473–480.

110. Shimizu T, Yamato M, Kikuchi A, Okano T, Two-dimensional manipulation of cardiac myocyte sheets utilizing temperature-responsive culture dishes augments the pulsatile amplitude, *Tissue Eng*, 2001, 7: 141–151.

111. Kushida A, Yamato M, Kikuchi A, Okano T, Two-dimensional manipulation of differentiated Madin-Darby anine kidney (MDCK) cell sheets: the noninvasive harvest from temperature-responsive culture dishes and transfer to other surfaces, *J Biomed Mater Res*, 2000, 54: 37–46.

112. Shimizu T, Yamato M, Akutsu T, *et al.*, Electrically communicating three-dimensional cardiac tissue mimic fabricated by layered cultured cardiomyocyte sheets, *J Biomed Mater Res*, 2002, 60: 110–117.

113. Shimizu T, Yamato M, Isoi Y, *et al.*, Fabrication of pulsatile cardiac tissue grafts using a novel 3-dimensional cell sheet manipulation technique and temperature-responsive cell culture surfaces, *Circ Res*, 2002, 90: e40–e48.

114. Harimoto M, Yamato M, Hirose M, *et al.*, Novel approach for achieving double-layered cell sheets co-culture: overlaying endothelial cell sheets onto monolayer hepatocytes utilizing temperature-responsive culture dishes, *J Biomed Mater Res*, 2002, 62: 464–470.

115. Nishida K, Yamato M, Hayashida Y, *et al.*, Corneal reconstruction with tissue-engineered cell sheets composed of autologous oral mucosal epithelium, *New Eng J Med*, 2004, 351: 1187–1196.

116. Ohashi K, Yokoyama T, Yamato M, *et al.*, Engineering functional two- and three-dimensional liver systems in vivo using hepatic tissue sheets, *Nat Med*, 2007, 13: 880–885 (2007).

117. Williams D, On the mechanisms of biocompatibility, *Biomaterials*, 2008, 29(20): 2941–2953.

Chapter 2

Endovascular Treatments for Brain Attack Introduction

Interventional neuroradiological treatments were practiced for the first time in 1974 when Sebinenko blocked the blood flow of a cerebral artery using a balloon catheter [1]. This method was then expanded to aneurysm treatment but was not adopted worldwide until sufficient tools were developed for that purpose. The use of digital subtraction angiography (DSA) spread quickly worldwide after that the first stable aneurysm embolization was presented in 1991 by Guglielmi *et al.* The embolization was achieved by using electricity for delivering a coil inside the aneurysm [2].

Generalized use of 3D DSA is the consequence of the remarkable improvements in accuracy done in recent years. Such technology increased safety during endovascular treatments. Therefore, interventional radiology used is not only for cerebral aneurysm treatments but also for various endovascular tools research and development, such as the creation of new treatment techniques for stenosis in carotid arteries, acute strokes and cerebral arteriovenous malformation. In this chapter, we will explain conventional and endovascular treatments for cerebrovascular diseases such as cerebral aneurysms, carotid artery stenosis, acute stroke and cerebral hemorrhage. Additionally, we will present how recently, from other minimally invasive treatments, the use of endoscopes has been adopted for neurosurgery.

Microsurgery: Advances, Simulations, and Applications
Edited by Toshio Fukuda and Carlos Tercero

ISBN 978-981-4364-69-0 (Hardcover), 978-981-4364-70-6 (eBook)
www.panstanford.com

2.1 Cerebral Aneurysms

2.1.1 Ruptured Cerebral Aneurysm

The principal cause of subarachnoid hemorrhage comes from rupture of cerebral aneurysms. It has been reported that this disease frequency varies from 1.04/100,000 to 20/100,000 persons and there are big differences on this frequency between regions and countries [3–5]. Even if the occurrence probability of aneurysm rupture is extremely low, its death probability is high compared with other types of cerebral hemorrhage and varies from 10% to 67% [6–8].

This high disease rate is result of different complications such as big size of initial rupture, recurrent hemorrhage and hydrocephalus, delayed cerebral vasospasm, and especially the prognosis of recurrent hemorrhage due to deterioration [9]. The medical treatments for that purpose are aneurysm clipping by microsurgery and embolization through an endovascular treatment, for both of them are needed to develop a treatment to prevent recurrent hemorrhage.

2.1.2 Unruptured Cerebral Aneurysm

The prognosis failure of subarachnoid hemorrhage motivated the development magnetic resonance angiography (MRA) and computer tomography angiography (CTA). With this development, intracranial blood vessels' depiction became easier. Moreover, the treatment of unruptured aneurysms became possible. However, deaths were reported as a result of the prognostic symptoms of the treatment, and complications were reported for 13.5% of the cases [10].

Aneurysms are classified into the following:

- *Group 1*: the unruptured aneurysms without medical records of subarachnoid hemorrhage.
- *Group 2*: the ruptured aneurysms.

Those groups are divided in two groups, *A* and *P*:

- *Group A*: aneurysms with less than 7 mm localized at internal carotid artery (ICA), anterior communicating artery (A-comA) and middle cerebral artery (MCA).
- *Group P*: aneurysms located between ICA and posterior communicating artery (P-com A), and in vertebrobasilar artery.

The rupture probability in the first 5 years is

- *Group 1A*: 0%.
- *Group 1P*: 2.5% with an increasing rate of 0.5% per year.
- *Group 2A*: 1.5% with an increasing rate of 0.3% per year.
- *Group 2P*: 3.4% with an increasing rate of 0.7% per year.

The difference between groups does not change as the size of the aneurysm increases. For sizes of 7–12 mm, the rupture rate in five years for *Group A* is 2.6% increasing at 0.5% per year; for *Group P* it is 14.5% increasing at 2.9% per year; for sizes of 13–24 mm *Group A* is 14.5% increasing at 2.9% per year and *Group P* at 18.4% increasing at 3.7% per year; and for 25 mm and more *Group A* has 40% at 8% and Group *P* has 50% at 10% [10]. Rupture risk factor relates with the size of the aneurysm, high blood pressure, smoking, multiple aneurysms, rupture medical records, family records, and aspect ratio. Genetic reproduction and inherent diseases increase that factor too [12–22].

2.1.3 Treatments

2.1.3.1 Clipping

Clipping surgery is achieved using a microscope and is the most conventional treatment for cerebral aneurysms. The patient receives general anesthesia, then the skin is cut and the skull open, after cutting the dura mater and using a microscope for surgery the subarachnoid is separated. Then the aneurysm is delicately exposed to place a clip on its neck.

Figure 2.1 shows the intra-operative appearance of an unruptured aneurysm and the DSA image of the cerebral artery before treatment. A large-size aneurysm is visible at MCA bifurcation and marked with an asterisk. In the intra-operative photograph, the large-size aneurysm is exposed and marked with an asterisk. Also an intra-operative photograph after the clipping is shown as well as the corresponding DSA image of the cerebral artery where the aneurysm has disappeared. Figure 2.2 shows 3D-CTA images of a ruptured aneurysm, the arrow points to a bleb that grew into a right middle cerebral artery (Rt. MCA) aneurysm. Intra-operative photographs show that subarachnoid hemorrhage produces the red color in the surgical field. The arrow points at a clot in the bleeding part of the aneurysm. Post-surgery image shows the clip attached to the aneurysm.

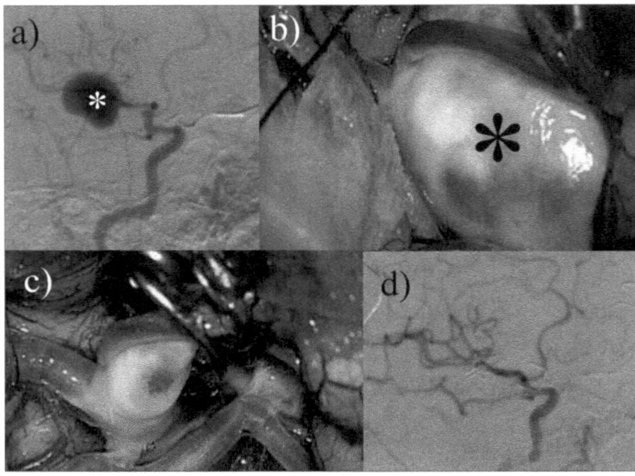

Figure 2.1 Clipping surgery for large unruptured Rt. MCA aneurysm: (a) Pre-operative digital subtraction angiography image. A large-size aneurysm is visible at Rt. MCA (*). (b) Intra-operative photograph, the aneurysm is exposed (*). (c) Intra-operative photograph, a clip was attached to the aneurysm neck. (d) DSA image of the cerebral artery after the treatment.

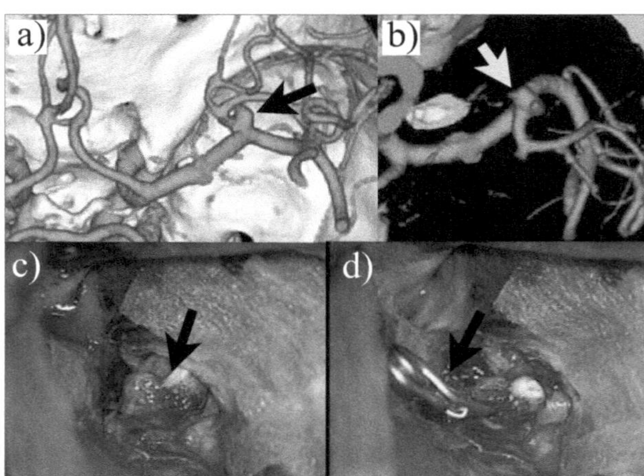

Figure 2.2 Clipping treatment for ruptured aneurysm in Rt. MCA (a,b) Pre-operative 3D-CTA images. The arrow points to a hemorrhage at the aneurysm in Rt. MCA. (c) Intra-operative photograph of a subarachnoid hemorrhage. Red appearance is given by the hemorrhage, the arrow points at a clot in the bleeding part of the aneurysm. (d) Intra-operative photograph, the clip attached to the aneurysm neck is pointed by the arrow.

2.1.3.2 Cerebral aneurysm embolization

From the development of electrically deployable coil in 1991 by Gugliemi *et al.*, the cerebral aneurysm embolization technique rapidly spread worldwide [2]. In conventional mechanically detachable coils, immediately after the detachable section of a coil leaves the micro-catheter used for the medical treatment, a mechanical lock is released and separates it from the micro-catheter. On the other hand, for electronically deployable coils, the detachment will not happen immediately after the detachable section leaves the catheter but will be safely released after it has been verified that the coil is installed in the target location. In those days, coils were made from platinum with helical shape (Fig. 2.3a) and three-dimensional shapes. A large number coil types were commercialized by different companies; for example, coil varieties with surface polymer coating may have characteristics such as enlarging its volume with moisture absorption or promotion of tissue organization (Fig. 2.3c) (by Boston Scientific Japan Co.). The development of DSA made possible to register 3D data accompanied with the development of rotational DSA, and flat panels were big progress for improving the medical treatments quality.

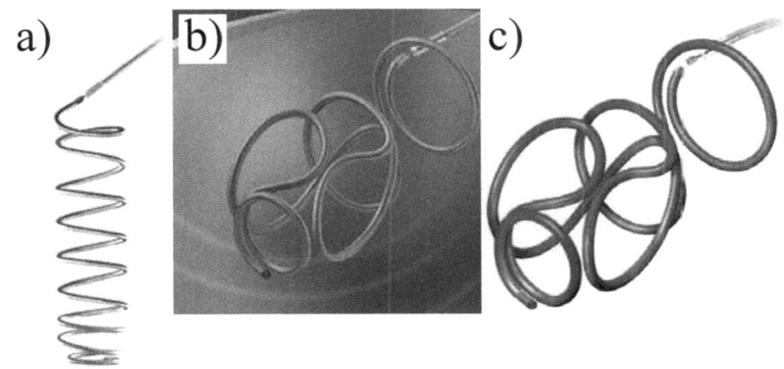

Figure 2.3 Coils (by Boston Scientific Japan Co.) (a) Helical type is the most common use type. (b) 3-D shape coils (c) Coil with macromolecule polymer coating for promoting tissue organization.

2.1.3.2.1 Medical treatments

Figure 2.4 shows a system for aneurysm embolization. The white arrow points at guiding catheter, the black one at a micro-catheter and the yellow to a balloon catheter. From those catheters, the micro-catheter (arrow head) is used for leading the system to the target location.

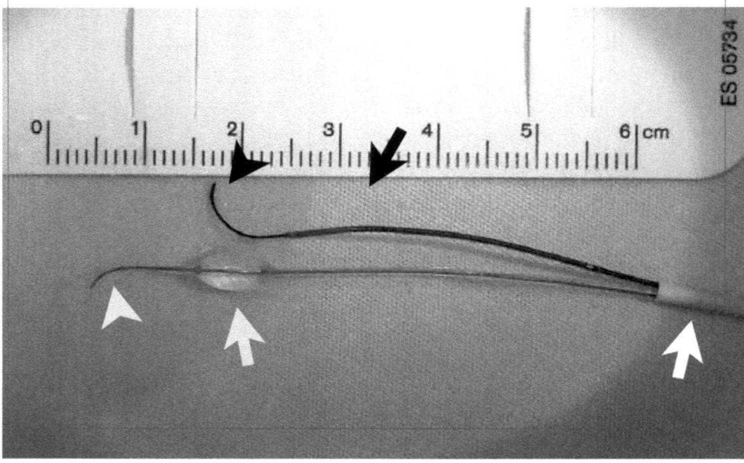

Figure 2.4 Aneurysm embolization system: guiding catheter (white arrow), micro-guide wire (black arrow head) for leading a micro-catheter (black arrow), micro-guide wire (yellow arrow head) for leading balloon catheter (yellow arrow).

The most basic treatment method for embolization using coils is to place a micro-catheter inside the aneurysm and deliver through it the coil for filling the aneurysm (Figs. 2.5 and 2.6). However, such treatment was less successful for wide-neck and large-size aneurysms [23]. The most conventional treatment for wide-neck aneurysms is the balloon-assisted technique (Figs. 2.7 and 2.8). For some cases it is difficult to use this technique; those treatments are completed assisting the coil deployment with a stent (stent assist coil) (Fig. 2.9a,b [by Johnson and Johnson]). In recent years aneurysm treatment using tight stent mesh became possible, such as pipeline and silk-stent (http://www.balt.fr/) (Fig.2.9c). That way, minimally invasive treatment results and adaptability have improved by the development of many instruments.

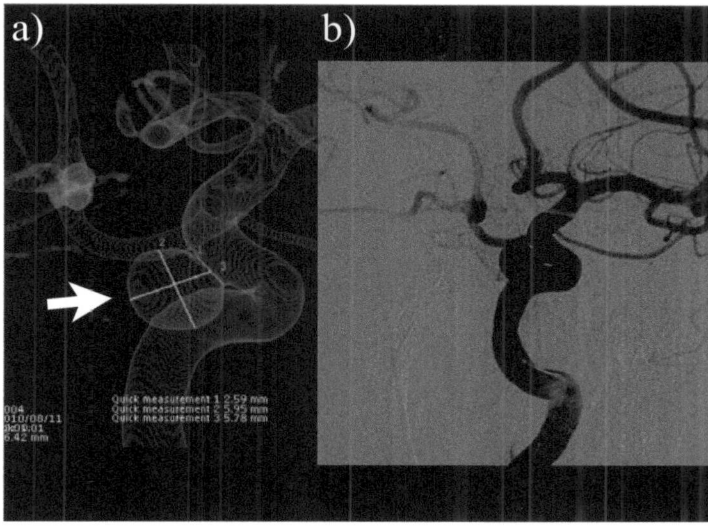

Figure 2.5 Unruptured aneurysm in left internal carotid artery (Lt. ICA): (a) 3D-DSA image of aneurysm with comparatively narrow neck in the Lt. ICA (arrow). (b) 3D-DSA and DSA registered from the same angle.

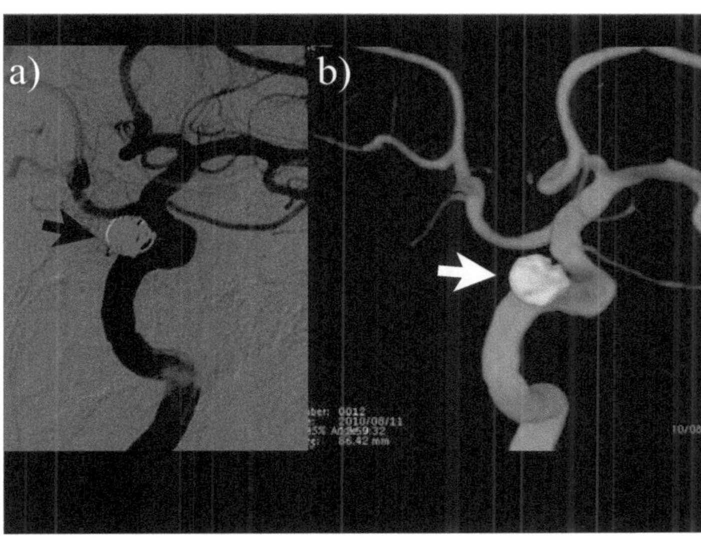

Figure 2.6 Unruptured aneurysm in Lt. ICA after embolization with coil: (a) DSA image, (b) 3DSA image. Embolized aneurysm is pointed with an arrow in both images.

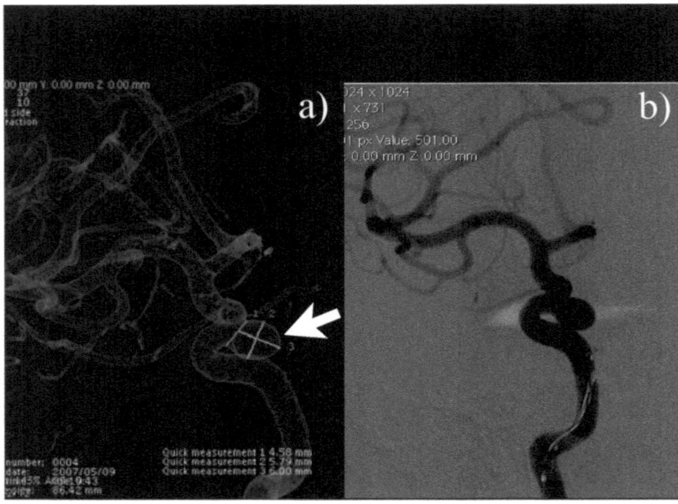

Figure 2.7 Unruptured aneurysm in Rt. ICA: (a) 3D-DSA image, aneurysm in with comparatively wide neck in the Rt. ICA artery (arrow). (b) 3D-DSA and DSA registered from the same angle.

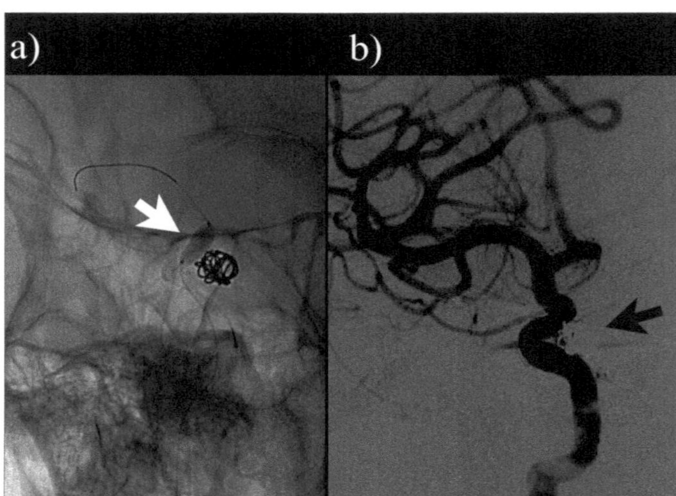

Figure 2.8 Balloon-assisted coil embolization of aneurysm in Rt. ICA: (a) Intra-operative simple x-ray image. The coil deployment within the aneurysm is protected using a balloon outside the aneurysm (arrow). (b) The coil remains stable inside the aneurysm (arrow).

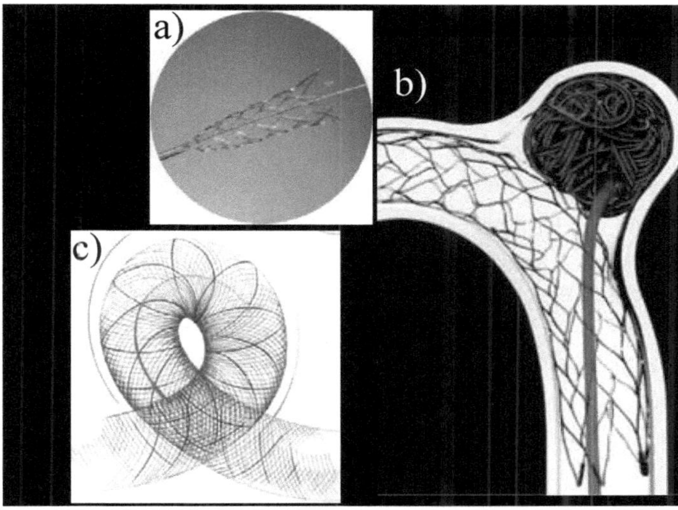

Figure 2.9 Stent for cerebral aneurysm. (a) Self-expandable-type stent (by Johnson and Johnson Co.) affixed to the neck of a wide-neck-type aneurysm. (b) Stent deployment for aneurysm coil embolization in a wide-neck-type aneurysm model (by Johnson and Johnson Co.) (c) The blood flow inside the aneurysm is restrained using a tight mesh stent placed in the blood vessel near the aneurysm location, promoting the coagulation of blood inside the aneurysm. This treatment is applied where aneurysm size makes difficult the use of conventional treatments (http://www.balt.fr/).

2.2 Carotid Artery Stenosis

The region located after carotid artery bifurcation has the tendency to develop atherosclerosis. About the half of strokes and transient ischemic attack (TIA) caused by transient insufficiency of cerebral blood flow are caused by stenosis as a consequence of atheroma growth in carotid artery. A total obstruction of the blood vessel by the stenosis is reported in 10% of the cases [24–26]. Noninvasive examination such as MRA, 3D-CTA and carotid artery ecograph enabled easy and comparative diagnosis. Consequently it is necessary to establish routine examination of carotid arteries in the neck of patients suffering from strokes and spasms caused by transient insufficiency of cerebral blood flow.

2.2.1 Treatment

The treatment for those patients includes prescription of anti-platelets agents; additionally it is needed to prescribe synthetic internal medicine treatments such as high blood pressure, anti-hyperlipidemia and diabetes control medicaments. However, for blood vessels with narrowing ratio above 70%, surgery complications occur in 6% of cases. The addition of carotid artery endarterectomy (CEA) to the best internal medicine treatments satisfactory improves prognosis [27]. Even if carotid artery stenosis lacks symptoms, surgical complications occur in less than 3% of the cases. For 60% of cases, the prognosis of treatments relying on CEA is better than for internal medicine treatments [28].

Similar to CEA, another treatment for carotid artery stenosis is the carotid artery stenting (CAS). The treatment with CEA has to adapt to difficult cases (acute lesion) and treatment failure factors (age, cardiac infarcts precedents); for those cases, it was verified that the use of protection devices improves the treatment effectiveness and safety [29–30]. Within the CAS study, CAS superiority over CEA within the group with low CEA treatment failure factor has not been verified yet [30–32].

2.2.2 Carotid Artery Endarterectomy

The most common treatment for carotid artery stenosis is surgical treatment. Under general anesthesia, the skin of the patient neck is cut. After peeling off the sternocleidomastoid anterior border and internal jugular vein, the common carotid artery, internal jugular vein and external carotid artery are exposed. After interrupting the blood flow in those arteries, an incision is made at the region where the stenosis is present, opening from the common carotid artery to the internal carotid artery. Then the peeling of the plaque thickening in the internal membrane starts. In some cases, the ablation of the diseased artery is needed. To reduce the time for which the blood flow is interrupted during the surgery, an internal shunt is used (Fig 2.10).

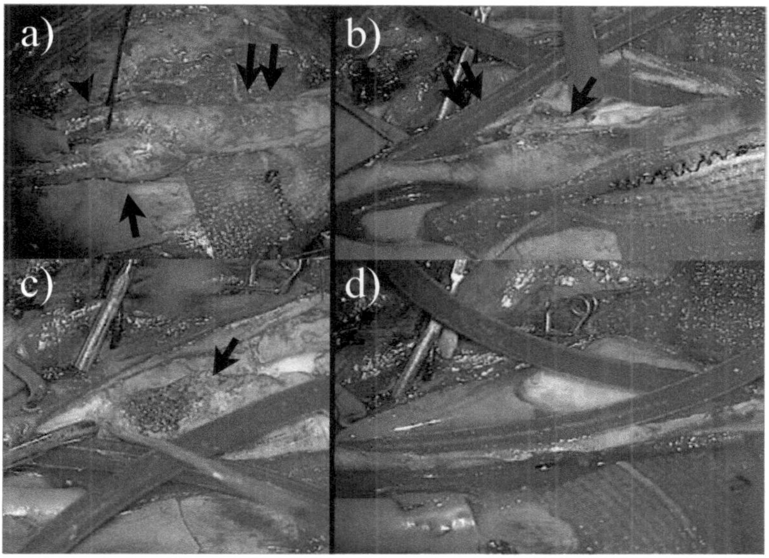

Figure 2.10 Carotid artery endarterectomy intra-operative photogram: (a) Exposure of common carotid artery (double arrow), internal carotid artery (arrow), and external carotid artery (arrow head). (b) Blood vessel after clamping, incision from common carotid artery to internal carotid artery is shown with an arrow and the internal shunt attached to them with a double arrow. (c) Peeling off the plaque (arrow) from the internal membrane. (d) The blood vessel lumen smoothening after the peeling and the recovery of sufficient lumen are confirmed.

2.2.3 Carotid Artery Stenting

In the upstream section of the blood vessel where the stenosis is located, a protection device such as a filter or a balloon is deployed (Fig 2.11a,b). After, with the assistance of the inflation of a small balloon, a self-expandable stent is deployed covering all the plaque from internal carotid artery to the common carotid artery (by Boston Scientific Japan Co. and Johnson and Johnson Co.). Finally after a large-size balloon is inflated and the protection device is retrieved (Figs. 2.12 and 2.13).

Figure 2.11 Sorts of carotid artery stents and protection devices: (a) Filter wire EZ (arrow) protection device and carotid wall stent carotid artery stent (double arrow) by Boston Scientific Japan Co.). (b) Angioguard protection device by Johnson and Johnson Co. (c) Precise carotid artery stent by Johnson and Johnson Co.

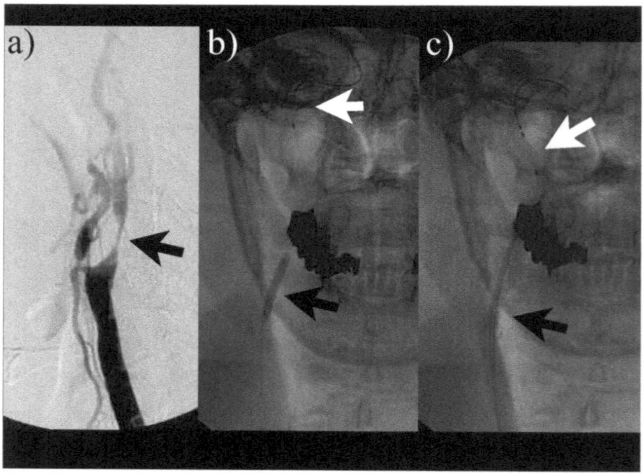

Figure 2.12 Right carotid artery stenosis: (a) Severe stenosis of right carotid artery (arrow). (b) Small-size balloon inflates from the front (black arrow). The Angioguard protection device is deployed upstream (white arrow). (c) Catheter delivering the stent is inserted within the blood vessel with stenosis (black arrow).

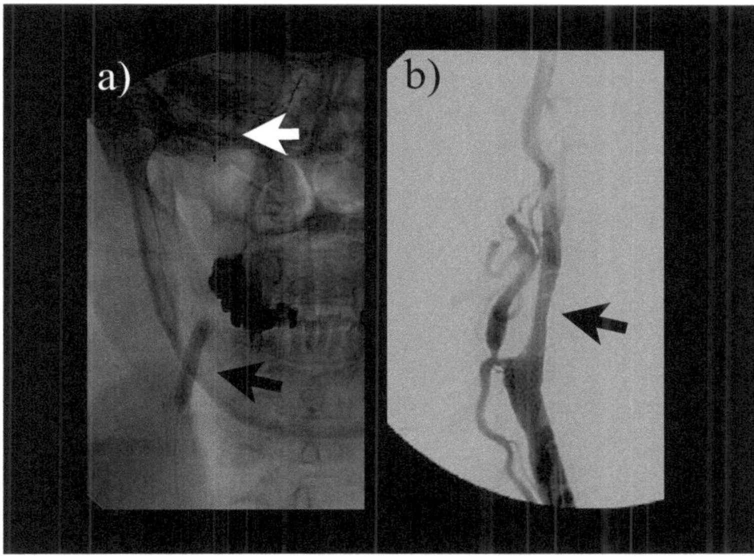

Figure 2.13 Right carotid artery stenosis. (a) After the stent deployment, a large-size balloon is inflated (arrow). The Angioguard protection device is deployed upstream (white arrow). (b) Blood vessel narrowing reduction after the treatment (arrow).

2.3 Cerebral Infarction

Within brain attacks, cerebral infarction is the most frequent type of diseases. There are divergences about the causes and the disease itself, and also there are different classifications for this kind of diseases. In this section, we will present the comparatively used classifications TOAST (Trial of ORG 10172 in Acute Stroke Treatment) and NINDS-CVD-3(1990) [35].

2.3.1 Cerebral Thrombosis

Atherosclerosis occurs in most cases in aorta and secondarily in other arteries' intima. This lesion originates from the multiplication of cells and lipid accumulation [36]. The identified causes of atherosclerosis are mechanical stimulation produced by high blood pressure, hyperglycemia, low-density lipoprotein (LDL), free radicals, toxins, inflammatory material from viral infections and smoking [37]. Therefore strokes originated by atherothrombosis, thrombus composed principally of platelets detached from plaque accumulated in neck and brain principal arteries. Anti-platelets agents such as aspirin and clopidogrel are used as general therapy.

2.3.2 Cardiogenic Cerebral Embolization

Cerebral arteries embolism occurs as a consequence of fibrin clots produced by coagulation activation triggered by blood accumulation due to low blood flow rates inside the heart due to arterial fibrillation and valve diseases [38]. In rare cases, cerebral embolism is produced by fibrin clots as a result of deep vein thrombosis due to patent foramen ovale and left to right shunt. Therefore, the treatment consists in antithrombotic agents and warfarin for the prevention of thromboembolism.

2.3.3 Lacunar Infarction

Lacunar stroke is in many cases asymptomatic. It is believed that silent brain infarction (SBI) occurs due to microscopic plaque penetrating branch proximal [39]. The progress of resistance symptoms to lacunar infarction treatments is called branch atheromatous disease (BAD); it is thought that BAD is originated by plaque from full-length penetrating branch occlusion [40].

2.3.4 Others

Moyamoya disease was named by the research group of Suzuki Jirou in 1963. For this disease, an abnormal capillary network is observed in the neighborhood of a stenosis or obstruction that occurs in the region near the terminal part of both internal carotid arteries and near the proximal part of anterior or middle cerebral artery. This disease is more frequent in Japanese people than in populations of Europe and the United States. It was found in all groups indistinctively of age; however, the ages with more probability of development of such diseases are at the age of 5, followed by population with ages between 30 and 40 years. It was observed that generally it produces ischemia when developed in young populations and hemorrhage in adults [1].

In European and US populations, cerebral artery dissection occurs frequently in the extracranial section of the internal carotid arteries; in Japan it occurs in the extracranial section of the vertebral artery. For extracranial cerebral artery dissection, in almost all cases brain attack is produced by cerebral ischemia. For intracranial cerebral artery dissection, brain attack is caused by ischemia or subarachnoid hemorrhage. In recent years, with the popular use of CT and MRI, it has been frequently diagnosed from headaches and asymptomatic condition.

2.3.5 Treatments

2.3.5.1 rt-PA (Tissue plasminogen activator: alteplase) intravenous therapy

Principally, until now the basic treatments for stroke have been the reduction of cerebral edema at ischemic regions, preventing the growth of ischemic regions by improving the microscopic circulation and at chronic stage rehabilitation for function recovery and reoccurrence prevention. Actually there is no change in the basic treatments. The National Institute of Neurological Disorders and Stroke (NINDS) rt-PA Stroke study proved in 1995 the effectiveness of rt-PA intravenous treatments for ischemic cerebrovascular diseases within 3 hours of occurrence [44], making possible intravenous treatments based on active revascularization for acute cerebral infarction. Before interventional neuroradiology, the effectiveness of revascularization was suggested by researchers [45, 46]. This treatment was not generally adopted quickly because of the limited number of physicians and facilities capable to provide it. Regardless of the time constraints and various treatment adaptation rules of rt-PA intravenous therapy, this treatment spread quickly as it enabled intravenous treatments.

2.3.5.2 Revascularization

In recent years some limitations of rt-PA were discovered bringing interventional neuroradiological treatments to the spotlight again. Merci is a guiding catheter with a balloon attached to it. When inserted it enables momentary blood flow interruption for thrombus extraction using a specialized coiled wire (Fig. 2.14). This treatment captured interest because it gives an example of external adaptation and effectiveness [48]. Good results of revascularization cases were reported in [49] using a system named Penumbra that relies on a specialized catheter and wire for treating cerebral artery occlusion by destruction and aspiration of thrombus (Fig 2.15). Figures 2.16 and 2.17 show a patient with Lt.MCA superior trunk occlusion. Figure 2.16a shows a 3D-CTA image of the obstruction in the Lt. MCA superior trunk. Cerebral blood flow (CBF) reduction at part of the Lt. MCA superior trunk area, duration of mean transit time (MTT) and time to peak (TTP), cerebral blood volume (CBV) and Penumbra are visualized using CT Perfusion (Fig. 2.16b–e). Vascular reconstruction by thrombus resection (Fig. 2.17) eliminates right hemiparesis and aphasia symptoms. From basilar artery through the left vertebral artery a near occlusion and thrombus with blood vessel are visible in Fig. 2.18. After the suction of the thrombus, a stent was deployed from basilar artery through the left vertebral artery (Fig. 2.18b). After the treatment the blood flow to the brain was reestablished (Fig. 2.18c).

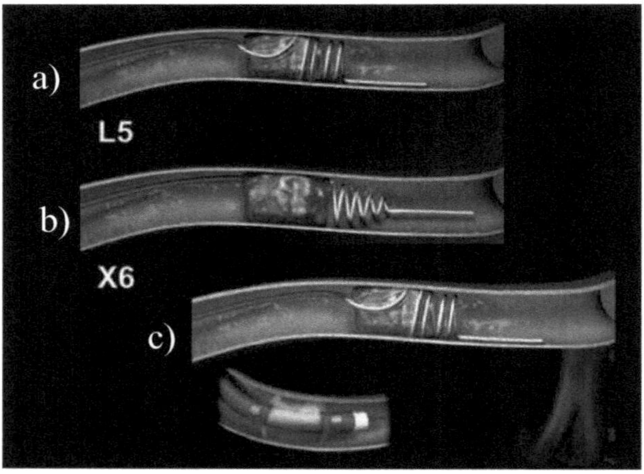

Figure 2.14 Merci retrieval system presented in (mechanical thrombectomy for acute ischemic stroke: final results of the Multi MERCI trial. *Stroke*, April 2008; 39(4):1205–1212). (a, b) A thrombus is caught using a specialized coiled wire that is inserted into the obstructed section. (c) After the blood flow is temporary stopped by inflating the balloon installed around the guiding catheter; the thrombus is suctioned to the inside of the guiding catheter for its extraction [48].

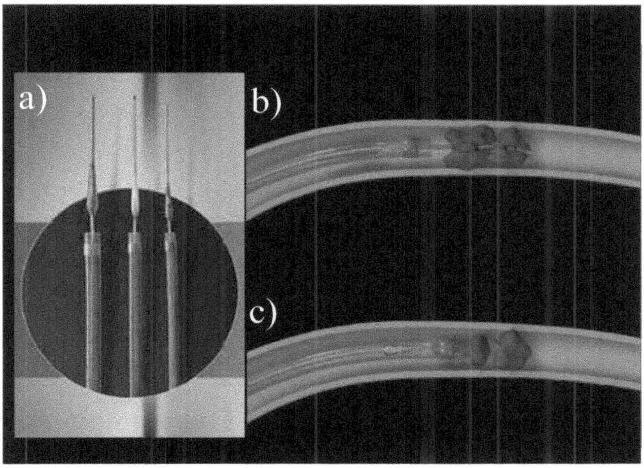

Figure 2.15 Penumbra system (http://www.penumbrainc.com/). (a) A specialized catheter is placed in front of the thrombus. (b) A specialized wire is placed inside while breaking the thrombus. (c) Suction of the thrombus through the catheter.

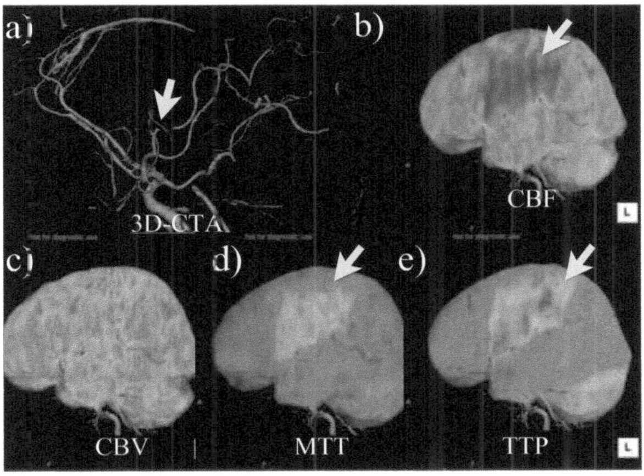

Figure 2.16 Lt. MCA superior trunk occlusion 3D-CT angiography (3D-CTA), CT perfusion (CTP). (a) 3D-CTA Lt. MCA superior trunk occlusion is shown with an arrow. (b) Cerebral blood flow (CBF) at the Lt. MCA superior trunk area (arrow) is reduced. (c) No visible change is seen for cerebral blood volume (CBV) in the region where the blood flow is obstructed. (d) Mean transit time (MTT) duration is confirmed at the Lt. MCA superior trunk area (arrow). (e) Time to peak (TTP) duration is confirmed at the Lt. MCA superior trunk area (arrow). See also Color Insert.

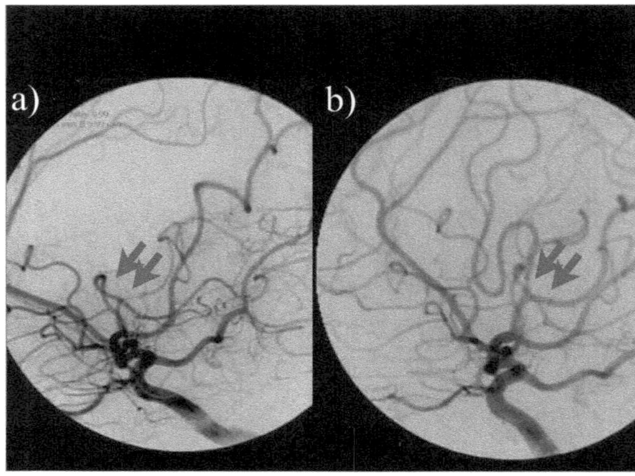

Figure 2.17 Lt.MCA superior trunk occlusion DSA internal carotid angiography (ICAG). (a) DSA before treatment, occlusion at the Lt. MCA superior trunk is indicated with an arrow. (b) DSA after treatment, the occluded artery was completely disobliterated (arrow).

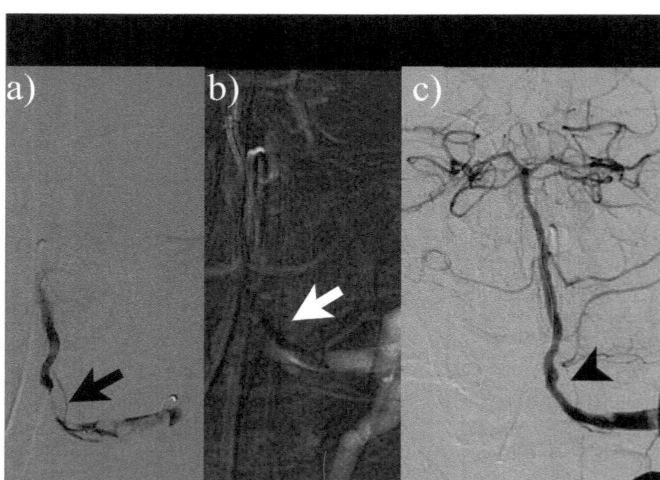

Figure 2.18 Near occlusion from basilar trunk artery (BA) to Lt. vertebral artery. (a) The near occlusion in BA is shown with an arrow; the accumulation of thrombus is appreciated in the peripheral side. (b) Stent (arrow) placed in near occlusion section of the BA after removing the thrombus. (c) Blood flow improvement after the treatment.

2.4 Intracerebral Brain Hemorrhage

The principal cause of intracerebral hemorrhage is hypertension (especially uncontrolled hypertension) followed by cerebral amyloid angiopathy. The place with more frequent occurrence is the putamen followed by the thalamus, sub-cortex, cerebellar dentate nucleus and pons. It is thought that a cause of intracerebral hemorrhage in patients with hypertension is the ruptured micro-aneurysm that appears as a consequence of the swelling up of cerebral blood vessels affected by fibrinoid necrosis [49, 50]. Large hematoma has low survival rate due to early bleeding. Conventionally emergency surgery to remove hematoma was carried out under the microscope. In the recent years, endoscopic systems have been used. For that, an incision of about 5 cm is done in the skin and a bar hole is opened to the skull (Fig. 2.20a), within a transparent sheath (Fig. 2.19a) an

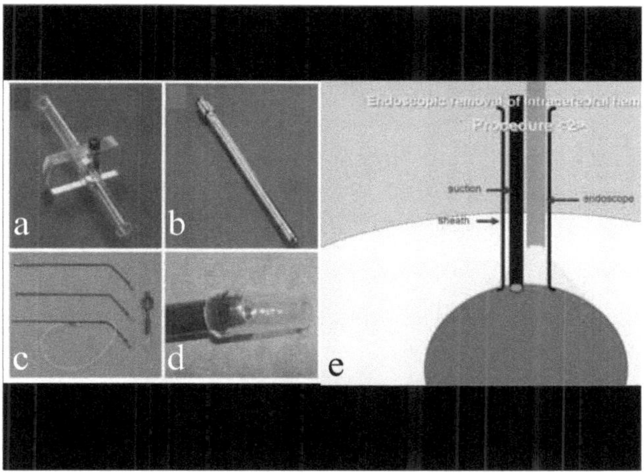

Figure 2.19 Schematic explaining the endoscopic system and treatment [51]. (a) Transparent sheath used as trocar for the endoscope and aspiration tube. (b) Inner cylinder, inserted within the transparent sheath for guiding the sheath insertion into the hematoma. (c) Aspiration tube, bent in its middle body for facilitating the endoscopic surgery. (d) Rigid endoscope with the light source and lens attached. (e) Schematic for treatment explanation: The sheath is inserted into the hematoma for extracting the hematoma. The endoscope is used for inspection within the hematoma.

inner cylinder is inserted and implanted into the cerebral hematoma. After that, the inner cylinder is removed and the aspiration tube (Fig. 2.19c) and the endoscope (Fig. 2.19d) are inserted. While inspecting with the endoscope, the hematoma is suctioned (Fig. 2.20b) [51]. A case of left putaminal hemorrhage is shown in Fig. 2.21a. The complete extraction of hematoma by endoscopic surgery is shown in Fig. 2.21b.

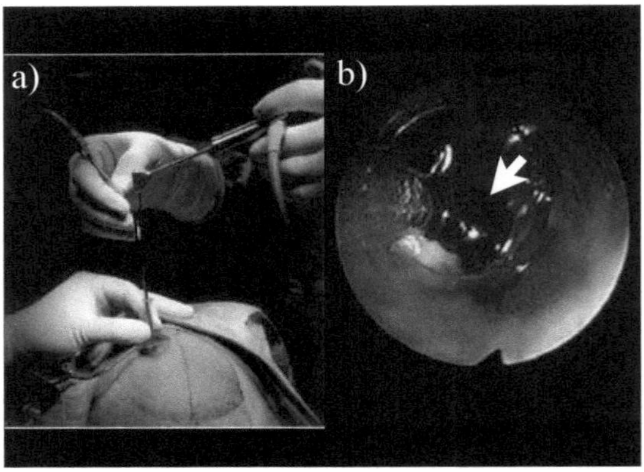

Figure 2.20 Surgery: (a) Intra-operative environment appearance (b) Endoscopic field of view: In the back of the transparent sheath a brown hematoma is pointed out with an arrow.

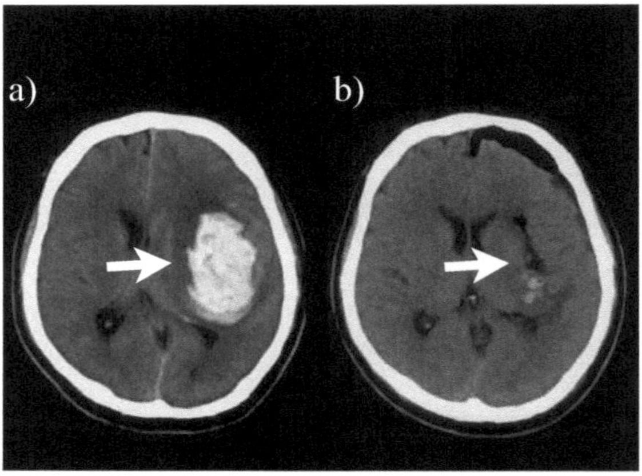

Figure 2.21 Head CT of left putaminal hemorrhage. (a) Before surgery a relatively large hemorrhage is visible at the left putaminal (arrow) (b) After surgery, almost all the hematoma was extracted (arrow).

References

1. Hopkins LN. Balloon catheterization. *J Neurosurg*, 2007, 107: 684.
2. Guglielmi G, Viñuela F, Sepetka I, *et al.* Electrothrombosis of saccular aneurysms via endovascular approach. Part 1: Electrochemical basis, technique, and experimental results. *J Neurosurg*, 1991, 75: 1–7.
3. Nogueira GJ. Spontaneous subarachnoid hemorrhage and ruptured aneurysm in the Middle East. A myth revisited. *Acta Neurochir (Wien)*, 1992, 114: 20–25.
4. Ingall T, Asplund K, Mahonen M, Bonita R. A multinational comparison of subarachnoid hemotthage epideminology in the WHO MONICA stroke study. *Stroke*, 2000, 31: 1054–1061.
5. Inagawa T, Tokuda Y, Ohbayashi N, Takaya M, Moritake K. Study of aneurysmal subarachnoid hemorrhage in Izumo City, Japan. *Stroke*, 1995, 26: 761–766.
6. van Gijn J, Rinkel GJ. Subarachnoid haemorrhage: diagnosis, causes and management. *Brain*, 2001, 12(4 Pt 2): 249–278.
7. Taylor B, Harries P, Bullock R. Factors affecting outcome after surgery for intracranial aneurysm in Glasgow. *Br J Neurosurg*, 1991, 5: 591–600.
8. Neil-Dwyer G, Lang D, Smith P, Iannotti F. Outcome after aneurysmal subarachnoid haemorrhage: the use of a graphical model in the assessment of risk factors. *Acta Neurochir(Wien)*, 1998, 140: 1019–1027.
9. Roos YB, de Haan RJ, Beenen LF, Groen RJ, Albrecht KW, Vermeulen M. Complications and outcome in patients with aneurysmal subarachnoid haemorrhage: a prospective hospital based cohort study in the Netherlands. *J Neurol Neurosurg Psychiatry*, 2000, 68: 337–341.
10. Wiebers DO, Whisnant JP, Huston J 3rd, *et al.* Unruptured intracranial aneurysms: natural history, clinical outcome, and risks of surgical and endovascular treatment. *Lancet*, 2003, 362: 103–110.
11. Raaymakers TW, Rinkel GJ, Limburg M, Algra A. Mortality and morbidity of surgery for unruptured intracranial aneurysms: a meta-analysis. *Stroke*, 1998, 29: 1531–1538.
12. Asari S, Ohmoto T. Natural history and risk factors of unruptured cerebral aneurysms. *Clin Neurol Neurosurg*, 1993, 95: 205–214.
13. Ishikawa T, Nakayama N, Yoshimoto T, Aoki T, Terasaka S, Nomura M, Takahashi A, Kuroda S, Iwasaki Y. How does spontaneous hemostasis occur in ruptured cerebral aneurysms? Preliminary investigation on

247 clipping surgeries. *Surg Neurol,* 2006, 66: 269–275 (discussion 275–266).

14. Morita A, Fujiwara S, Hashi K, Ohtsu H, Kirino T. Risk of rupture associated with intact cerebral aneurysms in the Japanese population: a systematic review of the literature from Japan. *J Neurosurg*, 2005 (April), 102(4): 601–606.
15. Nahed BV, Bydon M, Ozturk AK, *et al.* Genetics of intracranial aneurysms. *Neurosurgery*, 2007, 60: 213–225.
16. Rinkel GJ, Djibuti M, Algra A, van Gijn J. Prevalence and risk of rupture of intracranial aneurysms: a systematic review. *Stroke*, 1998 (January), 29(1): 251–256.
17. Ruigrok YM, Rinkel GJ. Genetics of intracranial aneurysms. *Stroke*, 2008, 39: 1049–1055.
18. Taylor CL, Yuan Z, Selman WR, *et al.* Cerebral arterial aneurysm formation and rupture in 20,767 elderly patients: hypertension and other risk factors. *J Neurosurg*, 1995, 8: 812–819.
19. Ujiie H, Tamano Y, Sasaki K, *et al.* Effect of size and shape (aspect rate) on the hemodynamics of saccular aneurysms: a possible index for surgical treatment of intracranial aneurysms. *Neurosurgery*, 1999, 48: 495–502.
20. Wermer MJ, van der Schaaf IC, Algra A, Rinkel GJ. Risk of rupture of unruptured intracranial aneurysms in relation to patient and aneurysm characteristics: an updated meta-analysis. *Stroke*, 2007 (April), 38(4): 1404–1410.
21. Wiebers DO, Whisnant JP, Huston J 3rd, Meissner I, Brown RD Jr, Piepgras DG, Forbes GS, Thielen K, Nichols D, O'Fallon WM, Peacock J, Jaeger L, Kassell NF, Kongable-Beckman GL, Torner JC. International study of unruptured intracranial aneurysms investigators. Unruptured intracranial aneurysms: natural history, clinical outcome, and risks of surgical and endovascular treatment. *Lancet*, 2003 (July 12), 362(9378): 103–110.
22. Yasui N, Suzuki A, Nishimura H, *et al.* Long-term follow-up study of unruptured intracranial aneurysms. *Neurosurg*, 1997, 40: 1155–1160.
23. Hayakawa M, Murayama Y, Duckwiler GR, Gobin YP, Guglielmi G, Vinuela F. Natural history of the neck remnant of a cerebral aneurysm treated with Guglielmi detachable coil system. *J Neurosurg*, 2000, 93(4): 561–568.
24. Flaherty ML, Flemming KD, McClelland R, Jorgensen NW, Brown RD Jr. Population-based study of symptomatic internal carotid artery

occlusion: incidence and long-term follow-up. *Stroke*, 2004, 35(8): 349–352.

25. Timsit SG, Sacco RL, Mohr JP, Foulkes MA, Tatemichi TK, Wolf PA, Price TR, Hier DB. Early clinical differentiation of cerebral infarction from severe atherosclerotic stenosis and cardioembolism. *Stroke*, 1992 (April), 23(4): 486–491.

26. Zhu CZ, Norris JW. Role of carotid stenosis in ischemic stroke. *Stroke*, 1990 (August), 21(8): 1131–1134.

27. Barnett HJ, Taylor DW, Eliasziw M, *et al.* Benefit of carotid endarterectomy in patients with symptomatic moderate or severe stenosis. North American symptomatic carotid endarterectomy trial collaborators. *N Engl J Med*, 1998, 339: 1415–1425.

28. Endarterectomy for asymptomatic carotid artery stenosis. Executive Committee for the Asymptomatic Carotid Atherosclerosis Study. *JAMA*, 1995, 273: 1421–1428.

29. Yadav JS, Wholey MH, Kuntz RE, *et al.* Protected carotid-artery stenting versus endarterectomy in high-risk patients. *N Engl J Med*, 2004, 351: 1493-1501.

30. Gurm HS, Yadav JS, Fayad P, *et al.* Long-term results of carotid stenting versus endarterectomy in high-risk patients. *N Engl J Med*, 2008, 358: 1572-1579.

31. Ringleb PA, Allenberg J, Bruckmann H, *et al.* 30 day results from the SPACE trial of stent-protected angioplasty versus carotid endarterectomy in symptomatic patients: a randomized non-inferiority trial. *Lancet*, 2006, 368: 1239–1247.

32. Mas JL, Chatellier G, Beyssen B, *et al.* Endarterectomy versus stenting in patients with symptomatic severe carotid stenosis. *N Engl J Med*, 2006, 355: 1660–1671.

33. Kastrup A, Groschel K, Krapf H, Brehm BR, Dichgans J, Schulz JB. Early outcome of carotid angioplasty and stenting with and without cerebral protection devices: a systematic review of the literature. *Stroke*, 2003, 34: 813–819.

34. Adams HP Jr, Bendixen BH, Kappelle LJ, Biller J, Love BB, Gordon DL, Marsh EE 3rd. Classification of subtype of acute ischemic stroke. Definitions for use in a multicenter clinical trial. TOAST. Trial of org 10172 in acute stroke treatment. *Stroke*, 1993 (January), 24(1): 35–41.

35. Special report from the national institute of neurological disorders and stroke. Special report from the national institute of neurological

disorders and stroke. Classification of cerebrovascular diseases III. *Stroke*, 1990 (April), 21(4): 637–76.

36. Yoshida Y. Pathophysiology of atherthrombotic infarction. *Nippon Rinsho*, 2006 (October 28), 64 Suppl 7: 97–101.
37. Ross R. The pathogenesis of atherosclerosis: a perspective for the 1990s. *Nature*, 1993 (April 29), 362(6423): 801–809.
38. Yoneda Y, Mori E, Uehara T, Yamada O, Tabuchi M. Referral and care for acute ischemic stroke in a Japanese tertiary emergency hospital. *Eur J Neurol*, 2001 (September), 8(5): 483–488.
39. Weksler BB. Platelet function and antiplatelet therapy in ischemic cerebrovascular disease. In: *Stroke: Pathology, Diagnosis and Management*, 2nd ed. (ed Barnett HJM, *et al.*), p 1041–1059, Churchill Livingstone, New York, 1998.
40. Caplan LR. Intracranial branch atheromatous disease: a neglected, understudied, and underused concept. *Neurology*, 1989 (September), 39(9): 1246–1250.
41. Fukui M. Current state of on moyamoya disease in Japan. *Surg Neurol*, 1997, 48: 395–400.
42. Yamaura A, Yoshimoto T, Hashimoto N, *et al.* Higai shousei zugainai kairisei doumyaku byouhen no zenkoku chousa (dai 1hou). Nousocchu no geka, 1998, 26: 79–86.
43. Schievink WI. Spontaneous dissection of the carotid and vertebral arteries. *N Engl J Med*, 2001, 344: 898–906.
44. The National Institute of Neurological Disorders and Stroke rt-PA Stroke Study Group. Tissue plasminogen activator for acute ischemic stroke. *N Engl J Med*, 1995, 333: 1581–1587.
45. del Zoppo GJ, Higashida RT, Furlan AJ, Pessin MS, Rowley HA, Gent M. PROACT: a phase II randomized trial of recombinant pro-urokinase by direct arterial delivery in acute middle cerebral artery stroke. PROACT investigators. Prolyse in acute cerebral thromboembolism. *Stroke*, 1998, 29: 4–11.
46. Furlan A, Higashida R, Wechsler L, *et al.* Intra-arterial prourokinase for acute ischemic stroke. The PROACT II study: a randomized controlled trial. Prolyse in acute Cerebral thromboembolism. *JAMA*, 1999, 282: 2003–2011.
47. Smith WS, Sung G, Saver J, Budzik R, Duckwiler G, Liebeskind DS, Lutsep HL, Rymer MM, Higashida RT, Starkman S, Gobin YP; Multi MERCI Investigators, Frei D, Grobelny T, Hellinger F, Huddle D, Kidwell C, Koroshetz W, Marks M, Nesbit G, Silverman IE. Mechanical

thrombectomy for acute ischemic stroke: final results of the Multi MERCI trial. *Stroke*, 2008 (April), 39(4): 1205–1212.

48. Bose A, Henkes H, Alfke K, Reith W, Mayer TE, Berlis A, Branca V, Sit SP. Penumbra phase 1 stroke trial investigators. The penumbra system: a mechanical device for the treatment of acute stroke due to thromboembolism. *AJNR Am J Neuroradiol*, 2008 (August), 29(7): 1409–1413.

49. Ogata J, Yamanishi H, Yukitani C, Ueda H. Mechanisms and pathology of hypertensive intracerebral hemorrhage. *Nippon Rinsho*, 2006 (October 28), 64 Suppl 7: 107–110.

50. Fisher CM. Cerebral military aneurysm in hypertension. *Am J Pathol*, 1972, 66: 313–330.

51. Nishihara T, Nagata K, Tanaka S, *et al.* Newly developed endoscopic instruments for the removal of intracerebral hematoma. *Neurocrit Care*, 2005, 2(1): 67–74.

Chapter 3

Patient-Specific Vascular Modeling

The morphology of cerebral arteries and cerebrovascular diseases vary greatly among individuals. Therefore, it is helpful to construct a vessel model of cerebral artery for each patient based on individual information and apply it for preliminary simulation of endovascular intervention. In this chapter, we propose a technique for constructing a vasculature model by using individual information obtained with computed tomography (CT) and magnetic resonance imaging (MRI). Several vasculature models of actual cerebrovascular diseases are constructed by using the patient-specific modeling method. Then the proposed vascular model is evaluated by simulating endovascular intervention procedures within this model, and the performance and utility are discussed from the viewpoint of clinical use. Finally we present a simulator including major vasculature models that achieves the comprehensive simulation of endovascular intervention by integrating the technologies proposed in this chapter, and then its utility is evaluated by a survey.

3.1 Introduction and Background

Recently established endovascular intervention is a potent treatment modality for various vascular diseases, which reduces the invasiveness for patients. Many devices for endovascular intervention have been developed since late 1980. The most prominent advancement was

Microsurgery: Advances, Simulations, and Applications
Edited by Toshio Fukuda and Carlos Tercero

ISBN 978-981-4364-69-0 (Hardcover), 978-981-4364-70-6 (eBook)
www.panstanford.com

the introduction of a micro-catheter into neurovascular intervention. So far, various types of micro-catheters have been developed with regard to their operating function to treat various diseases, which have been treated mainly with craniotomy in the past.

However, cerebral vasculature tortuosity makes difficult to drive intravascular devices to achieve a treatment. The blood vessels' morphology is different for each patient and it may contribute to make neuro-endovascular intervention more difficult. Endovascular intervention requires advanced surgical skills and experience. However, most of current environments for simulating endovascular intervention are less similar to the actual artery as to its morphology and physical characteristics. And this fact makes its surgical skills more difficult to acquire.

Traditionally, interventionalists have solely depended on images projected on displays or papers as the only means to get or to provide medical information obtained by angiography or other imaging modalities. Although the recent development of imaging processing technology has made possible the reconstruction of three-dimensional images from two-dimensional sliced images. Even so, it is still not easy to understand the accurate arterial shapes because the reconstructed three-dimensional images are still projected onto two-dimensional planes and it may lead to misinterpretation of the real morphology.

As a solution to these difficulties, we proposed the fabrication of an *in vitro* vasculature model that hollowly reproduces the three-dimensional configuration of the patient's vascular lumens within an elastic transparent silicone. The lumen of the model is obtained by CT, MRI, and other imaging modalities. With this platform, it becomes possible for young trainees to practice surgical skills of endovascular intervention. These models enable to practice surgical treatments for actual cases reproduced in silicone. And unlike the conventional image-based diagnoses, with this model it becomes possible to preliminary simulate neurovascular procedures. In addition, it is useful for diagnosis and informed consent. Moreover, this kind of model also provides evaluation platform for medical studies and surgical instruments.

In an earlier research, Guilloud *et al.* [1] presented techniques to reconstruct vessel lumens using modeling method-based dissection. This method utilizes human vessel as casting mold and is now largely utilized for various purposes. This modeling method allows precisely reconstructing three-dimensional configuration of vessel lumens and provides good platform for surgical trainings and preclinical testing. However, the dissection method is not applicable for clinical purposes and requires rather high time cost and labor cost for respective modeling. On the other hand, the reproduction of physical characteristics of vessel tissue have been less discussed in earlier works despite its importance, although the reproduction of delicate feel of operating catheters and the dynamic behavior of surgical instruments (e.g., its slip/stick motion and unwinding motion) are both important for simulating neurovascular procedures.

3.2 Required Properties for the Vascular Model

The cerebral arterial model that materializes (1) patient-tailored modeling with sub-millimeter modeling resolution and (2) the reproduction of physical characteristics of artery tissues should serve as an innovative platform for simulating neurovascular intervention before individual procedures and also helpful for young trainee to practice endovascular intervention procedures. In this section, we describe the conditions required for such an *in vitro* surgical simulator for endovascular intervention.

3.2.1 Patient-Specific Reproduction

Figures 3.1 and 3.2 show some typical configuration of the cerebral artery and aneurysm, respectively. As shown in these figures, the configurations of cerebral vasculature are quite different for each case, and this diversity is an outstanding feature of endovascular intervention, which contributes to making neurovascular procedures difficult. From this point of view, although the conventional surgical simulator that imitates the general configuration of cerebral artery with simplified configuration (such as glass simulator) is helpful to experience intervention procedures, it is insufficient for acquiring comprehensive surgical skills, since this kind of generalized vasculature model does not reflect the diversity of vascular configuration, especially the diversity of cerebrovascular diseases.

Moreover, since micro-catheter is used for neurovascular intervention, the diameter of the vascular lumen that can be treated by the endovascular approach was reduced to lumens with 1 mm of diameter. Therefore, the vessel model for simulating intervention is now required to reproduce that thin-vessel lumen too. However, other available fabrication methods do not satisfy that requirement.

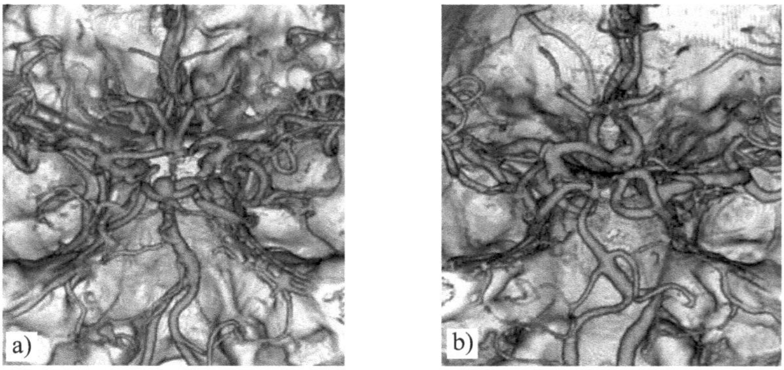

Figure 3.1 Typical configurations of cerebral artery: (a) Case A: 75 years old, male. b) Case B: 54 years old, male.

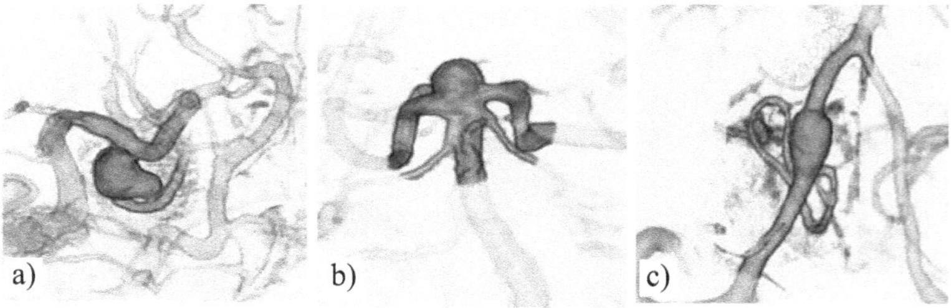

Figure 3.2 Typical configurations of cerebral aneurysm: (a) Middle cerebral artery (MCA) aneurysm, (b) basilar top (BT) aneurysm and (c) basilar artery (BA) aneurysm.

Hence, it is very helpful to materialize the fabrication method that allows constructing patient-tailored vessel model with sub-millimeter modeling resolution. And in this patient-tailored modeling, it is desirable to use medical imaging information such as CT, MR and ultrasonic diagnosis, which are commonly available for every case. This kind of patient-tailored high-precision vessel model should provide a helpful medical platform for diagnoses, preliminary simulations, pathological (hemodynamic) studies, surgical training, and testing of medical instruments.

3.2.2 Reproduction of Physical Characteristics

Two-dimensional monochromatic digital subtraction angiography (DSA) image and the subtle feel of operating medical instruments are the only information that interventionists commonly utilize in neurovascular procedures. Therefore, the operational feel plays an important role in treating cerebrovascular diseases safely without applying excessive piercing stress on the vessel structure. Hence, to reproduce the physical characteristics of vasculature, such as its elasticity and surface friction, becomes an essential requirement for a vasculature model for simulating neurovascular procedures, since the feel of operating medical instrument is never reproducible without the reproduction of physical characteristics. Moreover, with the reproduction of physical characteristics, dynamic behaviors of medical instruments, such as slip/stick motion and unwinding motion of micro-catheter, are also reproducible. Consequently the difficulties of manipulating medical instruments in neurovascular procedures could be reproduced on simulator in this way. Meanwhile, it is also important for a vasculature model to be compatible with medical imaging modalities (especially DSA), to allow simulating neurovascular procedures under practical IVR (interventional radiology) condition.

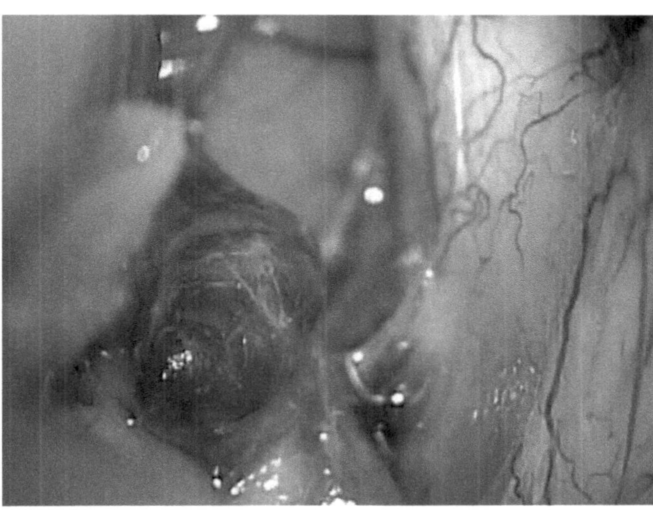

Figure 3.3 Configuration of actual cerebral artery and its locating condition inside the subarachnoid space.

3.2.3 Reproduction of Membranous Vascular Configuration

The deformation of vessel structure when catheter or other medical tools are inserted is related to its elastic membranous structure and the existence of surrounding soft tissue. Its soft structure even allows dynamic deformation against pulsation and blood streaming. Some neurological treatments are realized relying on this property. The expansion of narrowed artery (stenosis) with balloons and stents and aneurismal clipping are the major instances of these medical treatments. However, this property simultaneously results in technical difficulties. Therefore, the reproduction of this structural deformation is also important to allow simulating various medical treatments and experiencing technical difficulty accrued from the deformation of arterial structure.

3.2.4 Summary of Required Conditions

The required conditions for an *in vitro* vasculature model for endovascular treatments simulation are as follows:
The model must reproduce

1. Precise configuration of cerebral artery
2. Patient-specific vascular structure
3. Elastic characteristics of arterial tissue
4. Friction characteristics of arterial tissue
5. Thin membranous vascular structure
6. Visco-elastic vascular deformation

The model must be

7. Compatible with medical imaging modalities
8. Transparent to provide good visibility

3.3 Medical Image Processing

It is possible to acquire the information inside human body using various medical imaging modalities, such as X-ray CT and MRI. In these medical imaging systems, two-dimensional cross-sectional images of human body are obtained. It is possible to reconstruct with high resolution the three-dimensional structure of the scanned area from a set of cross-sectional images. As a result, doctors are able to recognize three-dimensional shapes of a disease and to provide an adequate diagnosis and treatment policy. Moreover, three-dimensional data obtained by this re-composition can be used for evaluations of dynamics of the living organ and blood stream and for the development of VR (virtual reality) simulators.

Here, the following method is used to reconstruct the three-dimensional shape of the object from these two-dimensional data: First, extracting the area of interest by applying a specific threshold to each two-dimensional image. Then, the obtained images are supplemented in a three-dimensional data set of the region of interest. Moreover, the blood vessel accurate shape acquisition relies on pursuing the continuousness of the CT value. We adopted the method of composing the shape of the brain blood vessel by designating a threshold on each individual fluoroscopic image taken by CT and MRI.

3.3.1 Medical Imaging Modalities

The following sections detail the methodology for reconstructing the three-dimensional shape of targeted vasculature from CT/MRI or other medical imaging modalities as a fundamental procedure for the patient-specific vascular reproduction. This section outlines the major medical imaging modalities, which are applicable to the proposed modeling.

Single-slice CT: X-ray CT was introduced by Hounsfield in 1973 [2]. X-ray CT of the first generation required a few minutes to take one picture, and the reconstruction operation required more time. As a result, it was applicable only for head, which is able to remain stationary while taking a picture. In the second generation, the time required was shortened to 10–20 seconds by adopting fan beam instead of the pencil beam, and taking a picture with two or more detectors, although it adopts translate/rotate modality as well as the first generation. In 1985, the introduction of the slipping ring mechanism made it possible to rotate the X-ray emitter. Thus the X-ray CT of the third and fourth generations were clinically introduced. As a result, a high-speed, continuous imaging became possible, and such improvement spread the CT system to various clinical applications.

Helical CT: Helical (spiral) was announced by Kalender in 1990 [3]. Helical CT continuously irradiates X-rays. This method collects the projection data while the table on which the patient lies slides through the slipping ring, at the same time the X-ray emitter continuously rotates in this ring. It is called helical CT, because the tracks of an X-ray beam draw spiral. This method allows (1) drastically shortening the imaging time, (2) reconstructing slice image of arbitrary position, (3) displaying three-dimensional information and MRP (multi-planar reconstruction).

Multi-slice CT: Multi-slice CT (MSCT), which has a multi-row detector, appeared in 1999. Technical improvements, such as high-speed scanning and reduction of slicing pitch, were made by the MSCT detector. As a result, it realized (1) drastic reduction of scanning time, (2) sagittal section imaging, (3) isotropic voxels, (4) high-resolution 3D image and (5) improvement of time resolution. Moreover, since the drastic reduction of scanning period widens the imaging range, it becomes possible to obtain sagittal section image that has the same resolution as the axial section image, by adopting isotropic voxel. Consequently, it becomes possible to reduce the partial volume effect (PVE) that originated in the slice interval. The resolution of CT has improved up to about 0.2 mm (about several millimeters in CT) by the appearance of MSCT.

Magnetic Resonance Imaging: The clinical application of MRI started in the 1980s. An excellent image can be obtained now because it is possible to obtain a strong and steady magnetic field by the development of super-conductivity and permanent magnet modality and because of the advancement of image reconstruction method. MRI has the following features:

(1) Arbitrary section image is attainable including sagittal/coronal direction.
(2) No X-ray irradiation.
(3) Excellent tissue resolution (easy separation of vascular region).
(4) Relationship between vasculature and disease can be recognized since the vasculature is detected as signal area.

On the other hand, the disadvantages of MRI are

(1) Low spatial resolution
(2) Long imaging period

The range of the use of MRI is limited because X-ray CT with high spatial resolution and time resolution are widespread, but MRI is useful to acquire the shape of blood vessel and organization information. The resolution of MRI is above 0.7 mm.

3.3.2 Three-Dimensional Vessel Shape Reconstruction from CT

In this section, we describe the method for reconstructing the three-dimensional shape of the vascular lumen structure from the cross-sectional image of human head, which was taken by the above-mentioned multi-slice CT modality. The fault image of the CT image is given as two-dimensional scalar place where the density value is assumed to be a scalar. The extraction of the blood vessel area from each fault image is achieved by giving a threshold processing to this density value. The extraction of the blood vessel area from each fault image is achieved by giving a threshold processing to this density value. This processing is useful for separation of a desired object with specific density value from the surroundings within a fault image. The density value in the CT images is given using the Hounsfield unit. In general, the density value changes according to the tissue type and organization.

Because blood is principally composed of water, the CT value is adjacent to the soft tissue. Therefore, in fluoroscopic images the distinction between a blood vessel area and the surrounding area is difficult. Therefore, to improve the CT value of blood in fluoroscopic images, a contrast medium is administered from the cervix part in the blood stream. The blood vessel image is taken by fluoroscopy of the mixture liquid of blood and the contrast medium flows into the brain blood vessel. In this case, it is about 150–250 HU, though the density value of the blood vessel depends on the kind of the contrast medium.

However, the CT value is variable because the density of the contrast medium is not uniform on the entire brain blood vessels. Moreover, when taking a CT image, the phenomenon that is called PVE appears, making the edge of the image duller.

As a result, when the blood vessel is extracted by the threshold processing, a thin blood vessel is recognized thinner. Partial volume effect makes the density value to decrease and to become a value almost similar to the blood vessel in an outline part of the bone tissue that has a high CT value (+1000 HU). Especially, the separation of both areas becomes difficult for vessels adjacent to the skull.

Figure 3.4 is a case where the blood vessel area is extracted by the threshold processing by using commercial medical diagnosis software. The shape of the brain blood vessel can be excellently recognized from CT imaging as confirmed from this figure. However, the separation of the blood vessels from the rest becomes difficult in the area where the blood vessel and the bone approach each other. This image confirms that blood vessels and bones have integrated in the brainpan bottom, and that the CT image is suitable for the areas where the blood vessel does not approach the bone. DSA, which excellently extracts the blood vessel area, is achieved by taking a fluoroscopic image before and after the contrast medium administration and operating the difference between them. It cannot be influenced by the existence of the bone area, and the shape of the blood vessel can be extracted easier.

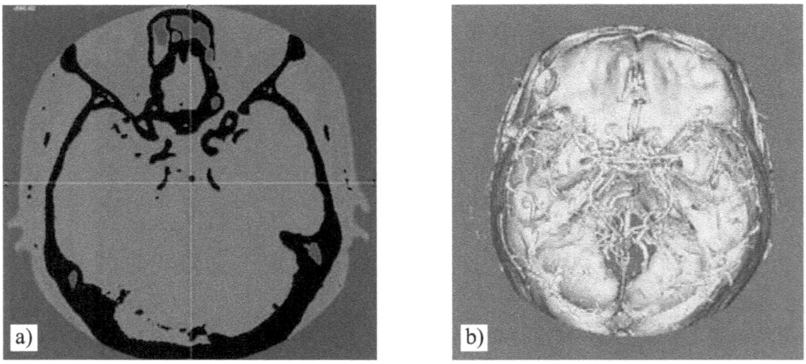

Figure 3.4 Reconstruction of three-dimensional vascular structure based on individual image obtained by multi-slice CT modality: (a) Extracted vessel area by threshold application. (b) Reconstructed three-dimensional structure.

3.3.3 Three-Dimensional Vessel Shape Reconstruction from MRI

In this section, we describe the re-composition of the three-dimensional shape of the blood vessel using the cross-sectional data of the head area obtained from MRI. In X-ray CT imaging, light and shade are only decided depending on absorption (penetration) rate of X-rays. On the other hand, for MRI, light and shade are generated based on a signal depending on the density of atomic nucleus of hydrogen; T1 and T2 are the longitudinal and transverse relaxation times of that signal which varies between tissue types. Therefore, various contrasts can be generated by selecting specific values of T1 and T2. When T2 is emphasized, water is a high signal, making the image whiter. When T1 is emphasized, water is a low signal and the pictures becomes darker, fat becomes a high signal. Bone tissue becomes a low signal in any case. Based on this property tissue characteristics and the blood vessel area can be differentiated from surrounding tissue in MRI.

Figure 3.5 is three-dimensional image of the blood vessel obtained by setting the threshold processing to 177 on the MRI image taken with emphasis on the T2 signal. The bone area can be completely separated compared with CT, and only the brain blood vessel area can be extracted as confirmed from this figure. Like this, MRI has an excellent merit in the parting of the blood vessel area and the bone tissue. However, it is a disadvantage that the resolution and the image registration speed are low compared with CT. Therefore, it is appropriate to limit its use for cases where the blood vessel is adjacent to the bone tissue.

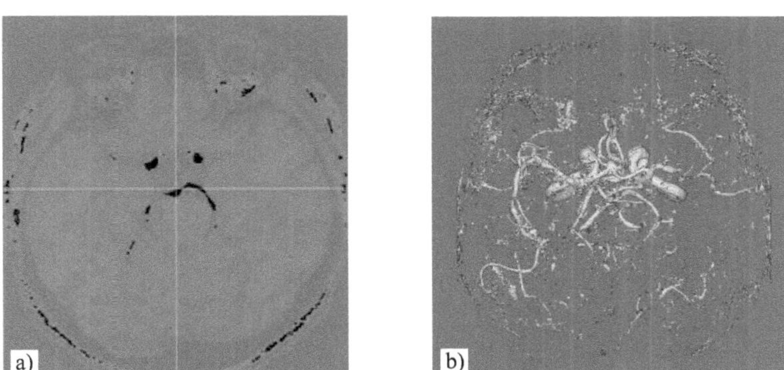

Figure 3.5 Reconstruction of three-dimensional vascular structure based on individual image obtained by MRI modality: (a) Extracted vessel area by threshold application. (b) Reconstructed three-dimensional structure.

3.4 Additional Vascular Shape Modification

By using CAD software, it becomes possible to make modifications on vascular shapes that are reconstructed based on angiographies or other medical imaging modalities. Vasculature lumen information obtained from CT, MRI and other imaging modalities often contains defects, such as distortion, partial loss, discontinuity and unifications. And it is usually difficult to completely avoid its occurrence since in many cases these defects are originated in the intrinsic characteristics of the imaging modalities; for example, the dispersion of X-rays on artifacts for CT, dependence of magnetic resonance on blood flow velocity for MRI, dispersion of contrast media for angiography and PVE.

Here the modification of vascular configuration with 3D-CAD operation plays an important role in restoring its original state. Figure 3.6 shows an instance of CAD modification. In this representative case, we restored the complete set of BT artery from two vascular shapes shown in Figs. 3.6a and 3.6b, respectively, reconstructed form CT and MRI information. Both of them lack some vascular portion due to the characteristics of imaging modalities. Vasculature in Fig. 3.6a is united with cranium due to PVE. Figure 3.6b lacks some aneurismal portion due to the flow velocity dependence. In those cases, the CAD operation allows complementary merging these two figures into one complete figure as shown in 3.6.c. This method is also helpful to construct cerebrovascular diseases on vascular tree with favorable configuration, such as wide-necked aneurysms and thin crooked stenoses, as a platform for comprehensive surgical training and for other various purposes.

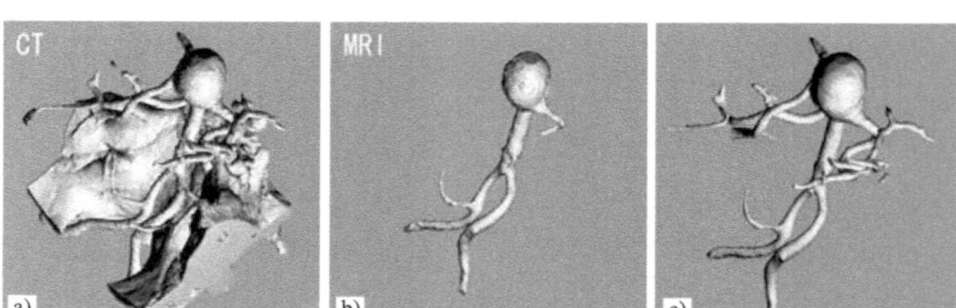

Figure 3.6 An instance of restoring original vasculature figure from two different data source, with the use of CAD: (a) 3D vasculature figure of BT artery reconstructed from CT angiographies, some vascular part is united with cranium due to PVE. (b) Three-dimensional vasculature figure of BT artery reconstructed from MRI lacks some aneurysm portion due to flow velocity dependence. (c) Complete figure of BT artery and aneurysm restored by merging two image sources into one with the use of CAD.

3.5 Patient-Specific Vascular Modeling

In this section, three-dimensional data of the blood vessel reconstructed from the individual information is applied to the fabrication method of a solid vascular model. A silicone model that reproduces individual cerebral vascular lumen inside its cubic structure is fabricated by using the CT and MRI information. The fabrication method of the model considering the previously stated requirements is as follows:

1. Reconstruction of 3D vascular configuration from CT/MRI information
2. Fabrication of vascular master mold by rapid prototyping (RP)
3. Injection molding using silicone elastomer
4. Lost wax reproduction of vascular lumen

In this modeling, we first reconstructed the three-dimensional configuration of cerebral artery using 100 digital slice images obtained with multi-slice CT scanning with regular 0.5 mm intervals and 0.3 mm/pixel resolution. Figure 3.7a shows one of the slice image used for the reconstruction. We took in each slice image, as two-dimensional scalar field comprising CT value, into three-dimensional space virtually created in computer according to the interval of 0.5 mm, and heaped up into three-dimensional scalar field. Then, by extracting areas that had identical CT value, which corresponds to the boundary region of artery, we extracted three-dimensional outline of the cerebral artery. Then interpolating this outline three-dimensionally, we generated a three-dimensional iso-surface composed of identical CT value. Then by eliminating unimportant branches and discontinuous segment from this structure, we simplified to leave only basilar tip artery and BT aneurysm. In this way, the three-dimensional structure of basilar tip artery was finally reconstructed from multi-slice CT information. Figure 3.8a shows the reconstructed image of this basilar tip artery (reconstructed area is indicated in Fig. 3.7b with dotted line; dimension: $50 \times 50 \times 50$ mm). Then we fabricated a tree-like lumen structure of the basilar tip artery by means of the above-mentioned RP modeling modality, using the reconstructed three-dimensional geometry of basilar artery (Fig. 3.8b). Laminating pitch for the RP modeling was set to 13 μm.

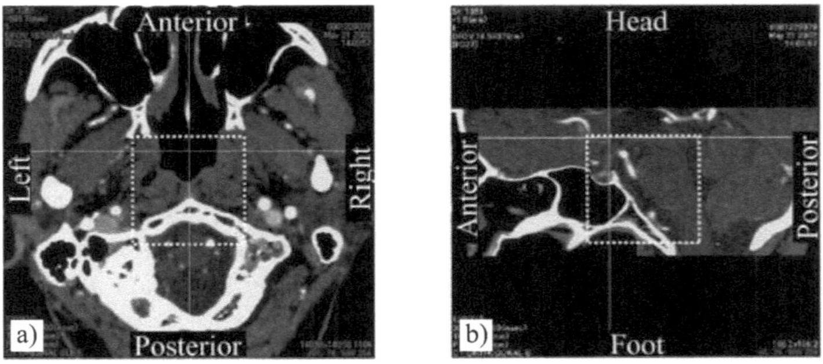

Figure 3.7 Head segment image obtained with CT angiography used for reconstructing three-dimensional vascular lumen configuration; dotted line indicates the modeling area of 50 × 50 × 50 mm: (a) Transversal plane and (b) sagittal plane CT slice images of the region of interest for vascular reconstruction.

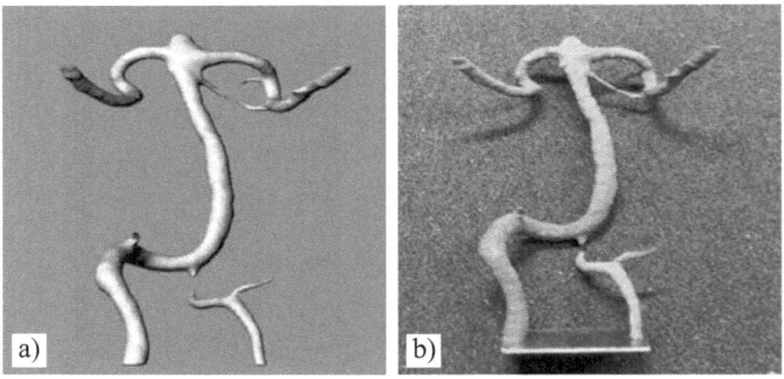

Figure 3.8 (a) Three-dimensional figure of basilar artery with BT (basilar tip) aneurysm, reconstructed from fluoroscopic information. (b) Vascular lumen model of basilar artery, fabricated by fused deposition modality rapid prototyping (laminating pitch: 13 μm).

Then we arranged the above-fabricated vascular master mold inside a cubic mold according to its orientation and embedded it within liquid-state silicone rubber, which solidifies into transparent elastomer through addition polymerization. Here, the physical characteristics (elastic modulus, Poisson's ratio and friction coefficient) of selected silicone are very similar to the arterial tissue, and its excellent transparency allows easily observing the endovascular intervention procedures. Here, molding the silicone into cubic structure has the following advantages: It allows maintaining vascular structure at correct position and eliminates optical distortion allowing correctly recognizing vascular configuration.

After curing the silicone into cubic elastomer structure, we eliminated the inward RP vascular mold by means of lost wax method. This process was achieved by two independent manners: heating and dissolving. We evacuated most part of the lumen model outside the silicon structure by heating whole the structure at 120°C, which is slightly higher than the melting point of the modeling wax and lower than the heatproof temperature of silicone elastomer (above 150°C). Then we completely dissolved the remaining quantity of lumen model, by injecting acetone inside lumens, which is hollowly constructed within the silicone structure through the former evacuation process. With this procedure, we could completely eliminate the lumen model within a very short period, and it allowed eliminating quantities in tight corners and dead ends.

Consequently, we finally constructed a cubic-shaped transparent elastic model of basilar tip artery that hollowly reproduces the vascular lumen of living patient with 13 μm modeling resolution (Fig. 3.9).

Figure 3.9 An *in vitro* patient-tailored vasculature model of basilar artery with BT aneurysm reconstructed from computed tomography angiography.

3.6 Reproduction of Membranous Vessel Structure

In a previous section, we showed a patient-specific model of cerebral artery within a cube structure and its potential for simulating endovascular intervention. Its rigidity restricts vascular deformation against medical treatment and blood streaming, and consequently limiting its applicability and reproducibility. From this reason, we propose an individual cerebral arterial model reproduced with elastic membranous configuration of vasculature and the physical characteristics of arterial tissue, expanding its potential for simulating endovascular intervention.

In this modeling, we reconstructed the three-dimensional structure of basilar artery with giant aneurysm with an approximate diameter of 15 mm by the hybrid method that combines multi-slice CT information (resolution: 0.3 mm/pixel, slicing pitch: 0.5 mm) and MRI information (resolution: 0.48 mm/pixel, slicing pitch: 0.7 mm) . Then we rapid-prototyped a vascular master mold using this information. In this procedure, we coated the RP master mold with the silicone elastomer. In this process, we fabricated a uniform thin membranous structure of 100 μm thickness by dipping the master mold in liquid-state silicone and drawing it up at constant velocity (1.0 mm/sec). We repeated this dipping, which is followed by additional polymerization until it attains the desired wall thickness. In this representative case, it yielded thin silicone membrane with uniform 300 μm thickness around the master mold. After the membrane was materialized, inward RP master mold was eliminated. In this way, we fabricated an *in vitro* patient-specific anatomical model of the human cerebral artery with a biologically accurate membranous configuration (Fig. 3.10).

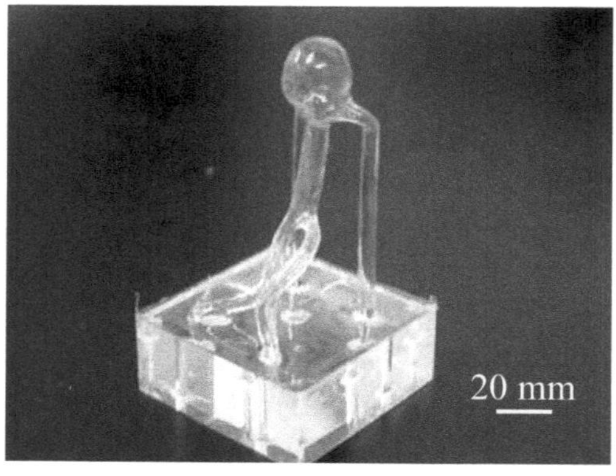

Figure 3.10 *In vitro* patient-specific anatomical model of the cerebral artery reproduced with a membranous configuration for simulating endovascular intervention.

3.7 Reproduction of Surrounding Brain Structure

Arterial model with soft membrane structure proposed in the previous section allows simulating variety of medical treatment including balloon PTA (percutaneous transluminal angioplasty) and aneurismal neck clipping. However, since its structure does not reflect the existence of surrounding soft tissues (mainly composed of very supple arachnoid tissue filled with cerebrospinal fluid), which actually supports most of the external force exerted on arterial wall, it is difficult to keep its anatomically accurate configuration in the atmospheric condition. In human body, the elasticity of vascular membrane and the existence of the surrounding soft tissue make vascular structure to deform when catheter or other medical tools are operated. And its soft structure even deforms against pulsation and blood streaming. Some neurovascular treatments are executed using this property, e.g., the expansion of narrowed artery (stenosis) with balloons and stents and aneurismal clipping. However, this property poses some technical difficulties, too. Therefore, it is necessary to reproduce the composite elastic structure composed of vascular membrane and surrounding brain structure.

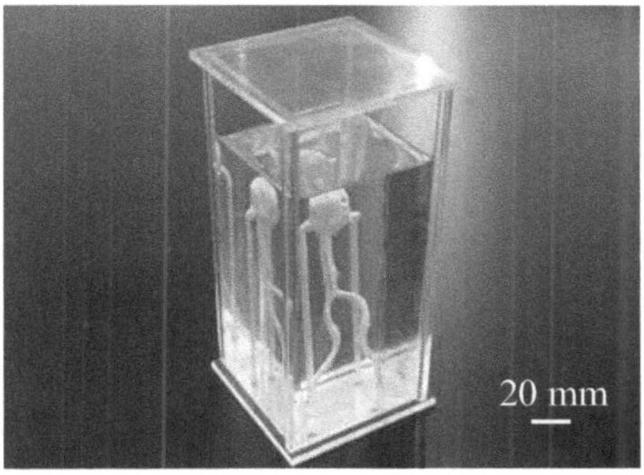

Figure 3.11 Arterial mode with composite structure reproducing membranous arterial structure and surrounding visco-elastic structure.

To reproduce these characteristics, we constructed a membranous cerebral arterial model surrounded by brain-like soft structure (Fig. 3.11). In this procedure, first we arranged the above-fabricated membranous vascular model inside a cubic mold according to its orientation and embedded it with liquid-state transparent silicone, which solidifies into brain-like soft gel structure through addition polymerization. Here, the physical characteristics (elastic modulus) of the selected silicone gel are adjusted to reproduce the physical characteristics of brain tissues (gray matter), and this material presents good transparency too. As the reflection indexes of silicone elastomer and silicone gel take almost identical value (silicone elastomer (nD25): 1.410, silicone gel (nD25): 1.404), no reflection and no refraction take place at its boundary and that realizes good visibility on its body.

3.8 Reproduction of Subarachnoid Space

The cerebral arterial model reproduced with a pliant membranous configuration presented in the previous section enables the simulation of a variety of medical treatments, including balloon percutaneous transluminal angioplasty (PTA) and aneurismal clipping. Its structure did not reflect the existence of circumferential pliant tissue that actually supports most external force exerted on the arterial structure, making it difficult to maintain an anatomically correct configuration in actual use, since its low rigidity simultaneously causes deformation against gravity. Much of the human cerebral artery is surrounded by a vacant subarachnoid space, formed by the wrinkles of the brain. This space contains cerebrospinal fluid and other pliant tissues that support the cerebral arterial structure inside tissues from external loads exerted by blood flow and medical treatment. Considering this, in this section, we propose a method to reproduce this subarachnoid space within a cube-shaped structure proposed in the previous section. This configuration enables the interventionalist to simulate the deformation of cerebrovascular disease and area of interest precisely against blood flow and surgical treatment, excluding the effects of distortion caused in non-applicable areas.

We fabricated the vascular model reproduced with subarachnoid space by the following procedure. First we reconstructed the three-dimensional structure of subarachnoid space using digital slice image obtained using MRI and then rapid-prototyped a lost-wax mold of this structure as disassembled state. Then we placed this mold around the vascular master mold covered by thin silicone membrane. After this, we removed both the molds (subarachnoid mold and vascular mold) through selective dissolution, leaving silicone structure. Figure 3.12 shows a representative of this model that partially reproduces the subarachnoid space around vascular membrane inside a cube-shaped silicone model.

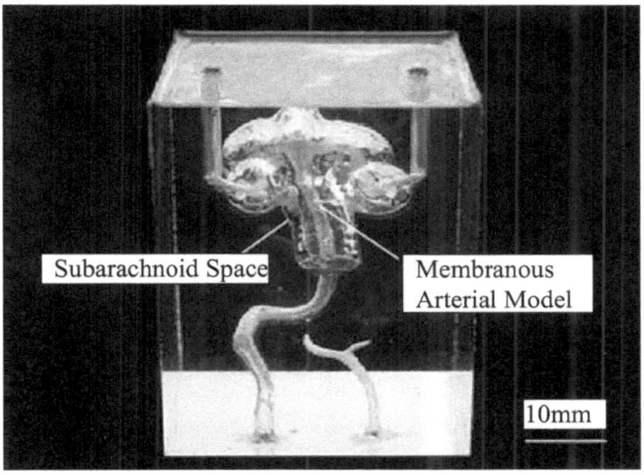

Figure 3.12 *In vitro* cerebral arterial model with a membranous vascular configuration reproduced with circumferential subarachnoid space.

Here, the presented vascular model presents unfavorable reflection at its subarachnoidal surface, making it difficult to observe the inward state. This reflection is eliminated by filling the subarachnoid space with a liquid having the same reflection indices as the silicone elastomer (nD25): 1.410; water solution (glycerine: 50wt%) and silicone oil serve aptly for such a fluid (Fig. 3.13).

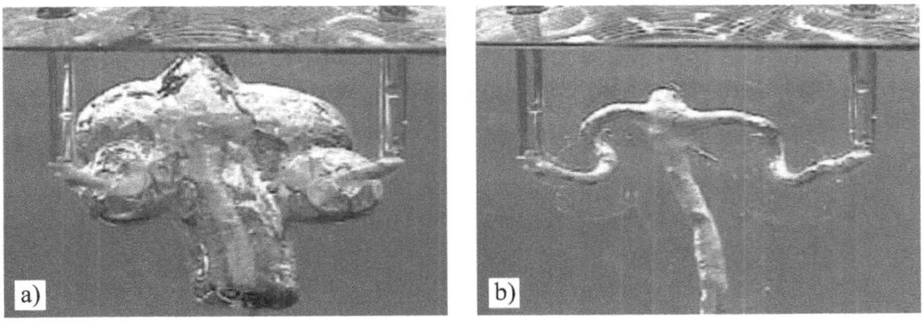

Figure 3.13 Elimination of reflection on subarachnoid space by filling the space with a liquid (silicone oil) having the same reflection indices as the model provides. (a) State before filling the space. (b) State after the space is filled with silicone oil.

3.9 Improvement of Visibility

Both the composite vascular model and the solid arterial model presented in this chapter realize good transparency on their bodies, but these models can cause deterioration in visibility under atmospheric condition with specific illuminating condition and specific observing direction. This visual deterioration is mainly caused by optical reasons and is classified into four categories according to the location of occurrence and the type of phenomenon causing deterioration:

1. Refraction at the external model surface: Makes the image of arterial structure distorted and reflects that image on several outward surfaces with deviation
2. Refraction at the internal arterial surface: Distorts or magnifies the image of inserted medical tools
3. Reflection at the external model surface: Deteriorates the visibility of arterial structure and inserted medical tools
4. Reflection at the internal arterial surface: Deteriorates the visibility of inserted medical tools

As it is evident, the deterioration of visibility mainly arises from refraction and reflection taking place at both internal and external surfaces of presented arterial model. The deteriorated visibility caused by these factors makes it difficult to observe the inward arterial structure and also the movement of inserted medical tools, and consequently this makes surgical simulation rather difficult.

However, as all the above categorized factors originate in the difference of refraction index between the presented arterial model itself and its circumferential environment (atmosphere), these unfavorable optical effects are eliminable with the use of optically adjusted liquid. Figure 3.14 shows the effectiveness of this liquid. In this case, we adopted glycerol solution (mixture of water (44%) and glycerin (56%), which has same the reflection index as silicone elastomer; refraction index (nD25): 1.403) as that liquid. Since the presented arterial model is made of silicone (refraction index (nD25): 1.404) too, these two material take similar refraction index for a wide wavelength range. A salt solution and a glycerol solution are also effective as this optically adjusted liquid; even water (refraction index (nD20): 1.333) can be great help to improve visibility.

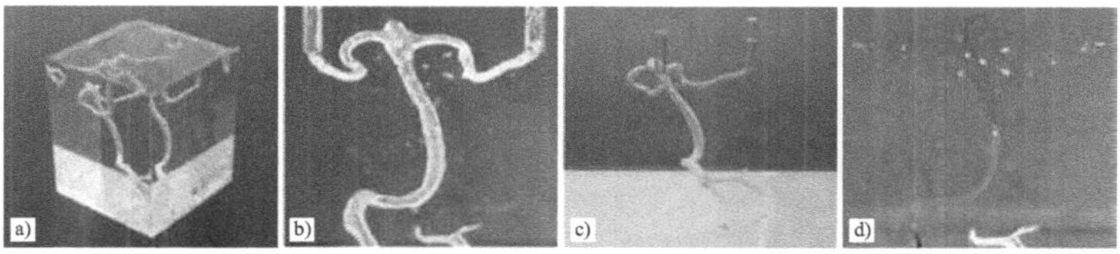

Figure 3.14 Elimination of reflection and refraction effects: (a) Original state, (b) after submerging in silicone oil, (c) original state, and (d) after injecting silicone oil.

3.10 Silicone Membrane Thickness Controllability

The production methodology presented makes possible to construct very thin membranous silicone structure with uniform wall thickness. The minimal attainable wall thickness was confirmed to be about 100 μm, and this thickness is sufficient to replicate the cerebral arterial structure treated in intravascular neurosurgery. Furthermore, its thickness is controllable from less than 50 μm to more than 250 μm by adjusting withdrawing velocity in dip-coating process. This also indicates that the proposed method enables layering several membranes with different thickness and different material property, allowing, for example, reproducing the three-layered general configuration of the artery (Figs. 3.15–3.17). We verified its validity for the neurosurgical treatments that affirmatively utilize the deformation of the arterial structure. Figures 3.18–3.20 show the deformation of the proposed membranous structure against expansion. As shown in these figures, the membranous structure deforms with moderate elasticity accrued from the elastic property and wall thickness close to actual artery, and this realistic deformation allows realistic surgical simulation and the simulation of various surgical treatments.

The proposed composite structure consists of a membranous arterial structure, and the circumferential soft structure showed good results for the reproduction of overall visco-elastic structural deformation. Figure 3.18 shows the deformation of arterial structure within the presented composite arterial model against the insertion of LDPE catheter (1.0 mm in diameter). As shown in this figure, the shape of the reproduced artery was widely transformed when catheter was inserted, and this transformation was accompanied by visco-elastic transition as seen in human artery. Furthermore, its structure even deformed against pulsation and blood stream dynamically (Figs. 3.19 and 3.20).

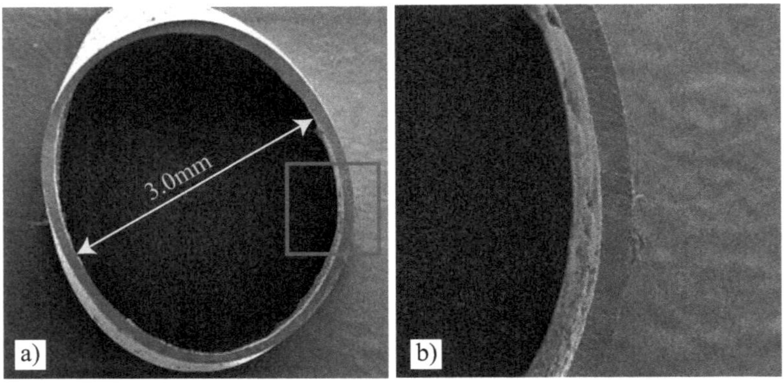

Figure 3.15 Micro-membranous structure fabricated with proposed RP-based production method: (a) SEM cross section. (b) Enlargement of wall cross section.

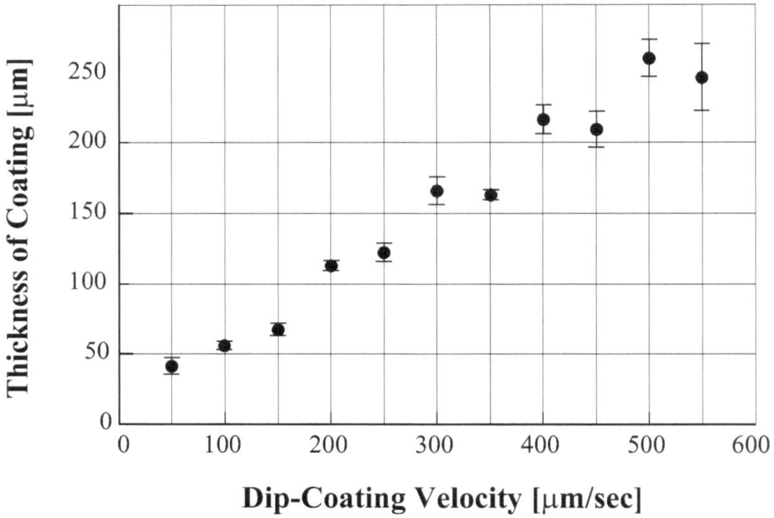

Figure 3.16 Correlation between thickness of micro-membrane and withdrawing velocity in dip-coating fabrication.

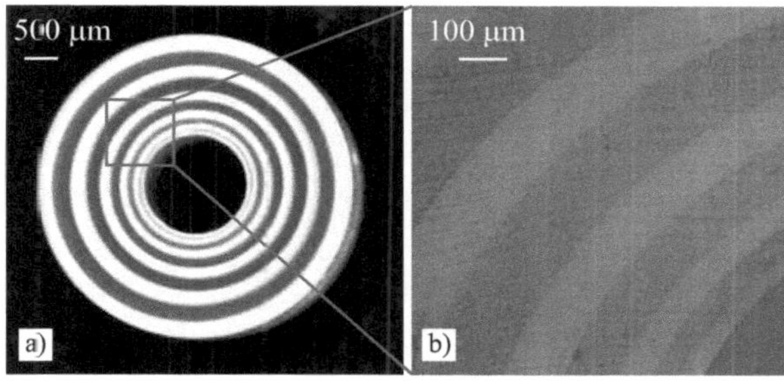

Figure 3.17 Micro-membrane cross-sectional thickness variation with dip-coating velocity: (a) Membrane cross section. (b) Enlargement of wall cross section.

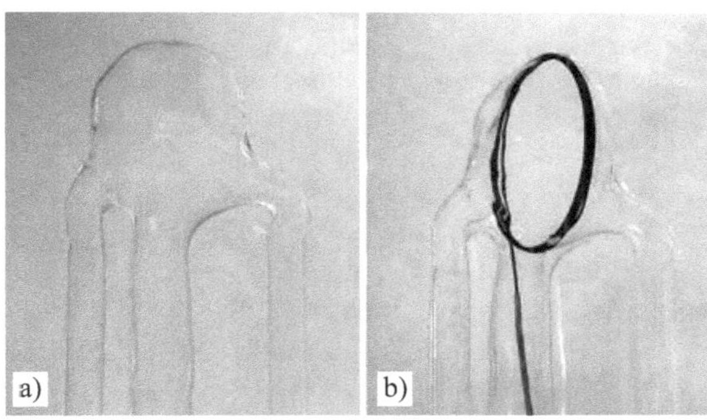

Figure 3.18 (a) Original aneurysm membranous model shape (diameter: 15 mm). (b) Deformation of the model after coil embolism simulation.

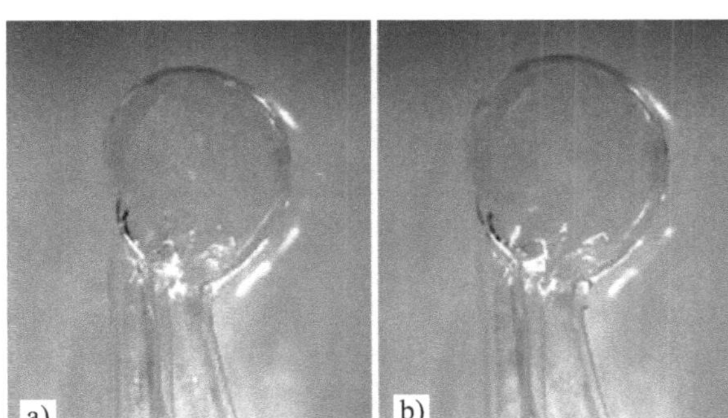

Figure 3.19 Deformation of aneurysm membranous model (original diameter: 15 mm) against pulsed blood flow with flow rate of 250 mL/min and beat rate of 75 bpm: (a) Systole, (b) Diastole.

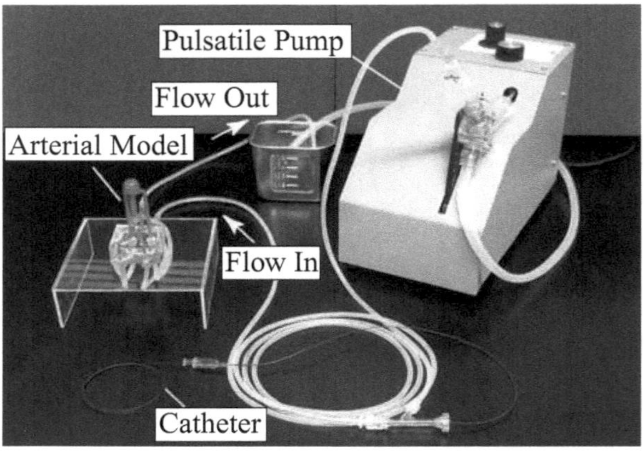

Figure 3.20 Hardware setup for simulating endovascular intervention with proposed vascular model in the existence of pulsed blood flow.

The proposed vessel model also reproduces the physical characteristics of artery tissue, such as elastic modulus, Poisson's ratio and friction coefficient, within 5% of errors. Consequently, it allows simulating neurovascular procedures with realistic dynamic characteristics: the delicate feel of operating catheters and the dynamic behavior of surgical instruments, such as slip/stick motion or unwinding motion. This modeling technique has sub-millimeter modeling resolution for patient-tailored models of blood vessels and reproduces physical characteristics of artery tissues; therefore it is useful for simulating neurovascular intervention for surgery planning and for simulator-based training.

3.11 Clinical Evaluation

In following section, several vessel models of actual cerebrovascular diseases are constructed by using the patient-specific modeling method presented previously. Then the proposed vascular model is evaluated by simulating endovascular intervention procedures with this model, and the performance and utility are discussed from the viewpoint of clinical use. Vascular models of five cases of conventional cerebrovascular disease are studied in the solid model type and membranous model type.

3.11.1 Preparation

Vascular models of five cases of conventional cerebrovascular disease were constructed for this modeling technique evaluation. All these representative cases have an aneurysm. To compare the difference in clinical, the following aneurysms, which have different size and generation part, were selected for modeling. For the comparison of vessel deformation reproduction and the necessary characteristics required for surgical simulation in models with solid and membranous structure, a model of each type was produced for each case. The medical imaging data used for the models construction, the solid structure models and the membranous structure models for all cases are shown in Figs. 3.21–3.25. Table 3.1 summarizes the dimensions of the aneurysms and other characteristics for each case.

Table 3.1 Case study description

	Case 1	Case 2	Case 3	Case 4	Case 5
Aneurysm type	BT	BT	BT	Left MCA	Left MCA
Mean diameter (mm)	4.3	6.8	3.1	3.8	6.9
Length (mm)	4.79	6.64	3.20	3.09	—
Breadth (mm)	3.85	7.01	3.02	4.61	—
Neck (mm)	3.10	4.15	3.73	2.91	—
Diameter 1 (mm)	—	—	—	—	6.68
Diameter 2 (mm)	—	—	—	—	7.14
Age (years)	55	45	73	78	59
Sex	Female	Female	Female	Male	Male

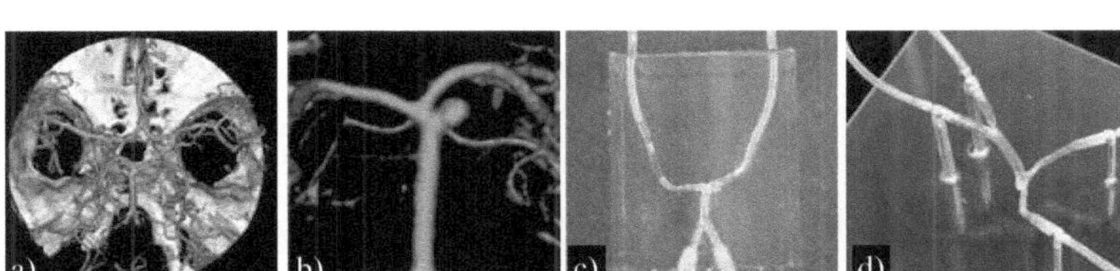

Figure 3.21 Case 1: BT aneurysm: (a, b) Medical imaging data. (c) Block model. (d) Membranous model.

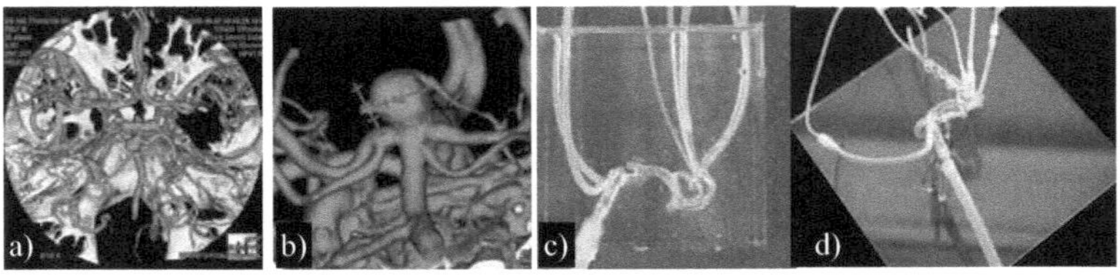

Figure 3.22 Case 2: BT aneurysm: (a, b) Medical imaging data. (c) Block model. (d) Membranous model.

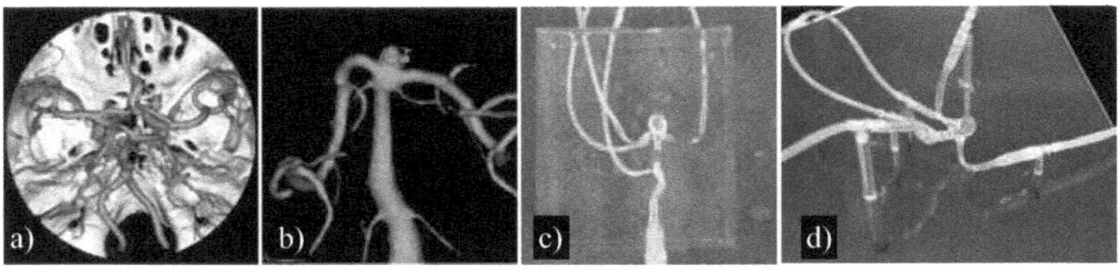

Figure 3.23 Case 3: BT aneurysm: (a, b) Medical imaging data. (c) Block model. (d) Membranous model.

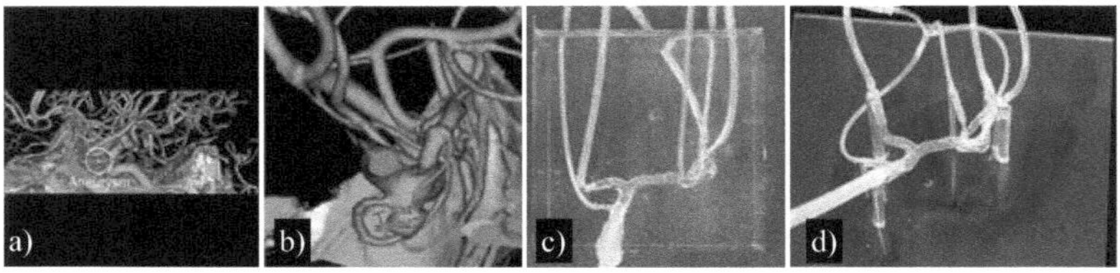

Figure 3.24 Case 4: Left MCA aneurysm: (a, b) Medical imaging data. (c) Block model. (d) Membranous model.

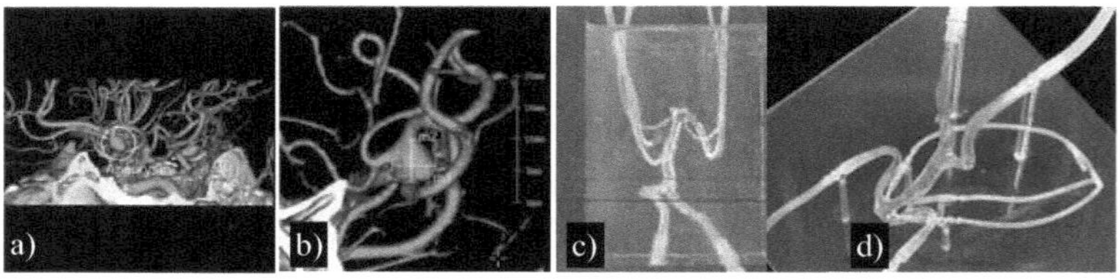

Figure 3.25 Case 5: Left MCA aneurysm: (a, b) Medical imaging data. (c) Block model. (d) Membranous model.

3.11.2 Flow Visualization

Blood flow visualization is executed with above-mentioned basilar tip aneurysm model of case No. 2. Figure 3.26 shows the composition of experimental setup for this flow visualization experiment. To remove the vascular deformation according to gravity and to secure the visibility inside vessel lumen, the vascular model was submerged under liquid, which has the same refraction index as the vessel model. Moreover, the lumen of the vascular model was filled with water solution that reproduces the viscosity of blood (glycerin: 33wt% and surface-active agent: 0.1wt%). Human blood stream was reproduced with pulsatile blood pump (Harvard Apparatus, model 1405), which is commonly utilized in medical treatment. Here, to reproduce the blood flow rate of the basilar artery, a pulsatile blood flow of 200 mL/min with a beat rate of 75 was generated with the pulsatile pump. And a pressure sensor (Fujikura Ltd., model-PSM) was installed on the entrance part of arteria basilaris, to measure the pressure transition during flow visualization. The hypertension of 140 mmHg was set with this pressure sensor. Figure 3.41 shows the flow velocity wave on the entrance of arteria basilaris measured with ultrasound sensor. This waveform indicates that human pressure fluctuation is satisfactorily reproduced.

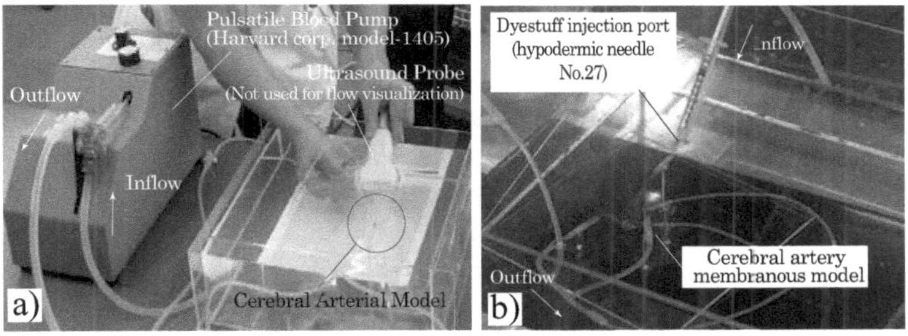

Figure 3.26 (a) Experimental setup for flow visualization. (b) Dyestuff injection into the vascular model for flow visualization.

Under this experimental setup, a hypodermic needle No. 27 was introduced into the vessel lumen from the entrance part of arteria basilaris, and its tip was fixed to the center of lumen. In this state, above-mentioned water solution including white dyestuff was injected into the flow, and by this way, the blood stream was made visible. Figure 3.26b shows the appearance at the dyestuff injection. The details of the flow in the BT aneurysm are shown in Fig. 3.27, where the flow is visualized for the solid vascular model and membranous vascular model constructed for case No. 2.

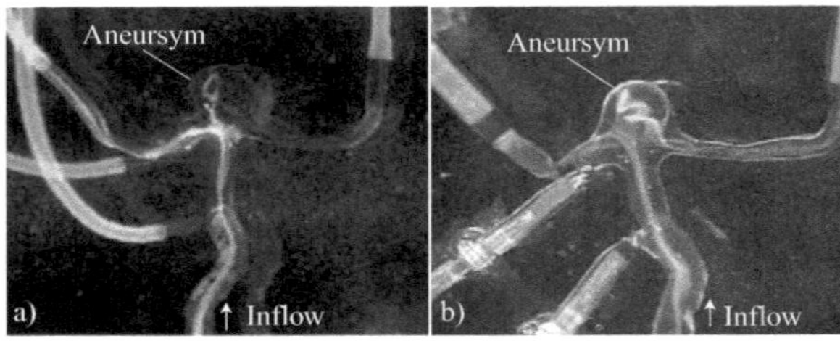

Figure 3.27 Visualized blood stream inside an aneurysm: (a) Block model. (b) Membranous model.

This experiment shows significant result for medicine, as it has been pointed out that the flow in artery changes by the vessel deformation as shown in Figs. 3.27 and 3.28. The membranous vascular model shown in Fig. 3.27b reproduces the flexible vascular wall structure of human. As a result, it reproduces the pulsatile vascular deformation similar to the living body along with the pulsatile blood streaming. Additionally it was confirmed that the membranous vascular model generates a high-speed swirling flow inside aneurysm. On the other hand, the lumen shape hardly deforms for the solid vascular model of Fig. 3.27a. Under such a condition, the high-speed swirling flow observed for the membranous vascular model has not appeared in its aneurysm. Thus, this result shows that the flow in blood vessel changes greatly by the existence of vascular deformation. Then, it is deduced from this result that it is indispensable to execute blood flow analysis to reproduce the pulsatile deformation of vascular structure. Consequently, it was confirmed that the proposed patient-specific cerebral arterial model with membranous configuration is the most suitable for hemodynamic studies and evaluation of medical treatment.

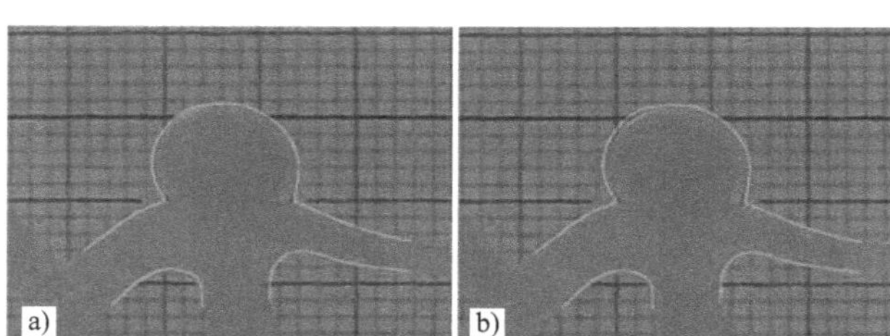

Figure 3.28 Deformation of membranous vascular model case No. 2 with the simulated pulsed blood streaming: (a) Diastole, (b) Systole.

3.11.3 Medical Treatment Simulation

This section reports on the result of simulating coil embolization, an endovascular treatment for aneurysm, by using the vascular models of the five representative cases shown in Section 3.11.1. Moreover, the simulation of aneurismal clipping, conventional craniotomy treatment for aneurysm, is also executed, and it experimental result is detailed. By this way, the utility of proposed vascular model for the simulation of clinical medical treatment is evaluated. To exert this experiment propriety, following experiments are executed by the interventionalist using an experimental setup similar to the one described in the previous section.

3.11.3.1 Aneurismal coil embolization simulation

Under above-mentioned composition, a micro-catheter was inserted through a 30 cm-long silicone tube connected to the entrance port of the vascular model, and coil embolization was executed to the aneurysm. Here, the Y-shaped connector, which is commonly used in actual surgery, was installed at the end of the silicone tube, and the medical equipment was operated by the similar procedure as the actual surgery.

Figure 3.29 shows the progress of the simulation of the surgical procedure for coil embolization executed for case No. 2. According to the interventionalist's criteria, the two-dimensional coil (spring shaped coil widely used for embolization treatment) deformed into intricate cage shape during the embolization treatment, and deploying in a similar way than actual surgery. Consequently, it was confirmed that the process of coil embolization was satisfactorily reproduced.

Figure 3.30 shows the visualized blood streaming after this treatment, which is managed by injecting minute air bubble from a Y-shaped connector port into the flow. As confirmed from this figure, the blood flow in the aneurysm was satisfactorily intercepted after the embolization treatment, and the interventionalist could confirm that coil embolization had been achieved. Consequently, it was confirmed that the proposed vascular model allows simulating embolization treatment at high-level reproduction.

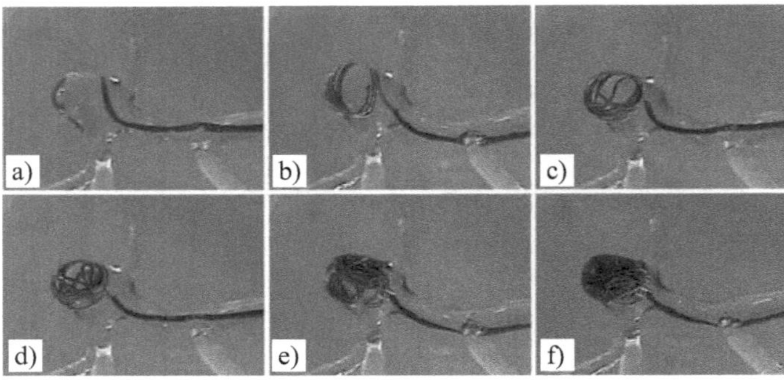

Figure 3.29 Progress of coil embolization executed for representative case No. 2. Embolism coil forms a cage-shape structure inside aneurysm as the treatment progresses. Time: (a) 0 s, (b) 23 s, (c) 5 min 54 s, (d) 6 min 7 s, (e) 6 min 36 s, and (f) 6 min 59 s.

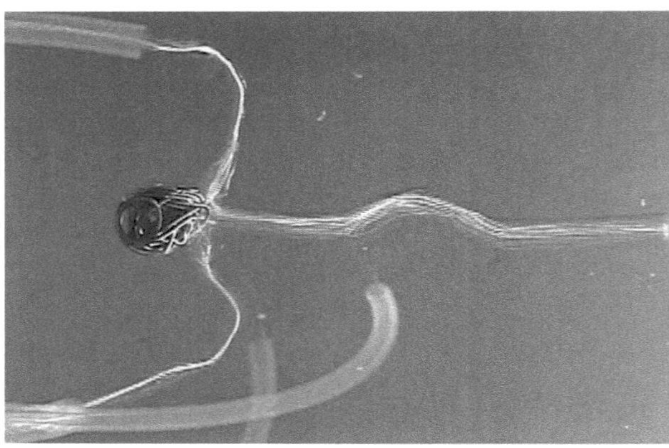

Figure 3.30 Visualized blood streaming after aneurismal embolization treatment for case No. 2. This visualization is executed by injecting a minute air bubble from a y-shaped connector port into the flow.

Coil selection: In embolization treatment, the selection of a coil with different appearance (with respect to shape, length, diameter, functionality, etc.) according to the morphology of aneurysm is required. Through this evaluation, it was confirmed that the proposed vascular model was useful for the selection of an appropriate coil for each case. The final state of coil embolization exerted with three different coils for a same aneurysm (case No. 1) is shown in Fig. 3.31. In an actual treatment, when the selection of the coil is improper, cage with appropriate shape may not be formed or a part of the coil may bulge out of the aneurysm. Figure 3.31 indicates that a similar condition is well reproduced by the proposed vascular model. In Fig. 3.31a, an appropriate cage shape is not formed because the 2D coil diameter is rather small. As shown in Fig. 3.31b, not the entire aneurysm was embolized because the selection of the coil was too short; as for Fig. 3.31c, the coil embolism was appropriately achieved. The main specification of the coil selected through the embolism simulation is summarized in Table 3.2. Consequently, it is confirmed that our vascular model is helpful for the evaluation of embolization treatment.

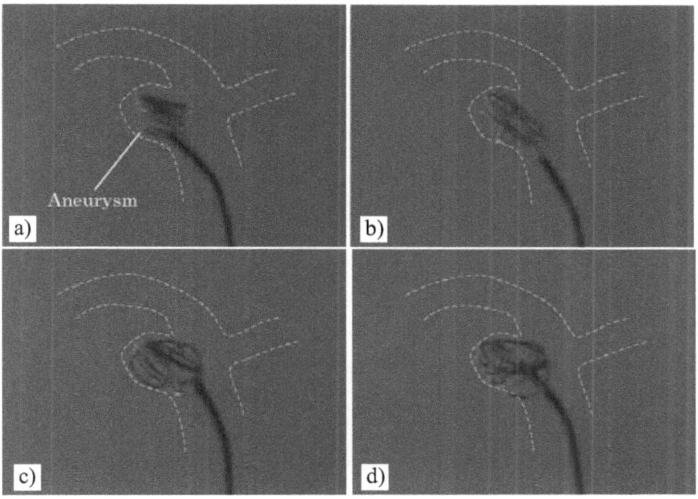

Figure 3.31 The final state of coil embolization exerted with three different coils for the aneurysm model of case No. 1: (a) Embolization with coil A. Appropriate cage shape is not formed because the 2D coil diameter is rather small. (b) Embolization with coil B. Not the entire aneurysm was embolized because the selection of the coil was too short. (c) Embolization with coil C. The coil embolism is appropriately achieved with 2D coil. (d) Embolization with coil D the coil embolism is appropriately achieved with 3D coil.

Coil disengagement: As stated above, the pulsatile flow similar to human blood circulation is reproduced in vessel lumen with presented simulation environment. As a result, it was confirmed that the swing and the inducement of the coil according to pulsatile blood flow, which is seen in actual surgery, were reproduced during surgical simulation. Figure 3.32 shows the swing of the coil after installation according to the blood streaming. In coil embolism, when the coil is appropriately applied to the aneurysm, the inflow of blood into the aneurysm becomes completely intercepted. Therefore, the swinging of the coil means that a part of the coil is bulging out of aneurysm and the coil is not appropriately installed. On the other hand, Fig. 3.33 shows a state where installed coil get disengaged from aneurysm during treatment, by the negative pressure caused by blood streaming. The result shown in these figures indicates that the phenomenon seen in actual surgery is well reproduced on this vascular model, and it validates the effectiveness of presented model as a simulator.

Table 3.2 The main specification of the coil selected through the embolization simulation for representative case No. 1

Trial ID	Dimension	Diameter (mm)	Length (mm)	Result
1st, Coil A	2D	3.0	60	Too small
2nd, Coil B	2D	4.0	60	Too short
3rd, Coil C	2D	4.0	100	Successful embolization
4th, Coil D	3D	5.0	100	Successful embolization

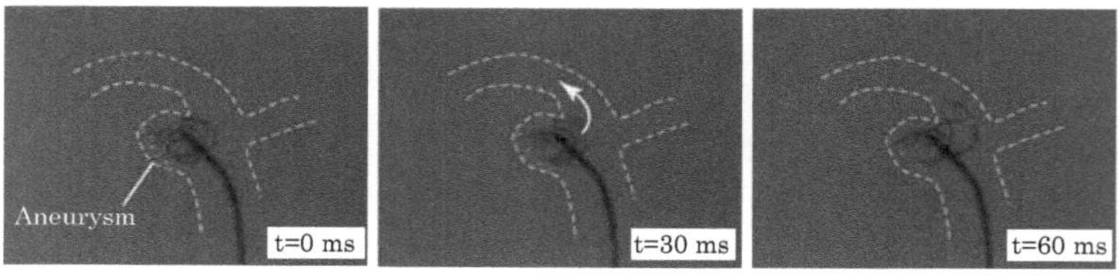

Figure 3.32 Swing of the installed coil after embolism treatment according to the blood streaming.

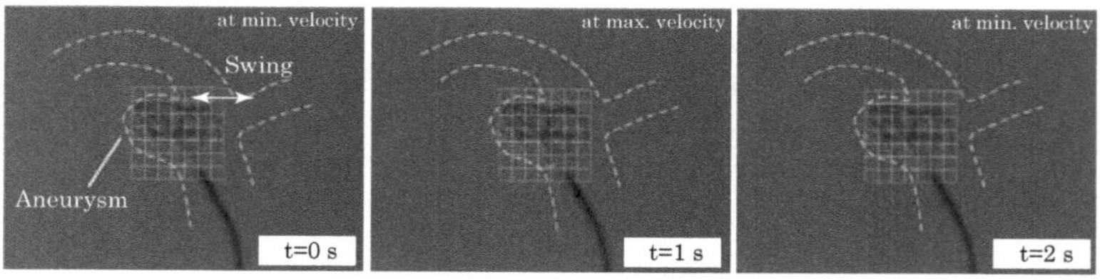

Figure 3.33 Observed state where installed coil gets disengaged from aneurysm during treatment, by the negative pressure caused by blood streaming.

Utility of membranous vascular model: Figure 3.34 makes the comparison between treatment results for both solid vascular model and membranous vascular model constructed for representative case No. 2. As a result of this evaluation, it was confirmed that the appropriate coil was different for these two models although these models reproduce the same aneurysm. The major specification of the coil appropriate to both models is summarized in Table 3.3. This result indicates that the reproduction of the vascular deformation is indispensable in embolization simulation, as well as the result of flow analysis. Here, the difference brought by these vascular models structure, solid and membranous, has a deep meaning. For instance, this result shows that it is improper to assume the rigid vascular structure for hemodynamic analysis, although researchers engage in numerical blood flow analysis reports analysis result made with this precondition.

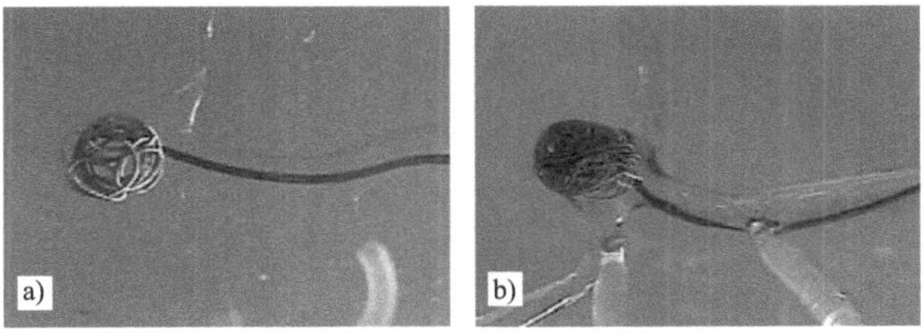

Figure 3.34 Comparison between the embolism treatment results for both solid vascular model and membranous vascular model constructed for representative case No. 2 model. (a) Treatment result for solid vascular model. (b) Treatment result for membranous vascular model.

Table 3.3 The main specification of the coil selected through the embolization simulation for representative case No. 2 model

Trial ID	Dimension	Suitable diameter (mm)
Solid Model	2D	5.0
Membranous Model	2D	8.0

3.11.4 Aneurism Clipping Simulation

Aneurism clipping treatment simulation performance was evaluated with prepared vascular model of case No. 4, under the experiment setup similar to that for coil deploy experiment. Figure 3.35a shows the applied Sugita aneurysm clip. By simulating aneurismal clipping with this clip, the applicability of craniotomy clipping treatment to the vascular model was confirmed. Figure 3.35b shows the appearance after clipping treatment. Figure 3.36 shows the transition of blood streaming condition before and after the treatment. It is confirmed that the inflow into the aneurysm is satisfactorily intercepted by this treatment, and appropriate clipping has been exerted. This result shows that the aneurismal clipping treatment is also applicable to the proposed vascular model.

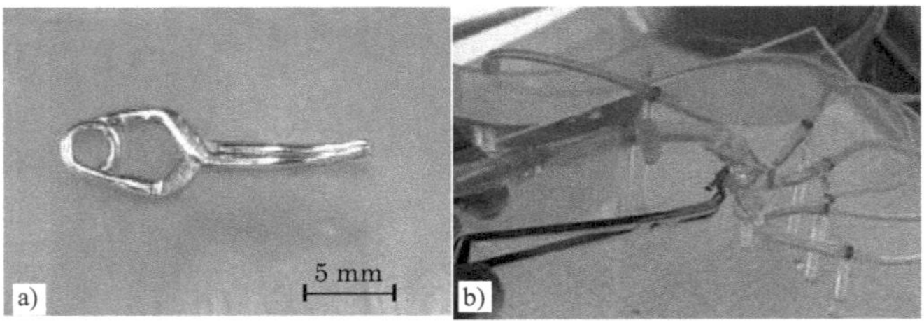

Figure 3.35 (a) Surgical simulation of aneurismal clipping treatment executed with a Sugita aneurysm clip. (b) Appearance after aneurism clipping treatment for vascular model No. 4.

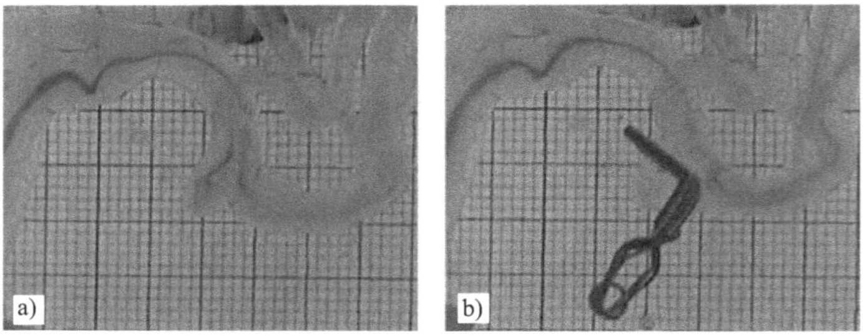

Figure 3.36 Transition of blood streaming condition before and after the aneurismal clipping treatment for representative case No. 2. (a) Before clipping. (b) Treatment result for membranous vascular model.

3.11.5 Applicability for Medical Imaging Modalities

3.11.5.1 Fluoroscopic X-ray Imaging

In endovascular intervention, X-ray fluoroscopy is irradiated to patient's head throughout surgery. Doctors skillfully manipulate the catheter inside the cerebral artery while confirming the real-time image projected to an X-ray monitor and execute the treatment for cerebrovascular disease. It is a peculiar restriction to this surgery not to obtain a clear blood vessel image by originating in the contrast of an X-ray vision system and the limitation of administering the contrast medium. Figure 3.37a shows a DSA image taken during surgery. Since the endovascular intervention is managed under X-ray fluoroscopy as described above, it is preferable that a simulator for endovascular intervention should be applicable to X-raying and other medical imaging modalities utilized in this surgery. As a result, the simulation with the use of the contrast medium is practically achievable under a similar IVR environment, and then the simulation under above-mentioned restriction peculiar to this surgery is achievable. Figure 3.37b shows a fluoroscopic image of presented solid vascular model taken by a conventional 2D fluoroscope. It is confirmed from this figure that the vascular model composed of silicone elastomer gives a satisfactory image. However, it was confirmed that the imaging condition for obtaining good contrast was somewhat different from the case of humans, because the X-ray permeability of silicone elastomer is rather lower than human body.

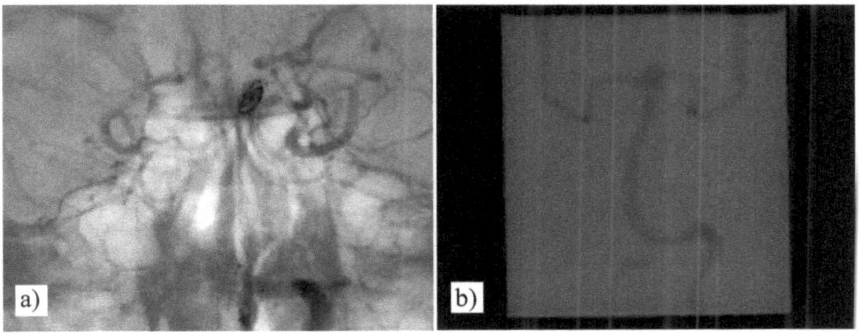

Figure 3.37 (a) DSA image utilized in endovascular intervention (b) Fluoroscopic image of presented solid vascular model taken by conventional 2D roentgenography.

In Fig. 3.37b, as imaging condition, G value and S value were set to 4.0 and 486, respectively. On the other hand, the result of applying multi-slice CT imaging to the proposed vascular model is shown in Fig. 3.38. It is confirmed from this figure that the proposed model is also applicable to this imaging modality and gives good contrast. Moreover, it was confirmed that the CT value of silicone elastomer was 180 HU from this experiment. From this result, when the vascular model with solid structure is applied for IVR surgical simulation, it is necessary to separate the CT value between the silicone elastomer and contrast media, for instance, by adjusting the density of the contrast medium, since these materials likely to take close CT value.

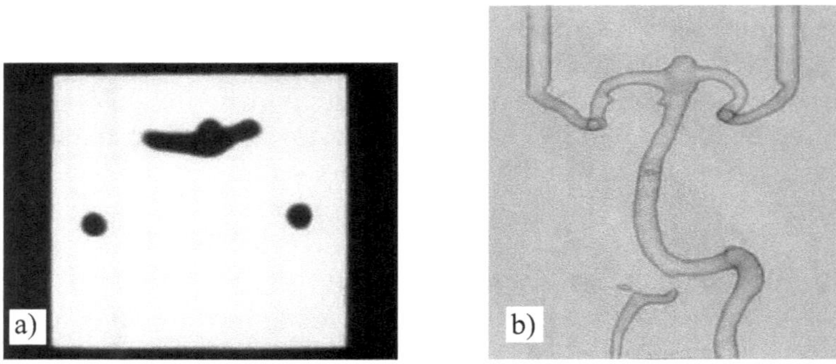

Figure 3.38 Fluoroscopic image of presented solid vascular model taken by multi-slice CT imaging: (a) Cross-sectional image. (b) Three-dimensional structure obtained by the reconstruction based on a set of cross-sectional image.

3.11.5.2 Ultrasound imaging

Using the same experimental setup, the applicability of the presented vascular model to ultrasound imaging was tested. Ultrasound imaging was executed under this experiment composition by placing the ultrasound probe tip near the vascular model, which is submerged under water. To improve reflection strength of the ultrasound wave to the fluid that is circulating inside vascular model, a small quantity of aluminum powder was mixed in the circulating fluid. Also to help the transmission of the ultrasound wave inside the model, a special jelly was spread over the model. Figure 3.39 shows the appearance of the experiment. It was confirmed that the membranous vascular model is applicable for this ultrasound imaging; on the other hand, the solid vascular model is difficult applying this imaging modality because the reflection wave was very low as the transmission of ultrasound wave in silicone elastomer was not good. Posterior tests revealed that the single boundary offered by the solid configuration offers better imaging quality for intravascular ultrasound systems. Chapter 7 shows how these models combined with intravascular ultrasound systems are used for *in vitro* imaging of vasculature.

Figure 3.40 shows the ultrasound image obtained for the membranous model of representative case No. 2. Figures 3.40a,

3.40b, and 3.40c are the ultrasonic data measured inside aneurysm, at the inlet and outlet of aneurysm, respectively. The colored area indicates the direction of the blood streaming obtained by ultrasound Doppler method. The red area indicates forward flow and the blue area indicates backward flow. Moreover, the flow velocity transition on some local domains obtained by the ultrasonic Doppler method is shown in Fig. 3.41. The domains for measurement are shown in Fig. 3.40 by "I" sign. The maximum and minimal flow velocities at the inlet of aneurysm were measured at 170.5 cm/sec and 34.3 cm/s, respectively. This value is found correctly measured, considering the inflow rate of the pulsatile pump. Through this experiment, it was confirmed that the vascular model with membranous structure is applicable to ultrasound measurement.

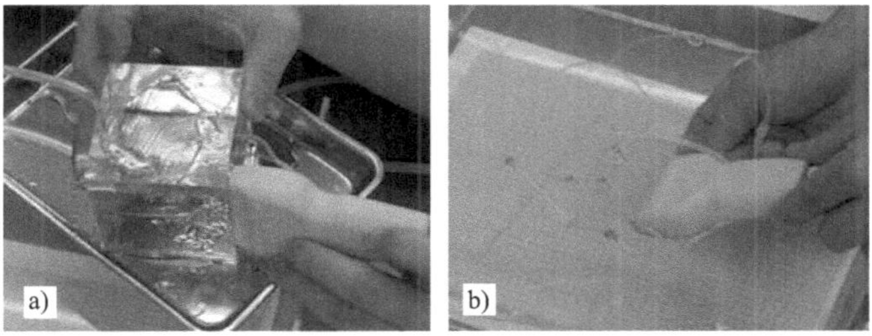

Figure 3.39 Appearance of ultrasound imaging experiment: (a) Ultrasound imaging for solid vascular model of case No. 2. (b) Ultrasound imaging for membranous vascular model of case No. 2.

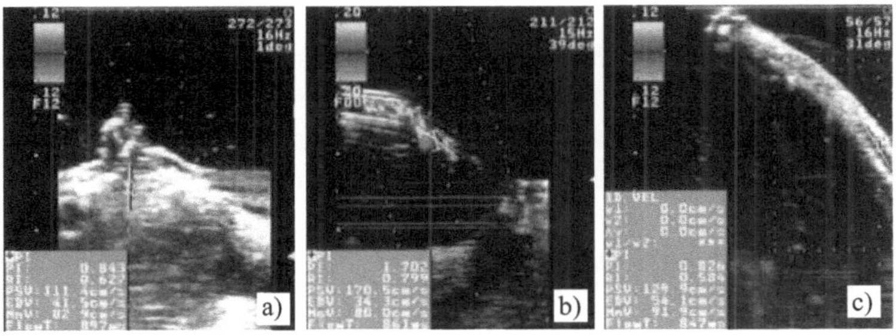

Figure 3.40 Ultrasound image obtained for the membranous model of representative case No. 2. Colored area indicates the direction of the blood streaming obtained by ultrasound Doppler method (red area indicates forward flow and blue area indicates backward flow). Ultrasound images: (a) Inside the aneurysm. (b) At the inlet of aneurysm. (c) At the outlet of aneurysm. See also Color Insert.

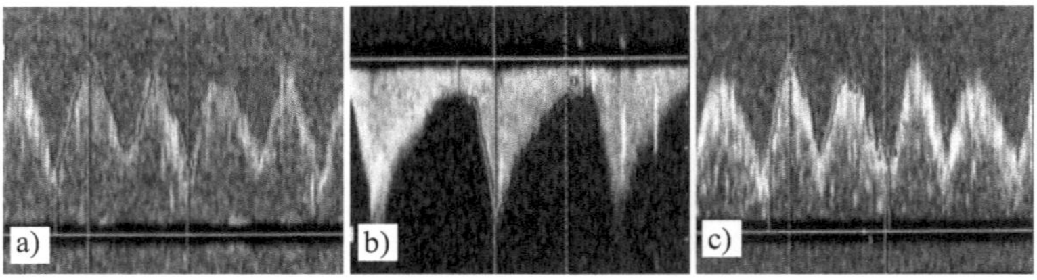

Figure 3.41 Flow velocity transition on some local domains (shown in Fig. 3.40 by "I" sign) obtained by the ultrasonic Doppler method: (a) Inside the aneurysm. (b) At the inlet of aneurysm. (c) At outlet of aneurysm.

3.11.5.3 Clinical evaluation summary

As the result of evaluations it was confirmed that proposed vascular model is at least useful in the simulation of endovascular intervention treatments intended for the following purposes.

1. Evaluation of blood streaming
2. Evaluation of aneurismal embolization treatment
3. Evaluation of aneurismal clipping treatment
4. Patient individual's diagnosis and surgical planning
5. Surgical evaluation under IVR environment
6. Surgical evaluation under ultrasound imagining environment

The proposed vascular model is useful for the individual patient's examination such as diagnosis, informed consent, surgical planning and preliminary simulation, and its usefulness was confirmed as the environment for medical research and studies such as hemodynamic study and pathological study. Major cerebrovascular diseases treated in endovascular intervention are stenosis and thrombosis in addition to aneurysm. Here, the simulation of stenosis treatment by the current vascular model is not complete, because it is necessary to reproduce the cholesterol and calcareous material that compose the stenosis in addition to the vessel reproduction. On the other hand, the simulation of arterial revascularization has not been evaluated because of the difficulty in forming (installing) blood clot (thrombosis) inside the proposed vascular model. However, the blood stream similar to that in humans is reproducible in the vascular model, and then it is predictable that the simulation of the thrombus treatment such as the administration of thrombolytic drug (e.g., urokinase and t-PA) through the catheter might be well reproduced. Meanwhile, it was confirmed that at least the following differences were found between the experimental results obtained with a solid vascular model and a membranous vascular model respectively, which reproduced the same case:

1. State of blood streaming (especially, flow in aneurysm)
2. Appropriate aneurismal embolism coil

This result shows that the vessel deformation during surgery according to blood pulsation and catheter operation is too big to be disregarded. Therefore, it was confirmed that the use of membranous vascular model, which is able to reproduce the vascular deformation, is appropriate in executing endovascular intervention simulation. Moreover, although most of researchers have been disregarding the existence of vessel deformation for hemodynamic analysis, this result strongly suggests that such a precondition would be improper. Thus, it was confirmed by this experiment that the proposed vascular model is also valuable for the researcher of science and engineering discipline besides the interventionalist and other medical researchers. Meanwhile, although this chapter did not mention, various medical equipment manufacturers (engaging in catheter, medical tools, medicine, imaging system, etc.) pointed out that the proposed vascular model is very useful for the performance evaluation of medical equipment.

To evaluate the clinical utility of the vascular model proposed in this research, five different actual aneurismal models were constructed from CT information. And, several clinical experiments (blood flow visualization, simulation of aneurismal embolization and clipping, application to surgical planning and application to medical imaging modalities) were executed, and the characteristic and the utility of the vascular model were evaluated from different viewpoint. Through these experiments, we confirmed that the proposed vascular model is useful for simulating endovascular intervention procedures, such as treatment for aneurysm and thrombosis, and valuable for diagnosis and surgical planning for individual patient. Especially, it was confirmed that the flow state and appropriate medical equipment become different according to this vessel deformation.

3.12 Comprehensive Surgical Simulator — EVE

From the previous analysis of vasculature modeling techniques, we propose an *in vitro* patient-specific vascular model of human major vasculature for the simulation of endovascular intervention procedures. The medical equipment, including the catheter and the embolism coil, is inserted from a femoral artery and introduced into the cerebral area passing through aorta and carotid artery. Therefore, it is desirable to reproduce not only the cerebral arterial structure but also the entire artery that operates as an insertion route of the catheter, in order to precisely reproduce the feature of this surgery. We propose the overall simulator that achieves the comprehensive simulation of endovascular intervention treatments by integrating the element technologies proposed in foregoing sections and then evaluate its utility. Almost all the medical equipment used in this surgery, such as catheters, coil and stent, has a flexible structure, and so, these tools are transformed and induced by the pulsatile blood flow and pulsed motion of vessel wall; therefore it shows a very complicated behavior. The vascular model proposed in foregoing sections cannot reproduce this characteristic peculiar to this surgery alone. No such systematization reproduction of all blood vessels related to endovascular intervention has been reported so far. Considering this situation, we propose an environment that achieves overall surgical simulation by integrating each element technology proposed in the preceding section. Afterward, the utility of such a simulation system is discussed in detail through a diversified evaluation.

3.12.1 Hardware Construction: Systematization

It is valuable for the surgical simulator for endovascular intervention to reproduce whole human arterial structure, from femoral artery to carotid artery in addition to the cerebral artery. And, it is necessary to reproduce the pulse blood stream similar to that in humans. On the other hand, for the cerebral part, where the scrupulous attention and the evaluation of given stress are required during surgery, it is useful to evaluate the stress given to the blood vessel by operating medical equipment inside the lumen. In this chapter, the method of constructing such a simulator is described by systematizing and developing each technology proposed in previous chapters. It is preferable that the comprehensive simulator reproduces the whole arterial area from femoral artery to carotid artery, of which diameter exceeds 1 mm, where catheters may be inserted by this surgical methodology. With this purpose, the high-resolution fluoroscopic image of the whole human body was acquired in this research by using 64-row multi-slice CT (subject: 53 year olds, male, no disease). Here, to acquire the high-resolution data of the cerebral artery, this imaging was executed separately for the head area (high-resolution) and the body area (normal resolution). For the head area, imaging resolution was set to 0.468 mm, and slicing pitch was set to 0.3 mm, in consideration of satisfactorily extracting three-dimensional shape of blood vessels with up to 1 mm internal diameter, and totally 1134 sheets of fluoroscopic images were taken. And for the body part, the resolution was set to 0.625 mm and the slicing pitch was set to 0.8 mm and we obtained totally 2930 sheets of images. Figure 3.42 shows the sagittal section image of the object human body reconstructed from this information. As confirmed from this figure, this data set contains the required entire vessel with satisfactory resolution and contrast.

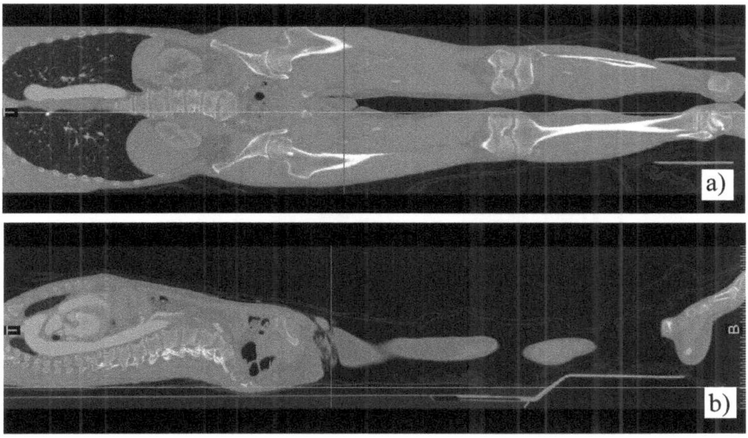

Figure 3.42 (a) Coronal and (b) sagittal cross-sectional image of the human body reconstructed from 2930 slices of CT images obtained with a 64-row multi-slice CT (resolution: 0.625 mm, slicing pitch: 0.8 mm; green area indicates the extracted artery).

The reconstructed three-dimensional blood vessel structure obtained from this scan is shown in Fig. 3.43. Figure 3.43a shows the reconstructed three-dimensional image of whole human artery related to endovascular intervention. Afterward, the three-dimensional image of Fig. 3.43b was generated by removing vessel branch with 1 mm or less in diameter because they are unrelated to this surgery. Then we constructed membranous vascular model with 1 mm in wall thickness of this structure using transparent silicone elastomer as shown in Fig. 3.44.

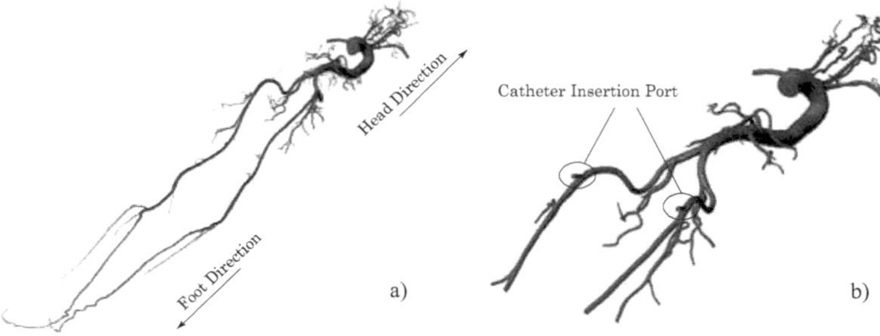

Figure 3.43 Reconstructed three-dimensional image of human arterial structure that is related to neuro-endovascular intervention: (a) State after three-dimensional reconstruction. (b) State after ablating small branches and prepared for vascular modeling.

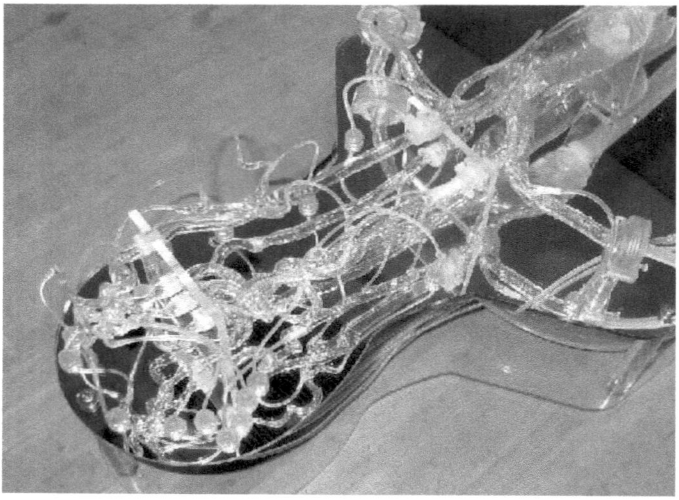

Figure 3.44 Membranous vascular model of overall human arterial structure with 1 mm in wall thickness made of transparent silicone elastomer.

For this large-scale vascular model, the structure was constructed by dividing the vasculature into five areas: femoral artery, abdominal artery, thoracic artery, aortic arch and carotid artery. This allows the exchange of each part with another vascular model that has a different shape and disease. And watertight structure was achieved by constructing rubber flange structure on each partial model and uniting them into one.

Meanwhile, a simulator body was constructed using transparent acrylic resin, imitating the posture of female patient (height: 172 cm) lying on surgical bed. By the interventionalist evaluation, we found it is appropriate for the overall simulator to have the human-like three-dimensional shape to reproduce surgery. The supporting structures were set up, and the above-constructed blood vessel model was anatomically accurately arranged inside the structure. As a result, a simulation system for the comprehensive surgical simulation of endovascular intervention was constructed. Moreover, an environment for the quantitative real-time stress evaluation was introduced for cerebral area by using photoelastic effect; the details of this stress analysis will be presented in Chapter 4.

On the other hand, the pulsatile blood flow influences the catheter during surgical treatment. Therefore, it is important for this simulator to reproduce the pulsatile human blood streaming in it. In this purpose, the pulsatile pump that is able to generate an arbitrary flow waveform based on a numeric setting was constructed in this research (range of flow rate: 0.0–10.0 L/sec; range of heart rate: 0–210). (Details about pumps for human blood pressure simulation will be presented later in this book.) The outlet of this pulsed flow pump was connected to the coronal area (ascending artery) of the vascular model locating inside simulator. Meantime, all the outlet tips of the arterial model were connected with inlet of this pulsatile pump via the accumulator which adjusts the max/min blood pressure, and thus, we reproduced the blood circulation similar to human inside the simulator. Here, water solution (glycerin: 33wt%, surface active agent: 0.1wt %), which reproduces the viscosity of human blood, was adopted as this circulation fluid. Figure 3.45 shows the comprehensive surgical simulator for endovascular intervention finally constructed by this research.

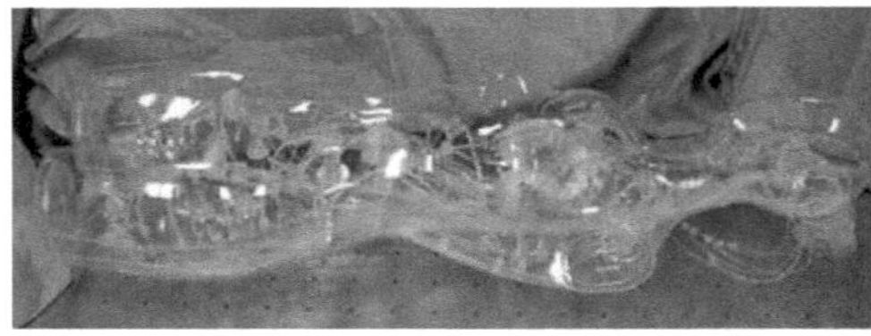

Figure 3.45 Comprehensive surgical simulator for endovascular intervention proposed and constructed by this research.

3.12.2 Surgical Simulation

To evaluate the performance of the comprehensive simulator, interventionalists made endovascular intervention simulation with it. The vascular model satisfactorily reproduces the elasticity characteristic and the friction characteristic of human blood vessel. As a result, operational feeling and the dynamic behavior of medical equipment is accurately reproducible with this vascular model. It was confirmed that this characteristic is similarly achieved in this overall simulator, and the operational feel and the behavior of medical equipment were reproduced. Figure 3.46 shows the state of the catheter passing through its abdomen artery.

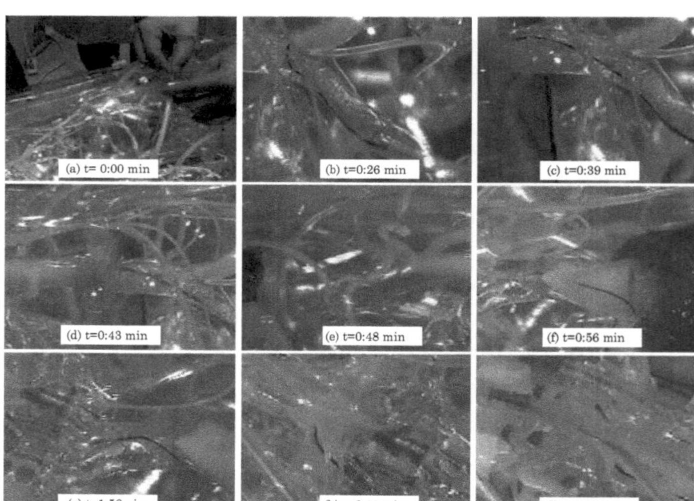

Figure 3.46 State of the catheter passing through the abdomen artery reproduced inside the constructed comprehensive surgical simulator.

The pulsatile motion of vessel wall similar to the human was reproduced by reproducing the pulsated blood circulation in this simulator; consequently, the swing and the inducement of the medical equipment according to the flow were satisfactorily reproduced. Consequently, the reproducibility of the surgical procedure has dramatically improved compared with the cerebral arterial model, in which reproduction is limited to local area, and, it becomes possible to know the procedures that have to be taken care of, such as the blood flowing out caused by forgetting to close a cork provided on catheters and other medical equipment. It is confirmed by interventionalists that the constructed simulator is very helpful for the acquisition of technical skills required for endovascular intervention. Meantime, the evaluation function of the stress applied to the cerebral arterial model was confirmed and was valuable. The stress can be directly observed and quantitatively analyzed (and displayed) by computer analysis as a real-time information, and it is confirmed to be very helpful in evaluating surgical procedures and in improving surgical methodologies. With respect to this stress evaluation, a lot of interventionalists gave us their opinion that points out the necessity of this function and the needs of materializing more diversified evaluation environment using this technology. On the other hand, it was also confirmed that the whole body reproduction is not essential for diagnosis purpose, including preliminary surgical simulation and surgical planning. It was confirmed that the application of patient-specific local vessel model to the corresponding part of overall simulator is appropriate, considering the time cost and labor cost needed for vascular modeling. Moreover, it was confirmed that the local cerebral arterial model is enough for making diagnosis for the skilled interventionalist. The constructed comprehensive surgical simulator was exhibited in the international exposition 2005 Ai-Chikyu-Haku, held in Aichi Prefecture, Japan. The simulator was evaluated by more than 150,000 visitors during this period (about 5000 of them experienced this simulator). During that time, it attracted attention from doctors and medical manufacturers as a new simulation tool. Due to its utility, the simulator received then a positive evaluation from the visitors (Fig. 3.47).

Figure 3.47 (a) The simulator EVE is evaluated by two interventionalists: Dr. Makoto Negoro and Dr. Ikuo Takahashi. (b) Appearance of the presented surgical simulator exhibited in the international exposition 2005 Ai-Chikyu-Haku, Televised by NHK "Yu-Gata Chance." (c) Appearance of the presented surgical simulator exhibited at the International Robot Exhibition 2005. (d) Appearance of the presented surgical simulator exhibited at the 21st Annual Meeting of the Japanese Society for Neuro-Endovascular Therapy.

3.12.3 Evaluation of Surgical Simulator

3.12.3.1 Evaluation by interventionalists

To obtain the evaluation about the reproducibility and the utility of this simulator, we exhibited it at the 21st Annual Meeting of the Japanese Society for Neuro-Endovascular Therapy for three days in November 2005. At this medical congress, the surgical simulator was evaluated by more than 100 interventionalists. The evaluation was made as follows:

Evaluation method: Leading the tip of a micro-catheter to the BT aneurysm fabricated on the cerebral artery, using widely utilized medical tools, which include catheter, guide-wire, micro-catheter, and micro guide-wire. The interventionalist inserts the tip of the micro-catheter into the one of the three branches (diameter: 1 mm; Fig. 3.48) artificially fabricated on the aneurysm. The treatment of aneurysm (coil embolism) is not executed in this evaluation for the convenience of the time required.

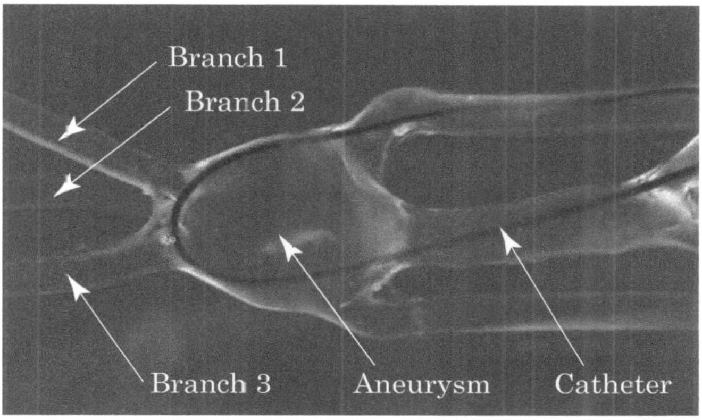

Figure 3.48 Three branches artificially fabricated on the aneurysm for evaluation purpose. In this figure, the proposed photoelastic stress visualization is applied.

Evaluation item: After exerting the above-mentioned surgical treatment, the comprehensive simulator is graded from the following viewpoints by four ranks (1: "not good," 2: "ordinary," 3: "satisfactory" and 4: "excellent").

Evaluation item:

1. Reproduction of vessel shape
2. Reproduction of vessel deformation
3. Reproduction of surgical feeling
4. Reproduction of surgical monitor
5. Reproduction of surgical room
6. Effectiveness of stress analysis
7. Effectiveness of proposed system
8. Affordable system

The data acquired as a result of this evaluation is summarized in Table 3.4. The number of evaluators who executed this questionnaire is 37. The evaluation result of this simulator made by the interventionalists was very good, as confirmed from Table 3.4. The score of 3.0 means that "satisfactory" is obtained for vascular shape reproduction, vascular deformation reproduction and surgical feeling reproduction. A score of 3.4 means that "almost excellent" is obtained for the effectiveness of the proposed system that gives the overall evaluation of the simulator. A score of 3.4 means that "strongly recommend" is obtained for the system affordability for the institute (university and hospital) to which the evaluator belongs. Thus, it was confirmed from this evaluation that the simulator constructed by this research is very useful as the simulation environment of endovascular intervention procedure. High appraisal, 3.3, on average means that "almost excellent" was obtained for the effectiveness of stress evaluation.

Table 3.4 Evaluation result made by a population of 37 interventionalists at the 21st Annual Meeting of the Japanese Society for Neuro-Endovascular Therapy

Item	Evaluation result
Reproduction of vessel shape	3.0/4.0
Reproduction of vessel deformation	2.8/4.0
Reproduction of surgical feeling	3.0/4.0
Reproduction of surgical monitor	2.8/4.0
Reproduction of surgical room	2.5/4.0
Effectiveness of stress analysis	3.3/4.0
Effectiveness of proposed system	3.4/4.0
Affordable system	3.4/4.0

After their evaluation, the interventionalist pointed out that the proposed stress analysis method is extremely important for the future development of endovascular surgery, since the method to know the stress on artery (aneurysm) during surgery had not existed so far. On the other hand, scores below 3.0 are obtained for the reproduction of the surgical monitor and the surgical room. The principal reason for this was that the foot pedal mechanism that enables to change the position of the surgical bed and X-ray direction freely has not been introduced into this system. In this research, patient-specific reproduction of patient body was the main objective in constructing the simulation environment; however, this result shows the utility of introducing the medical equipment used in this surgery into the surgical simulator.

3.12.3.2 Evaluation by layperson

The evaluation of the utility of the simulation system proposed in this research was exerted by a general person (layperson) from the standpoint of receiving medical treatments. In order to exert this evaluation, we exhibited the simulator at the International Robot Exhibition 2005 for five days in November 2005. The following evaluations were executed by more than 1000 visitors for this period. Figure 3.47 shows the appearance of this exhibition.

Evaluation method: Leading the tip of a micro-catheter to the BT aneurysm fabricated on cerebral artery, using widely utilized medical tools, which include catheter, guide wire, micro-catheter, and micro guide-wire. The layperson inserts the tip of micro-catheter into the one of the three branches (diameter: 1 mm, c.f. Fig. 3.48) artificially fabricated on the aneurysm.

Evaluation item: After exerting the above-mentioned surgical treatment, the comprehensive simulator is graded from the following three different viewpoints by four ranks ("excellent," "satisfactory," "ordinal" and "not good"). Here, we explained verbally that the vascular model is constructed by individual patient information and applicable for individual patient diagnosis and surgical planning.

Evaluation Aspect:

1. As patient receiving medical treatments
2. As the interventionalist who executes endovascular treatments

Evaluation item:
Aspect 1-1: Effectiveness of patient-specific vascular model
Aspect 1-2: Do you consider necessary to construct a vascular model for your treatment?
Aspect 2-1: Did you understand the simulated surgical method?
Aspect 2-2: Effectiveness of stress evaluation
Aspect 2-3: Effectiveness as a training environment

As a result of this evaluation, acquired data is summarized in Figs. 3.49 and 3.50. The number of evaluators who executed this questionnaire is 325.

First of all, as to the "evaluation from the standpoint as a patient who receive medical treatments," three-fourth (74%) of the entire population answered that "the proposed patient-specific vascular model is very useful for diagnosis" (average score calculated by supposing "excellent = 3," "good = 2," "not good = 1" was 2.7/3.0).

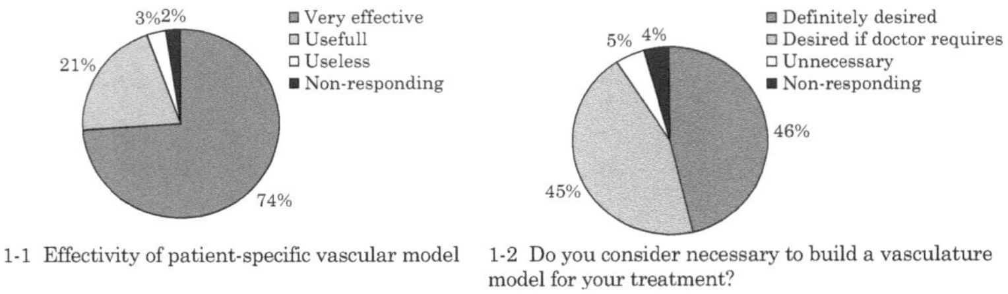

Figure 3.49 Evaluation 1: As a patient who receive medical treatments.

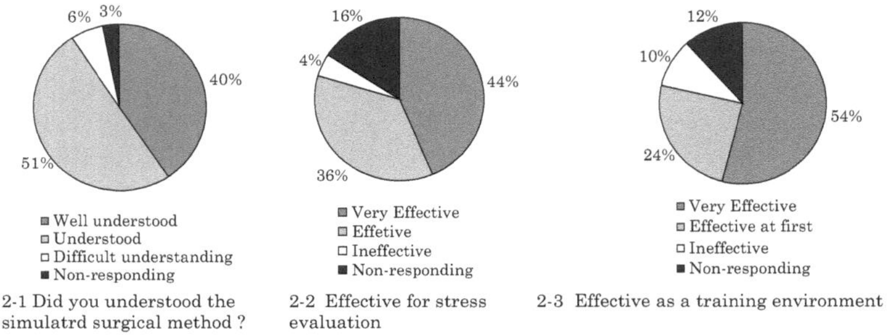

Figure 3.50 Evaluation 2: As interventionalist who execute endovascular treatments.

Moreover, the population who answered, "wish to make their own vascular model if the doctor want to use" had majority (45%) of the remainder, and the evaluator who had answered, "presented vascular model is unnecessary" stayed in 5% of the whole (average score was 2.4/3.0). From this result, the utility of the presented patient-specific vascular model was evaluated very high, from the standpoint as "a patient who is to receive endovascular treatment."

On the other hand, as to the "evaluation from the standpoint as interventionalist who execute endovascular treatments," 40% of the entire population answered that "was able to well understand the methodology of endovascular treatment by using presented comprehensive simulation system." The evaluator who answered that "he was able to understand the outline of endovascular treatment with the simulator" had majority (51%) of the remainder, and the evaluator who had answered "could not understand the methodology well" stayed in 6% of the whole (average score was 2.4/3.0). From this result, it is confirmed that the presented simulator is very helpful in understanding surgical methodology considering the evaluation time allowed less than 10 min. Especially, regarding the stress evaluation function, 80% of the evaluators termed it "useful," and the percentage of evaluator who judged it "useless" was only 4% of the whole (average score was 2.5/3.0). Moreover, comments such as "it was able to know the transition of stress on vascular wall applied by medical equipment, and dangerous operation and dangerous state was clearly recognized" and "presented simulator is extremely useful in acquiring surgical skills" were given a lot. For the question, "Do you want to use the presented simulator for surgical training, if he were an interventionalist," more than half (54%) of the evaluators answered "definitely yes," and they said, "I want to improve their surgical skill by continuously making simulation with this platform." Moreover, 78% of evaluator answered, "I want to use the simulator at all" (averaged score was 2.5/3.0).

Consequently, from these results, it was confirmed through this evaluation that the existence of the proposed simulator had been evaluated high, because malpractices had gained high attention recently in Japan. Moreover, we confirmed that it is possible to easily and accurately understand the correct medical procedures even for an inexperienced person by using the proposed simulator, and thus the presented simulator is valuable for surgical training purpose. It was confirmed that the comprehensive simulator presented in this chapter greatly improved the reproducibility of surgery (especially, behavior and operational feeling of medical equipment) compared with the cerebral arterial model presented in preceding chapters.

Moreover, it was confirmed that the stress evaluation method is progressive as an environment for surgical simulator. And, as a result of the evaluation made by actual interventionalist, it was confirmed that the proposed simulation system realize the "condition required for surgical simulator" at high standard, consequently, it is appropriate as the simulation environment for endovascular intervention. On the other hand, it was also confirmed that interventionalist hopes the reproduction of entire surgical room, in addition to the human body reproduction as described in the foregoing paragraph. The necessity of considering ideal simulator construction was found important. On the other hand, as a result of the evaluation made by layperson, it is confirmed that the surgical procedures is concretely and surely understandable even for the people who did not know the treatment method at all (layperson) by using this simulator. This result strongly indicates that the constructed simulator is suitable as the training environment for this surgery. From the result of a diversified evaluation above, it has been understood that the surgical simulator proposed by this research is advanced as a platform for simulating endovascular intervention procedures. We confirmed that this simulator is very useful in a wide variety of purposes, which include surgical training and diagnosis and evaluation of procedures, providing innovative platform for engineering researchers and science researchers in addition to medical doctors.

In this chapter, we developed the patient-specific vascular modeling technique and reproduced overall arterial area of diameter 1 mm or more, which can be treated by this endovascular approach. In this way, we constructed a comprehensive surgical simulation system for endovascular intervention. Meanwhile, human-like pulsed blood streaming was reproduced in this simulator by developing a pulsatile blood pump, and we confirmed that the behavior and the surgical feeling can be realistically reproducible by this simulator. As a result, it was confirmed that the proposed simulation system realize the "required condition" itemized in the beginning of the chapter and the system is progressive and valuable as a simulation environment. It was also confirmed that the system is useful as the environment for an individual medical treatment (diagnosis and rehearsal, etc.), and its practical use is strongly expected.

References

1. Gailloud P, Pray JR, Muster M, Piotin M, Fasel JHD, Rufenacht DA. An *in vitro* anatomic model of the human cerebral arteries with saccular arterial aneurysms, *Surg Radiol Anat*, 1997, 19: 119–121.
2. Hounsfield GN, Computerised transverse axial scanning (tomography) I. Description of system, *Br J Radiol*, 1973, 46: 1016–1022.
3. Kalender WA, Seissler W, Klotz E, Vock P, Spiral volumetric CT with single- breath-hold technique, continuous transport, and continuous scanner rotation, *Radiology*, 1990, 176: 181–183.

Chapter 4

Respect for Tissue Representation Using Photoelastic Stress Analysis for Endovascular Surgery Simulation

Silicone vasculature was presented in previous chapters for endovascular surgery simulation and built relying on the tomogram of human vasculature, and human arterial lumen was recreated in silicone with 13 μm of accuracy. Major vasculature models were built and integrated in the endovascular evaluator EVE. This silicone vasculature allows flow circulation and a maximum pressure of 200 mmHg, enabling practice of endovascular tools deployment in models simulating diseased vasculature (Fig. 4.1a). It is of particular interest for endovascular surgery research, medical training and endovascular tools industry to develop sensing technologies for quantitative evaluation of interaction between intravascular tools and silicone vasculature, to enable the profiling of human skills and device performance. For that purpose the use of photoelastic effect for measuring the stress in walls of urethane vasculature was proposed in 2005 [1]. Polarized light undergoes a phase shift when passing through photoelastic materials producing birefringence (Fig. 4.1b). In this chapter we will present how to measure stress, morphology changes, catheter shape and trajectory from photoelastic images during endovascular surgery simulation. Micro force sensors placed on the catheter tip [2] cannot replace photoelastic stress analysis as they modify

Microsurgery: Advances, Simulations, and Applications
Edited by Toshio Fukuda and Carlos Tercero

ISBN 978-981-4364-69-0 (Hardcover), 978-981-4364-70-6 (eBook)
www.panstanford.com

the catheter stiffness and lumen, reducing the range of tools that may be evaluated or used for evaluation.

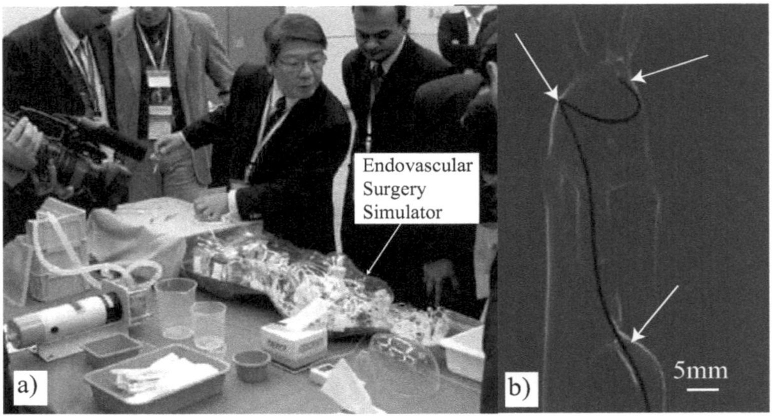

Figure 4.1 (a) Endovascular surgery simulator. (b) Birefringence consequence of stress applied to the vasculature model by a guide wire. See also Color Insert.

4.1 Photoelastic Stress Analysis Fundamental Equations

The resulting phase shift of polarized light while passing through photoelastic materials is called retardation and its variation is described by

$$\left(\frac{\mathrm{Re}}{cD}\right)^2 = \left(\begin{array}{l}(1-v_1^2)^2(\sigma_1-\sigma_2)^2 \\ \quad +(1-v_3^2)(\sigma_2-\sigma_3)^2 \\ \quad +2(v_2^2-v_3^2v_1^2)(\sigma_1-\sigma_2)(\sigma_2-\sigma_3)\end{array}\right) \tag{4.1}$$

where Re is the retardation and is measured in nanometers, D is the optical path length (thickness of the photoelastic material), c is the photoelastic coefficient of the photoelastic material, v_1, v_2 and v_3 are the components of the unitary vectors that describe the light beam direction in terms of the principal stresses σ_1, σ_2 and the radial stress component σ_3 as described in [3] (Fig. 4.2a).

When the principal stress plane is parallel to the light beam, $v_1 = -\cos\varphi = \sin\varphi\, v_2$ and $v_3 = 0$, then

$$\left(\frac{\mathrm{Re}}{cD}\right)^2 = \begin{pmatrix} (1-\cos^2\varphi)^2(\sigma_1-\sigma_2)^2 \\ +(\sigma_2-\sigma_3)^2 \\ +2\sin^2\varphi(\sigma_1-\sigma_2)(\sigma_2-\sigma_3) \end{pmatrix} \tag{4.2}$$

where φ is the angle between the light beam and σ_1 (Fig. 4.2). If the radial stress component σ_3 is considered zero, equation (4.4) is reduced to

$$\frac{\mathrm{Re}}{cD} = (\sigma_1 \sin^2\varphi + \sigma_2 \cos^2\varphi) \tag{4.3}$$

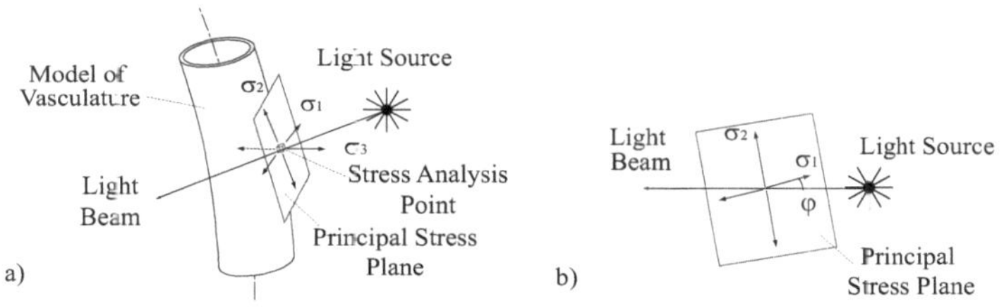

Figure 4.2 (a) Stress components in photoelastic stress analysis. (b) Angle φ between principal stress component σ_2 and light beam (right).

The difference between the principal stresses on each side of a biaxial stress field σ_1, σ_2 is obtained when the light beam is parallel to σ_3, $v_1 = v_2 = 0$ and $v_3 = 1$ [3, 4]:

$$\sigma_1 - \sigma_2 = \frac{\mathrm{Re}}{cD} \tag{4.4}$$

For stress analysis in model borders we consider that the model of vasculature has a cylindrical shape, and we set its axis perpendicular to the light source beam. Therefore in this case equation (4.1) can be reduced to

$$\sigma_2 = \frac{\mathrm{Re}}{cD} \tag{4.5}$$

Measurement of stress relying on photoelastic effect (equations 4.1–4.5) depends on the photoelastic coefficient of the material and the optical path length (amount of photoelastic material between the light source and the observation point). The method for measuring those parameters is explained in the following sections.

4.2 Vasculature Modeling for Stress Analysis

Silicone elastomer is a photoelastic material, with excellent mechanical properties for blood vessel modeling; however, its photoelastic coefficient is too low for enabling a practical measurement of stress produced by endovascular tools with blood vessel models. On the other hand, urethane elastomer has enough sensitivity to visualize stress produced by human blood pressure ranges or by micro-coil deployment. Single-layer models of urethane elastomer with low wall thickness are the best selection for measuring low-magnitude stress fields; however, single-layer models of urethane elastomer suffer plastic deformation for pressures in the range of 60–200 mmHg. This plastic deformation is drastically reduced by shielding the urethane layer with silicone elastomer layers. Transmittance of light for silicone elastomer is above 88% for 10 mm thickness and its photoelastic coefficient is several times smaller than that of urethane elastomer. Shielding layers of silicone layer are about 60 μm thick and colorless; the urethane elastomer layer is about 250 μm thick and dyed yellow [6]. Therefore, at the studied pressure ranges, the contribution of silicone elastomer layers to retardation and transmittance may be neglected.

4.3 Blue Light Transmittance Equation

The measurement of the optical path length is achieved relying on light attenuation while passing through the photoelastic material. Blue light is more sensitive to attenuation than red or green light when passing through yellow-dyed materials. For that reason, and its high photoelastic coefficient, we selected NIPPOLAN5120 urethane elastomer dyed yellow with I-01-001Y of Epoch Co. at 0.1w% as preferred modeling material. Thus, transmittance of blue light was used to calculate the optical path length in multilayered models of silicone and urethane elastomer.

For constructing a mathematical model of the blue light transmittance in urethane elastomer, a urethane elastomer membrane with variable thickness was built using a stair-case shaped casting mold. The depth difference between neighbor levels of the casting mold was 0.2 mm. A blue filter was placed over a white light source; an image of the model was captured. The thickness of the membrane was measured at 11 locations using a UV laser microscope. The relation between transmittance and thickness was then deduced (Figs. 4.3 and 4.4).

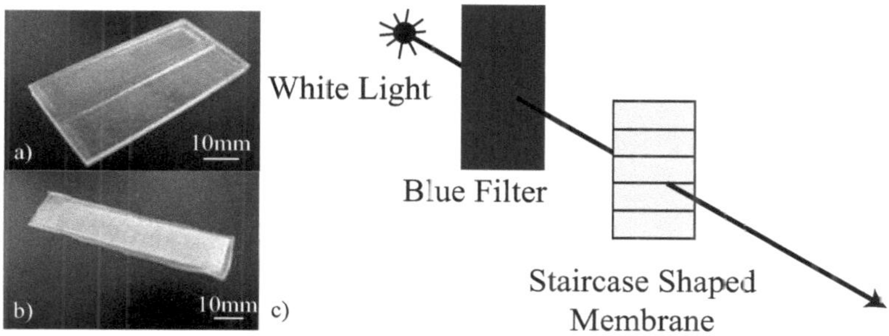

Figure 4.3 (a) Stair case-shaped casting mold. (b) Variable-thickness urethane elastomer membrane. (c) Blue light transmittance measurement optic system.

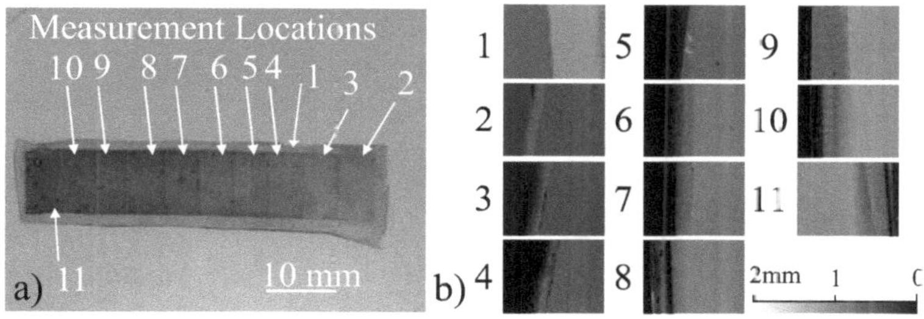

Figure 4.4 (a) Location of thickness and blue light intensity measurements in the variable thickness membrane. (b) UV Laser Microscope Images: Location 1 shows the depth difference between the base plate and the border of the membrane. Locations 2–11 show the difference between the border of the membrane and the level of the stair case shape.

The plot describing the relation between the measurements done with the UV laser microscope against the values of blue light transmittance is shown in Fig. 4.5. The obtained mathematical model of the relation is

$$D = -T_c \ln(I_B/I_{BMax}) \tag{4.6}$$

where I_B is the blue light intensity at the studied location, I_{BMax} = 132.08 is the average blue light intensity in the area where there is no membrane covering the filter, and T_c = 1666.66 is the transmittance coefficient. The robustness of equation (4.6) is increased by the normalization of I_B by I_{BMax} and has a correlation coefficient of R = 0.982 with the measurements [7].

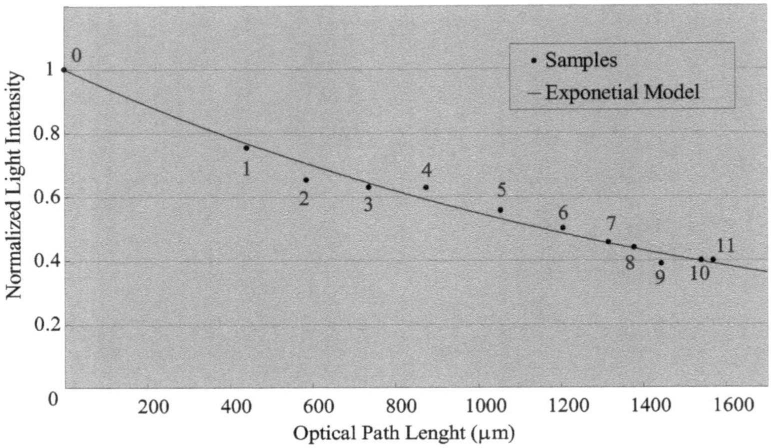

Figure 4.5 Plot of relation between blue light transmittance and membrane thickness. Locations names are shown for each measurement; location 0 is measured directly from the filter.

4.4 Polariscope for Stress Magnitude Analysis

As it is desirable to integrate the stress measurement system with the endovascular surgery simulator EVE, it is necessary to use film instead of glass for polarizers, and quarter-wave plates and color filters to reduce the space occupied by the optic system. Therefore, the polariscope configuration should be robust against the quarter-wave manufacturing accuracy. For that, we propose the polariscope shown in Fig. 4.6.

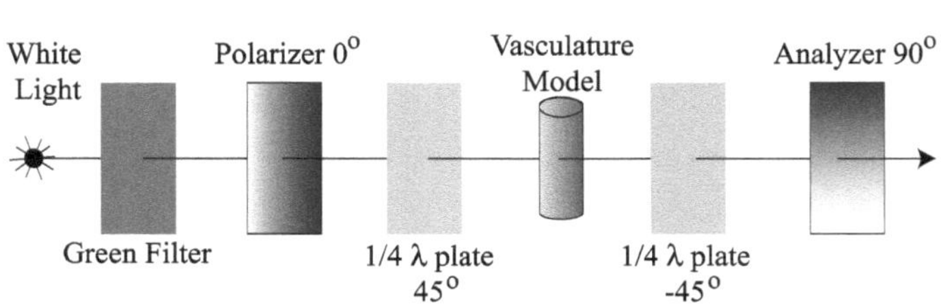

Figure 4.6 Polariscope configuration to reduce the effect of quarter-wave plates manufacturing precision in the system error.

Using Jones calculus, the components of the polariscope can be defined as follows [5]:

- **Polarizer**

$$J_{P(0)} = \begin{bmatrix} 1 & 0 \\ 0 & 0 \end{bmatrix} \quad (4.7)$$

- **Analyzer**

$$J_{P(\pi/2)} = \begin{bmatrix} 0 & 0 \\ 0 & 1 \end{bmatrix} \quad (4.8)$$

The total phase shift between the field components for quarter-wave retarder is represented by

$$\Phi_1 = \frac{\pi\lambda_{ex}}{2\lambda_G} \quad (4.9)$$

For this study, we selected wavelength retarders of λ_{ex} = 560 nm wavelength to study the retardation of the observed beam that has a wavelength λ_G. The matrix corresponding to a quarter-wave retarder rotated π/4 is

$$J_{QR(\Phi_1,+\pi/4)} = \begin{bmatrix} \cos\frac{\Phi_1}{2} & i\sin\frac{\Phi_1}{2} \\ i\sin\frac{\Phi_1}{2} & \cos\frac{\Phi_1}{2} \end{bmatrix} \quad (4.10)$$

And corresponding to a quarter-wave retarder rotated –π/4 is

$$J_{QR(\Phi_1,-\pi/4)} = \begin{bmatrix} \cos\frac{\Phi_1}{2} & -i\sin\frac{\Phi_1}{2} \\ -i\sin\frac{\Phi_1}{2} & \cos\frac{\Phi_1}{2} \end{bmatrix} \quad (4.11)$$

The matrix corresponding to the blood vessel phantom is then represented by

$$J_{M(\alpha,\theta)} = \begin{bmatrix} \cos\frac{\alpha}{2} + i\sin\frac{\alpha}{2}\cos 2\theta & i\sin\frac{\alpha}{2}\sin 2\theta \\ i\sin\frac{\alpha}{2}\sin 2\theta & \cos\frac{\alpha}{2} - i\sin\frac{\alpha}{2}\cos 2\theta \end{bmatrix} \quad (4.12)$$

where θ is the stress direction and α is the retardation in radians and Re in nanometers. For green light study,

$$\alpha = \frac{2\pi \,\mathrm{Re}}{\lambda_G} \quad (4.13)$$

The Jones vector for an incident beam is defined as

$$E_{in} = \begin{bmatrix} E_x \\ E_y \end{bmatrix} \quad (4.14)$$

The emerging beam is then obtained from

$$E_{out} = J_{P(\pi/2)} J_{QR(\Phi_1,-\pi/4)} J_{M(\alpha,\theta)} J_{QR(\Phi_1,\pi/4)} J_{P(0)} E_{in} \quad (4.15)$$

$$\Leftrightarrow E_{out} = \begin{bmatrix} 0 \\ E_x \sin\left(\frac{\alpha}{2}\right)(\cos(2\theta)\sin(\Phi_1) + i\sin 2\theta) \end{bmatrix} \quad (4.16)$$

The emerging beam components are defined with

$$E_{out} = \begin{bmatrix} E'_x \\ E'_y \end{bmatrix} \quad (4.17)$$

And its intensity using

$$I_{out} = E'_x E'^*_x + E'_y E'^*_y \quad (4.18)$$

$$I_{out} = E_x^2 \sin^2\left(\frac{\alpha}{2}\right)\left[\cos^2(2\theta)\sin^2\Phi_1 + \sin^2(2\theta)\right] \quad (4.19)$$

The maximum observed intensity in the emerging beam occurs when $\alpha = \pi$ if $\phi_1 = \pi/2$ or $\theta = \pi/2$.

$$I_{outMAX} = E_x^2 \quad (4.20)$$

Relying on this, we measure the maximum $I_{G0\lambda ex/2}$ and minimum $I_{G0\lambda ex}$ green light intensities from stacked quarter-wave plates placed within the polariscope and oriented with $\theta = 0$ and we normalize the observed green light intensity I_G as follows to compensate for external noise (Fig. 4.7):

$$I_{GN} = \frac{I_G - I_{G0\lambda ex}}{I_{G\lambda ex/2} - I_{G0\lambda ex}} \quad (4.21)$$

$$I_{GN} = \sin^2\left(\frac{\mathrm{Re}\pi}{\lambda_G}\right)\left[\cos^2(2\theta)\sin^2\left(\frac{\lambda_{ex}\pi}{2\lambda_G}\right) + \sin^2(2\theta)\right] \quad (4.22)$$

In this configuration, we expect λ_G to be as close as possible to λ_{ex} in order to reduce the influence of stress direction in magnitude measurements. For models of vasculature, θ is the direction of the projection of σ_2 in the polarizer plane, it is desirable to quantify its contribution into the stress magnitude measurements. For that purpose we placed a variable retardation device inside the polariscope shown in Fig. 4.7. This device is composed of stacked quarter-wave plates oriented in the same direction, and providing five retardation zones of $0\lambda_{ex}$, $\lambda_{ex}/4$, $\lambda_{ex}/2$, $3\lambda_{ex}/4$, and λ_{ex} = 560 nm. Changing the orientation of the variable retardation device inside the polariscope and measuring the green light intensity at each zone enables us to study the effect of θ in the measurement of Re. Green light intensity measurements were done in each zone for the orientation range 0–180° in 15° intervals.

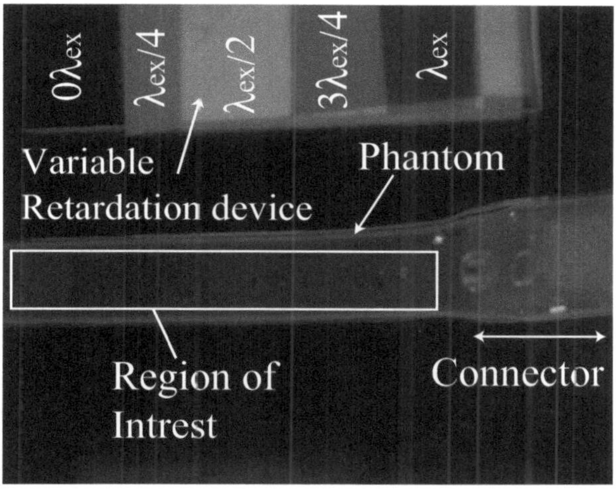

Figure 4.7 Image of light retardation using the polariscope. Quarter-wave plates stacked in a variable retardation device used for λ_G calibration and the blood vessel model.

The measured green light wavelength was λ_G = 510 nm, the quarter-wave plates provide a retardation of λ_{ex} = 140 nm with 10 nm accuracy. (The method for measuring the emerging beam wavelength will be presented in the next section.) Therefore equation (4.22) reduces to

$$I_{GN} = \sin^2\left(\frac{\mathrm{Re}\cdot\pi}{510}\right)\left[0.976 + 0.024\sin^2(2\theta)\right] \tag{4.23}$$

The variation of green light intensity with the direction of the variable retardation device is shown in Fig. 4.8. Equation (4.23) forecasts a maximum contribution of 2.4% of the device direction to the intensity measurements. The maximum contribution of stress direction in the intensity measurements was observed when retardation was 280 nm, where it contributed in 2.52%. The contributions for 0 nm and 140 nm are 0% and 1.93%, respectively. Considering the above error, equation (4.22) may be approximated by [8]

$$I_{GN} \approx \sin^2\left(\frac{\mathrm{Re}\cdot\pi}{\lambda_G}\right) \tag{4.24}$$

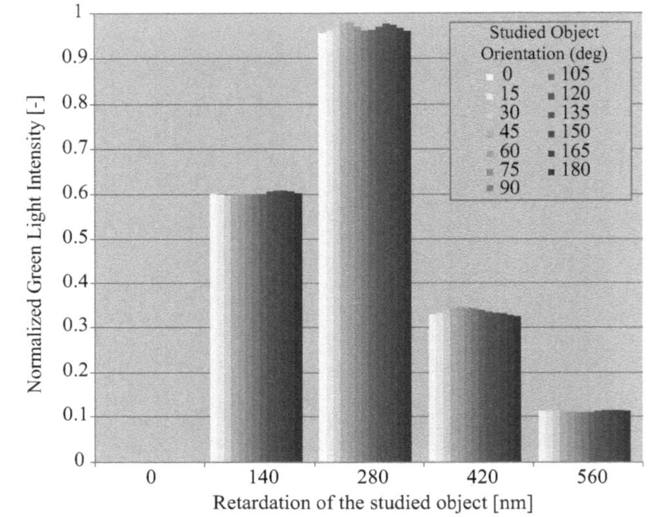

Figure 4.8 Variation of green light intensity with the direction of quarter-wave plates.

4.5 Camera Calibration

A Toshiba IK-TF2 camera was used for capturing the photoelastic images with field of view of 48.4 mm × 36.3 mm. When calculating the optical path length, blue light wavelength is not a parameter of interest; however, to calculate Re, green light wavelength λ_G is essential. A green filter reduces the effect of red and blue light in the green channel of the camera. λ_G is the principal wavelength value captured by the camera's green channel, changing with the environment illumination and lens configuration.

By placing the variable retardation device with $\theta = \pi/2$, shown in Fig. 4.7, normalized green light intensity can be deduced by reducing equation (4.22) to

$$I_{GN} = \sin^2 \frac{\pi}{\lambda_G} \mathrm{Re} \tag{4.25}$$

The green light intensity was measured for each region of the variable retardation device and normalized using equation (4.21). As the value of Re is known for each region, the value of λ_G may be calculated from equation (4.23).The normalized values of green were plotted against their corresponding retardation in Fig. 4.9. The value of λ_G is obtained by minimizing the error between the plot and the mathematic model of equation (4.25). The error minimizes when $\lambda_G = 510$.

The quarter-wave plates are provided with a precision of 10 nm; therefore, they may add in some cases a maximum error of 14.5% to the model proposed in equation (4.24). This error may be suppressed by configuring the camera to capture $\lambda_G = \lambda_{ex}$.

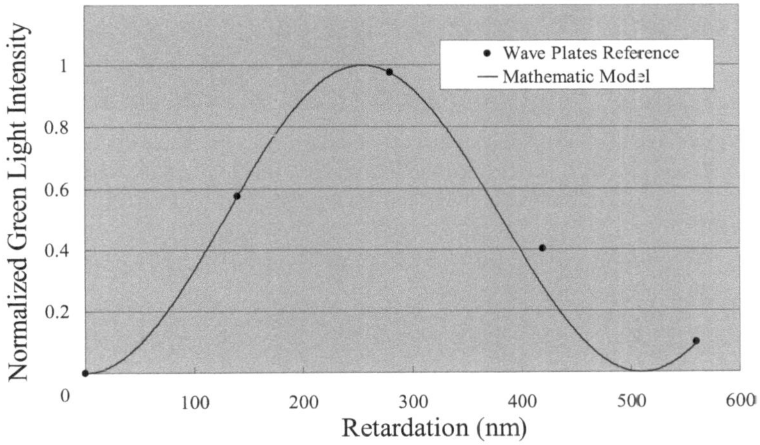

Figure 4.9 Plot of the retardation introduced by the quarter plates in the studied areas, and their respective normalized green light intensity. The mathematical model proposed by equation (4.25) with $\lambda_G = 510$ nm is also shown.

4.6 Photoelastic Coefficient of Urethane Elastomer

In this study, we used NIPOLAN5120 urethane elastomer-dyed yellow with I-01-001Y of Epoch Co. at 0.1w% for manufacturing all models. If we posit that the urethane elastomer membrane of rectangular cross section of Fig. 4.10 is under a tensile force F when placed inside the polariscope in Fig. 4.6 instead of the vasculature model, the longitudinal stress σ_1 is given by

$$\sigma_1 = \frac{F}{LD} \tag{4.26}$$

where L and D are the width and thickness of the membrane's cross section. Because in this case σ_2 equals zero, the stress in the membrane may be calculated using equation (4.4) or (4.26).

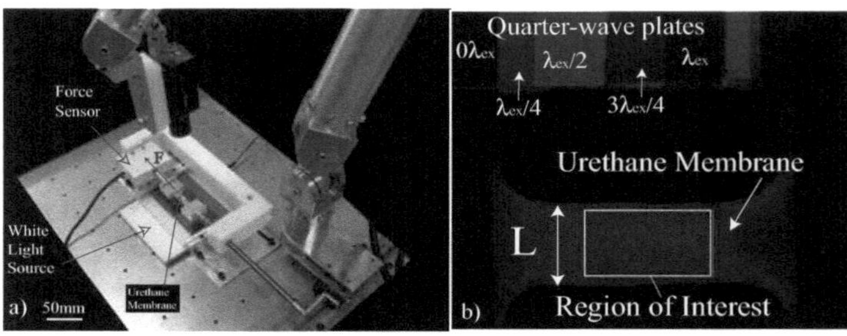

Figure 4.10 (a) Variable tension device for photoelastic coefficient calculation. (b) Membrane used for deducing the photoelastic coefficient of urethane elastomer.

Therefore, we are able to calculate the photoelastic coefficient of urethane elastomer C using

$$C = \frac{\mathrm{Re}\,L}{F} \tag{4.27}$$

We built the device shown in Fig. 4.10a, composed of a C-clamp and a force sensor for gradually applying and measuring tension to the membrane. Some of the elements of the polariscope are shown in this figure. A total of 19 samples were captured obtaining simultaneous measurements of Re, F and L. Figure 4.10b shows the membrane inside the polariscope when F = 2.1 N, Re is measured after calibrating the camera from the average of green light intensity at the region of interest and equation (4.22). Tension was applied to the membrane in a range of 0.3–3.7 N, while the photoelastic images were captured for visualizing the first fringe. L is obtained from the image captured when force is applied to the membrane. For this experiment, λ_G was found to be 521 nm. Figure 4.11 shows the plot of variation in the normalized green light intensity with applied force, showing that complete first fringe was captured, and the value of the photoelastic coefficient obtained for each sample. Their average value is 1.284×10^{-9} Pa^{-1}.

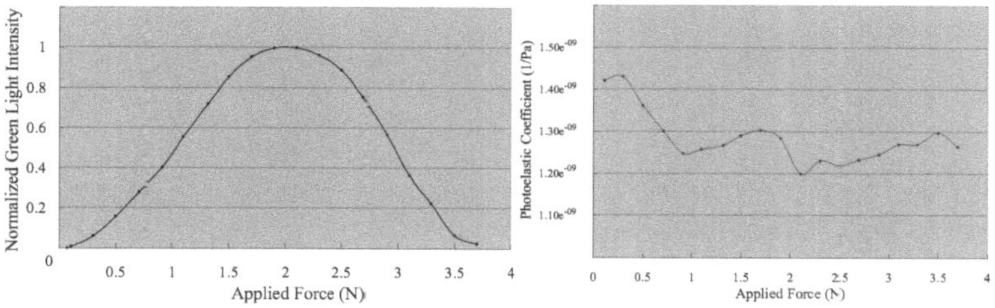

Figure 4.11 Variation of the green light intensity in urethane elastomer membrane with the applied force, showing the first fringe (left). Variation of the photoelastic coefficient of NIPOLAN 5120 dyed yellow with the applied force (right).

4.7 Photoelastic Stress Analysis Error Quantification

The difference between longitudinal and tangential stress for a cylinder wall with inner pressure *P* is given by the following equation [9]:

$$\sigma_1 - \sigma_2 = \frac{2rP}{D}\left(\frac{r-D}{2r-D/2}\right) \tag{4.28}$$

where the cylinder outer radius is *r* and its wall thickness is *D*/2. If $P > 40$ mmHg for the blood vessel model, the stress on the model wall is described by equation equation (4.28).

When a phantom is placed into the polariscope inside a glycerin bath, stress on the model wall may be calculated using equations (4.4, 4.22) or with equation (4.28) by measuring Re, *D* and *r*. Photoelastic measurement system error may then be quantified using the deformation measurements.

P was increased within the blood vessel model using a syringe and measured using a sensor for pressures below 200 mmHg (Fig.4.12). For a range from 40–189 mmHg average stress in the region of interest (ROI) was calculated using both methods for 15 samples (Fig. 4.7). Values of green were measured in ROI for each sample.

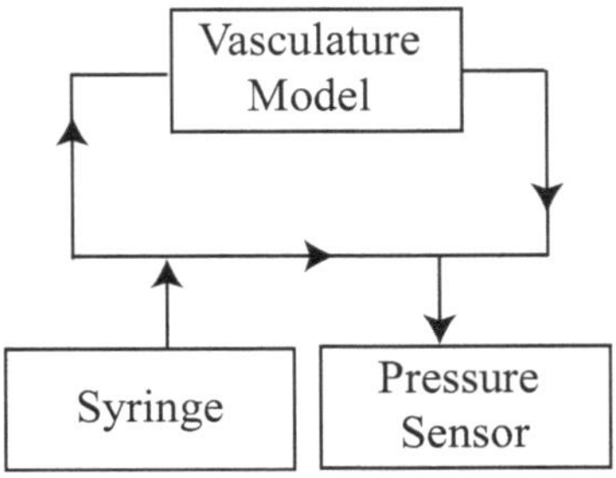

Figure 4.12 Fluid closed loop for pressure variation and measurement inside the blood vessel model.

The average green value of ROI I_{GP} for a pressure P is normalized using

$$I_{GNP} = \frac{I_{GP} - I_{GMin}}{I_{GMax} - I_{GMin}} \tag{4.29}$$

I_{GMin} and I_{GMax} are the minimum and maximum values of green, respectively, and are obtained from

$$I_{GMin} = I_{G40} - \varepsilon_{IG40} \tag{4.30}$$

$$I_{GMax} = I_{G200+} + \varepsilon_{IG200+} \tag{4.31}$$

where ε_{IG40} and ε_{IG200+} are the standard deviation of I_{G40} and I_{G200+}.

A sample of source images of photoelasticity and transmittance measurements is shown in Fig. 4.13. We deduced from I_{G40} that I_{GMin} is 23.32 and from I_{G200+} that I_{GMax} is 53.86. For this experiment the value of λ_G = 510 nm and was obtained as described in previous sections. The optical path length was calculated by applying equation (4.6) to the transmittance images set.

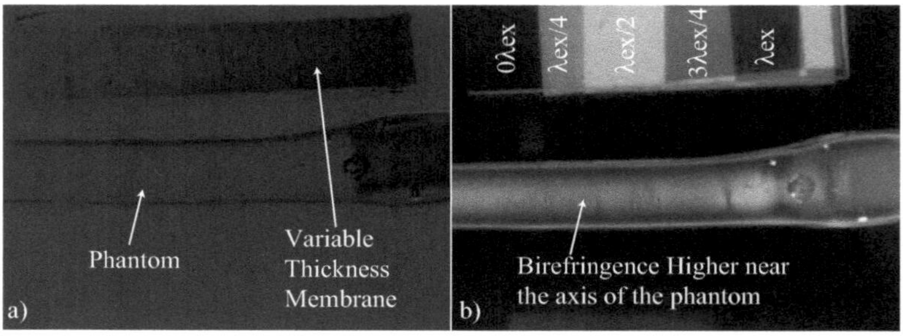

Figure 4.13 Source image example when P = 189 mmHg for: (a) optical path length calculation (b) and retardation measurement.

The resulting set is shown in Fig. 4.14; for eliminating noise a threshold of 200 mm was applied to the optical path length images, this enhanced the borders of both urethane objects. Average stress measurements were performed for the different values of *P* in the ROI of the photoelasticity source images, first using the photoelastic stress analysis and then by using equations (4.6, 4.28).

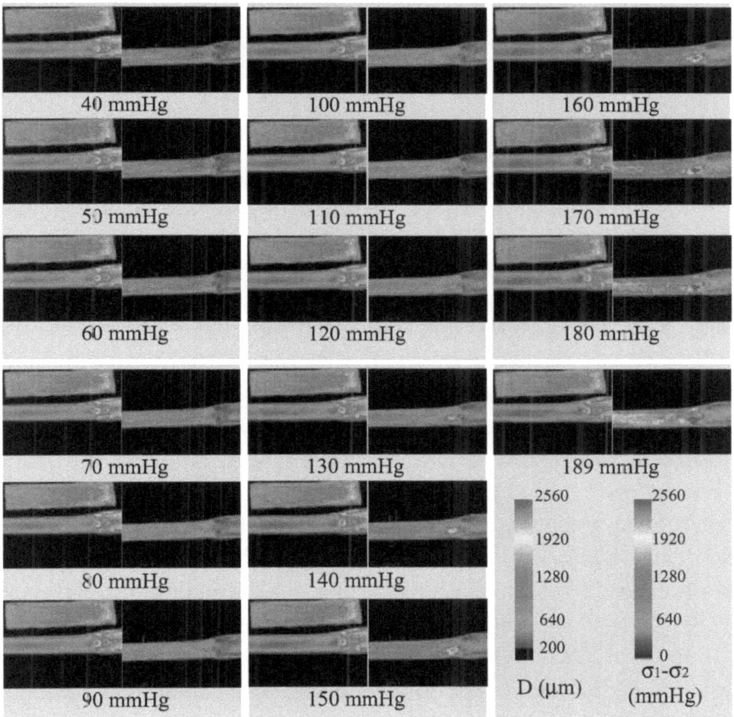

Figure 4.14 Result image of the 16 samples. Below each set of two images, the corresponding pressure is written. For each set, optical path length *D* (left) and the corresponding photoelastic stress analysis (right) are shown. A local maximum of stress appears on the right side of the phantom as pressure increases. See also Color Insert.

If $r >> D$, the wall thickness measurement influence in the final result is almost the same for both methodologies; then the stress measured using equations (4.6, 4.28) may be used as a reference for the photoelastic error quantification. The results for both methods and their difference are plotted in Fig. 4.15. An average error of 3.9%

is present for pressure above 60 mmHg. Error is large for pressure ranges below 60 mmHg [7].

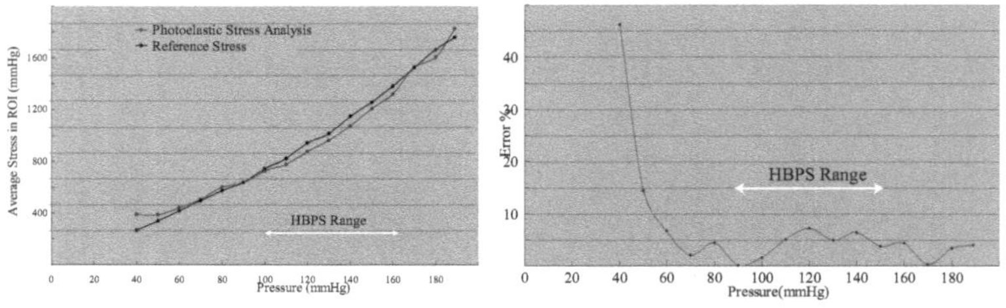

Figure 4.15 Stress in the model wall variation against pressure inside the model, calculated using photoelastic stress analysis and model radius variation as reference stress (left). Variation of error of photoelastic stress analysis with pressure changes inside the vasculature model (right).

4.8 Angular Distortion Correction

To obtain the measurements shown in Fig. 4.14, a correction method for angular distortion is needed. This distortion is produced as the photoelastic crystals orientation changes related to the camera orientation by the blood vessel model shape, producing more retardation near the cylinder's axis and less in the borders of the ROI (Fig. 4.16). Provided that the error obtained by applying the photoelastic stress analysis to the average green light intensity value in ROI is low, we could then apply equations (4.6, 4.26) to deduce the stress in $(\sigma_1-\sigma_2)_{(P,\,x,\,y)}$, a pixel located in row x and column y of a source image set corresponding to a pressure P inside the model. For correcting this, we used

$$(\sigma_1-\sigma_2)'_{(x,y)} = (\sigma_1-\sigma_2)_{(x,y)} + k\cos\left(\frac{2\pi y}{40}\right)\left[(\sigma_1-\sigma_2)_{AVG(x,ROI)} - (\sigma_1-\sigma_2)_{\min(x,ROI)}\right] \qquad (4.32)$$

where for each column of ROI, $(\sigma_1-\sigma_2)_{\min\,(P,\,y)}$ is the minimum and $(\sigma_1-\sigma_2)_{AVG(P,\,y)}$ the average stress value, and k = 0.8 is correction factor.

For the stress analysis in every pixel of the ROI, the applied correction eliminated the distortion introduced by the blood vessel shape, as the stress distribution is more homogeneous in Fig.

4.14 than in Fig. 4.16. As seen in Fig. 4.14, the correction method preserved local maxima and minima of stress. This characteristic is essential to quantify the stress produced by medical tools on the blood vessel wall. As we can see in Fig. 4.14, there is a weak area in the model wall near the right side of the ROI. At that point, the wall gets thinner and produces a stress local maximum, the size of this local maximum increasing with pressure. Stress local minima are visible where air bubbles were trapped between silicone layers, as air does not produce retardation of polarized light.

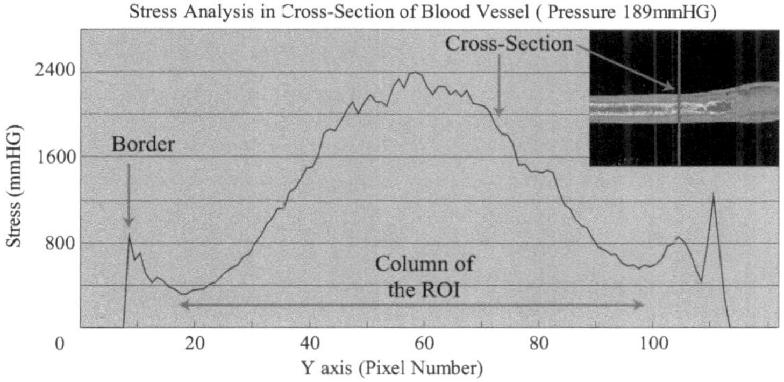

Figure 4.16 Distortion introduced by the model shape to the stress analysis and studied cross section of blood vessel. Pressure inside the model is 189 mmHG.

4.9 Stress Direction Measurements

The method proposed in the previous section allows quantitative stress analysis for neurovascular simulation, and it is very helpful for evaluating the performance of instruments. However, inserted medical instrument are difficult to perceive because generally these objects are opaque and in the circular polariscope, light penetrates only through photoelastic materials under stress. Additionally, as Fig. 4.8 shows, stress direction is not practically measurable. To solve this problem, we propose another polariscope construction shown in Fig. 4.17, adding an extra wave plate at slight inclination between second set of quarter-wave plates and polarizer. With this construction, the transit background color from black to violet, remarkably improves the visibility of inserted surgical instruments, improves the sensitivity to color transition (color changes from violet to red and blue at slight stress) and enables simultaneous stress magnitude and direction analysis. However, high-precision wave plates and quarter-wave plates are necessary to obtain more accuracy in measurements of stress magnitude and direction [1].

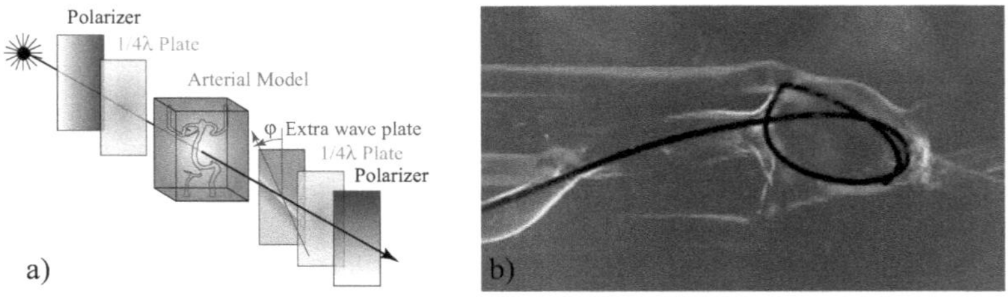

Figure 4.17 Circular polariscope for simultaneous measurement of stress magnitude and direction (left). High contrast between background and surgical instruments (right). See also Color Insert.

For this polariscope configuration, the phase shift introduced by the extra wave-plate wavelength is

$$\Phi_2 = \frac{4\pi\lambda_{ex}}{\lambda} \tag{4.33}$$

and the effect of wave-plate into the light beam is defined by

$$J_{(\Phi_2,\varphi)} = \begin{bmatrix} \cos\frac{\Phi_2}{2} + i\sin\frac{\Phi_2}{2}\cos 2\varphi & i\sin\frac{\Phi_2}{2}\sin 2\varphi \\ i\sin\frac{\Phi_2}{2}\sin 2\varphi & \cos\frac{\Phi_2}{2} - i\sin\frac{\Phi_2}{2}\cos 2\varphi \end{bmatrix} \tag{4.34}$$

where φ is the inclination angle of the extra wave-plate. The polariscope equation is then defined as follows:

$$E_{out} = J_{P(\pi/2)} J_{R(\Phi_2,\varphi)} J_{R(\Phi_1,-\pi/4)} J_{R(\alpha,\theta)} J_{R(\Phi_1,\pi/4)} J_{P(0)} E_{in} \tag{4.35}$$

Then the emerging beam intensity is given by

$$\begin{aligned} I = {} & 4c_1^2c_2^2\sin^2\left(\frac{\Phi_2}{2}\right)\cos^2\left(\frac{\alpha}{2}\right) + \sin^2\left(\frac{\alpha}{2}\right)\left\{c_1^4 + c_2^4 + 2c_1^2c_2^2\cos\frac{\Phi_2}{2}\right\} \\ & + c_1c_2\sin(\alpha)\left\{(c_1^2 - c_2^2)\sin 2\theta - c_1^2\sin\left(2\theta - \frac{\Phi_2}{2}\right) + c_1^2\sin\left(2\theta + \frac{\Phi_2}{2}\right)\right\} \end{aligned} \tag{4.36}$$

Figure 4.18 shows a color map calculated from above equation considering Φ_2 = 1060 nm and φ = 22.5°. The heterogeneous dispersion of RGB ratio (color) over the map is maximized by the polariscope design. This map indicates the correlation among observed color (RGB), magnitude of retardation Re and direction of the stress θ. From the matching the observed color with the color map Re and θ can be deduced. Then the corresponding intensity can be calculated with equation (4.36) and then stress magnitude is calculated.

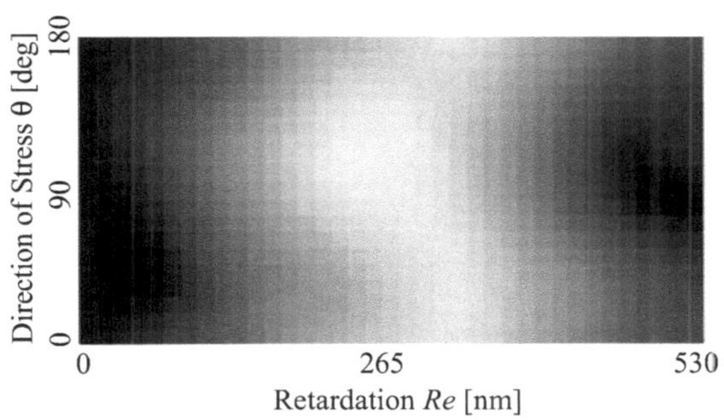

Figure 4.18 Color map calculated from equation 4.36 for RGB for retardations within the model below 530 nm and stress direction below 180°. See also Color Insert.

This method is extensible to real-time stress analysis and provides information for evaluating the medical instruments performance in terms of stress magnitude and direction, computational fluid dynamics validation for hemodynamic studies, and surgical skills evaluation.

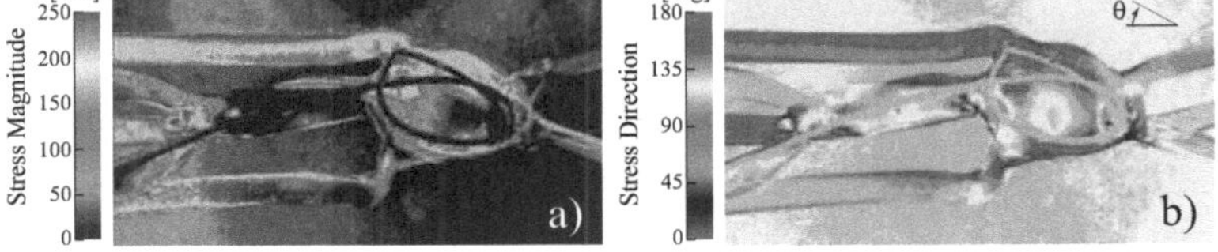

Figure 4.19 Simultaneous measurements of stress magnitude (a) and direction (b) for the single-layer model shown in Fig. 4.17. See also Color Insert.

4.10 Three-Dimensional Visualization of Stress

Three-dimensional stress visualization is necessary in endovascular surgery simulation for observation and measurement of stress distribution change in blood vessel model wall during human blood pressure simulation, health care specialist training and intravascular tools evaluation. ML-EM is an iterative method that reconstructs the cross section of an object from the sinogram of a slice of the object and an initial guess of the cross section. The sinogram is an image where each row *a* represents a direction, each column *b* a spatial coordinate of the studied cross section, and color represents the object thickness. Let us name the image B^0 the initial guess of the object cross section, and B^k the result image of the reconstruction process at iteration *k*. For recording the sinograms of vasculature phantoms a three-dimensional scanner was constructed. The polariscope shown in Fig. 4.6 and a camera were attached to a glycerine tank. This structure has its rotation axis aligned to the tank's vertical axis, enabling to register the sinograms by scanning in 360° a phantom submerged into the tank (Fig. 4.20). A multilayer phantom representing a straight segment vasculature was submerged into the tank, concentric to the tank rotation axis. For this experiment, pressure inside the model was set at three levels P_1 = 80 mmHg, P_2 = 120 mmHg and P_3 = 160 mmHg. For each pressure, two sets of images were recorded: A first scan was performed with a blue filter instead of the polariscope to obtain in a range of 180° 128 source images of the optical path length (OPL) for each slice.

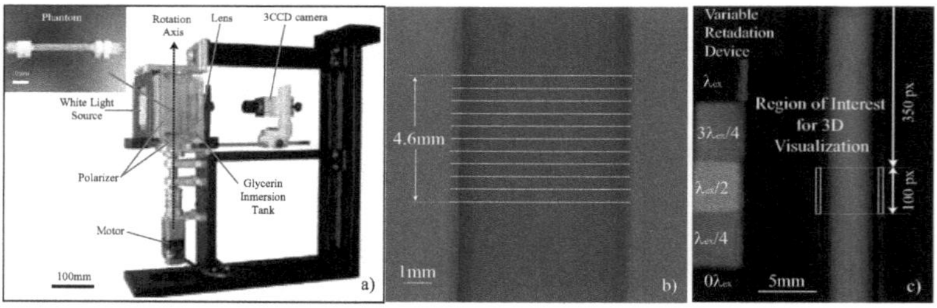

Figure 4.20 Scanner for building the sinogram of each slice of the model of vasculature submerged into the glycerin tank. (a) Studied slices for the three-dimensional visualization of stress over a source image sample for optical path length calculation. (b) Retardation of green light produced by stress in the model wall.

From them, the sinograms $A^n{}_P$ of slice *n* at pressure *P* were built for each pressure and slice. Then a second scan was done with the polariscope in place for the same directions to register the green light retardation. From them, the sinograms R^n_P of slice *n* at pressure *P* were built for each pressure and slice. For each image, the region of interest (ROI) is limited to the borders of the model from rows 350 to 449. In the same way, two additional scans were done with a guide wire inserted into the phantom and inner pressure of 80 mmHg. From them the sinograms A^n_{80GW} representing the OPL and the R^n_{80GW} representing retardation were obtained (Fig. 4.20). Optical path length and retardation images were captured for the slices within the region of interest at 128 different directions for pressures of 80, 120 and 160 mmHg. When the model had an inner pressure of 80 mmHg, an additional set of images was captured when a guide wire was inserted. The resulting sinograms of slice 400 are shown for the four samples in Fig. 4.21. For reconstructing the stress distribution from A^n_P and R^n_P, ML-EM method was used twice. First, the slices of the blood vessel thickness were reconstructed by using A^n_P images as source and using a white image with the same dimensions of A^n_P as initial guess of the cross section B^0_P. The result images are named $B^{n,0}_P$; best results were obtained after 20 iterations $B^{n,20}_P$. $B^{n,20}_P$ are grayscale images where the thickness is represented by the color value. A threshold level was applied to set to zero the pixels with a value below 128, and set to 255 those with a value above. The resulting image $B'^{n,20}_P$ is used as the first stress cross section approximation for using the ML-EM method a second time to reconstruct the stress distribution (Fig. 4.22).

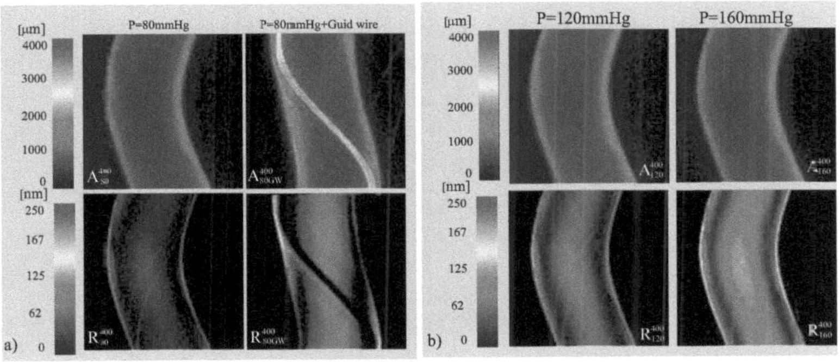

Figure 4.21 (a) Sinograms of optical path length and retardation with 80 mmHg of pressure inside the model and with guide wire inserted. (b) Sinograms of optical path length and retardation with pressures of 120 mmHg and 160 mmHg. See also Color Insert.

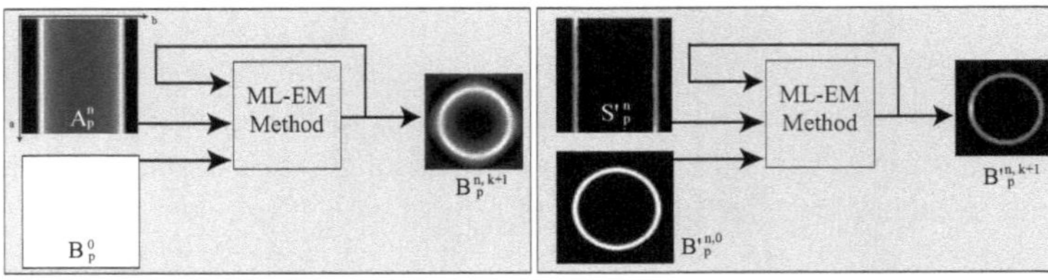

Figure 4.22 (a) Preliminary thickness slice reconstruction $B^{n,k}{}_p$ from thickness sinogram $A^n{}_p$ and first cross section initial guess $B^0{}_p$. b) Average stress slice reconstruction $B'^{n,k}{}_p$ from filtered average stress sinogram $S'^0{}_p$ and first cross section initial guess $B'^{n,0}{}_p$.

Let us name $A^n{}_P$ *(a, b)* the pixel located at row *a* and column *b* of $A^n{}_P$, the limits of the ROI for the row *a* are defined by the external wall location L_{oa} and the inner wall location L_{ia}. When $A^n{}_P$ *(a, b)* reaches 1200 µm, $L_{oa} = b$, and $L_{ia} = b$ when $A^n{}_P$ *(a, b)* reaches the first local maximum. The average stress sinogram $S^n{}_P$ is calculated by the boundaries defined by the ROI by

$$S^n_p(a,b) = \frac{\sum_{b=L_{oa}}^{L_{ia}} R^n_p(a,b)/cA^n_p(a,b)}{L_{ia} - L_{oa}} \tag{4.37}$$

Pixels outside the ROI were set as zero. Finally the average stress sinogram $S^n{}_P$ is filtered by multiplying it with the sinogram of $B'^{n,\,0}{}_P$ to obtain $S'^n{}_P$. Then the ML-EM was applied to images $S'^n{}_P$ and $B'^{n,\,0}{}_P$, after five iterations a slice of average stress was reconstructed $B'^{n,\,5}{}_P$. During the application of the ML-EM method if a pixel value of $B'^{n,\,0}{}_P$ or $S'^n{}_P$ was zero, the detection probability was brought to zero for that pixel. For accuracy evaluation, we used the model for longitudinal stress variation in a cylinder according to [9]; this model describes the stress on the blood vessel model edge:

$$\sigma_2 = \frac{(r-t)^2}{r^2 - (r-t)^2} P_{in} \tag{4.38}$$

where P_{in} is the pressure inside the cylinder, *r* the cylinder radius and *t* the wall thickness.

Slices of average stress corresponding to the sinograms of Fig. 4.21 were reconstructed relying on ML-EM method, as shown in Fig. 4.23. Ten stress distribution slices within the region of interest were loaded to the *Volume Viewer* of *ImageJ* to display the three-dimensional visualization of stress. Figure 4.23 shows three-dimensional views of the blood vessel model when its inner pressure was 120 mmHg and also shows views of the model with a guide wire applying stress to the wall and with inner pressure of 80 mmHg. Stress measurement error of images $B'^{400,5}{}_{80}$, $B'^{400,5}{}_{120}$ and $B'^{400,5}{}_{160}$ is 10.73%, 4.55% and 3.18%, respectively [8].

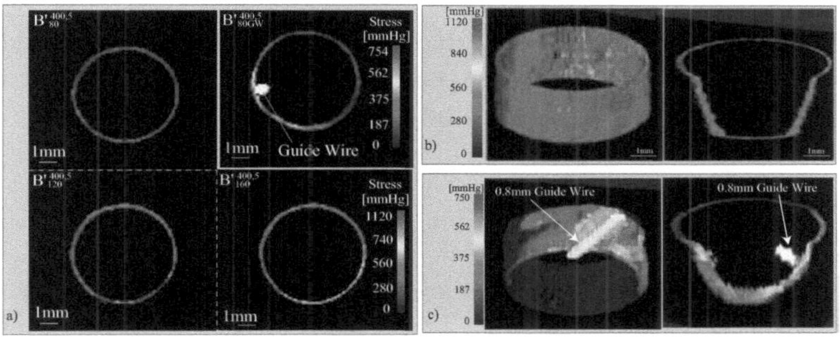

Figure 4.23 (a) Reconstruction stress distribution slices using ML-EM method in a model of vasculature slice at pressures of 80, 120 and 160 mmHg, and with guide wire inserted at 80 mmHg. (b) Three-dimensional visualization of stress and longitudinal stress variation within the model wall for a phantom with inner pressure of 120 mmHg. (c) Three-dimensional visualization of stress and longitudinal stress variation within the model wall for a phantom with a guide wire inserted and with inner pressure of 80 mmHg. See also Color Insert.

Slices of stress produced by a guide wire and inner pressure on a blood vessel model were reconstructed and three-dimensionally visualized; however, for this experimental setup and measurement accuracy, the lumen cross-sectional diameter of the model is limited to 11.8 mm. Image sizes of the reconstruction method are limited to powers of 2; for this case images of 128 × 128 pixels were used. The model lumen limit and measurement accuracy depend on the camera resolution and the size of the blood vessel model on the screen. The complete cross section of the model must appear into the sinogram to be reconstructed properly, and it is desirable to maximize the number of pixels belonging to the model wall. The membrane thickness must be at least three pixels to achieve the accuracy shown in this chapter. However, increasing resolution implies increasing the time required to reconstruct each slice and produce the three-dimensional visualization.

4.11 Complementary Image Processing for Real-Time Analysis

To quantify tissue respect during endovascular surgery simulation, it is desirable to measure in real time the stress within the blood vessel model wall and its deformation and then relate those measurements to the actual catheter tip location within the model. For that purpose, first it is necessary to diferentiate from the background of photoelastic images the parts belonging to blood vessel model and the parts belonging to the catheter and then locate the catheter tip into the image.

4.11.1 Vision System

For the demonstration of real-time catheter trajectory analysis including photoelastic effect, a multilayer model of a carotid artery filled and submerged in glycerin is used. We selected the polariscope shown in Fig. 4.24 due to the contrast quality and simplifications enabled by the mathematic model. The polariscope equation (4.39) simplifies to equation (4.24) if we consider $\lambda_G = \lambda_{ex}$. The pressure inside the model remained below 40 mmHg during the experiment.

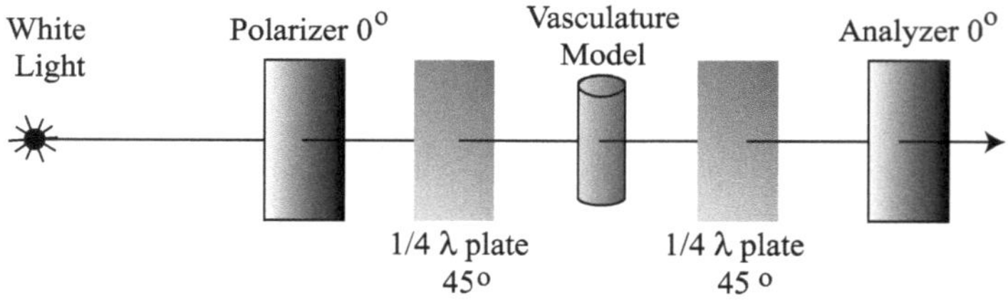

Figure 4.24 Polariscope for real-time analysis.

By aligning the model's principal branch axis parallel to the X-axis of the cameras in a bi-planar vision system, the progression of the catheter trajectory is principally perceived at the X-axis and branch selection at Y- and Z-axes. Respecting this principle, the cameras were aligned to the tank containing the blood vessel model. A polariscope was placed parallel to each camera plane and having the blood vessel model in between their quarter-wave plates. The beams from a single light source were separated to pass through each polariscope. Toshiba IKTF-2 cameras were used for the stereovision system and Micro Technica MTPCI-DC2 as video capture boards. This enabled to load video frames of 640 × 480 pixels to a dynamic memory array every 15 ms. The video frames were captured and processed in different computers for each plane, and they were synchronized through TCP/IP. For this analysis, we use a region of interest of 640 × 260 pixels containing the entire blood vessel model (Fig. 4.25).

$$
\begin{aligned}
I_{GN} = {} & \sin^2 \frac{\pi \mathrm{Re}}{\lambda_G}\left[\cos^2 2\theta + \sin^2 2\theta \sin^2\left(\frac{\pi\lambda_{ex}}{2\lambda_G}\right)\right] \\
& + \cos^2 \frac{\pi \mathrm{Re}}{\lambda_G}\left[2\cos^2\left(\frac{\pi\lambda_{ex}}{4\lambda_G}\right) - 1\right]^2 \\
& - \sin\frac{2\pi \mathrm{Re}}{\lambda_G}\sin 2\theta \sin\left(\frac{\pi\lambda_{ex}}{2\lambda_G}\right)\left[2\cos^2\left(\frac{\pi\lambda_{ex}}{4\lambda_G}\right) - 1\right]
\end{aligned}
\qquad (4.39)
$$

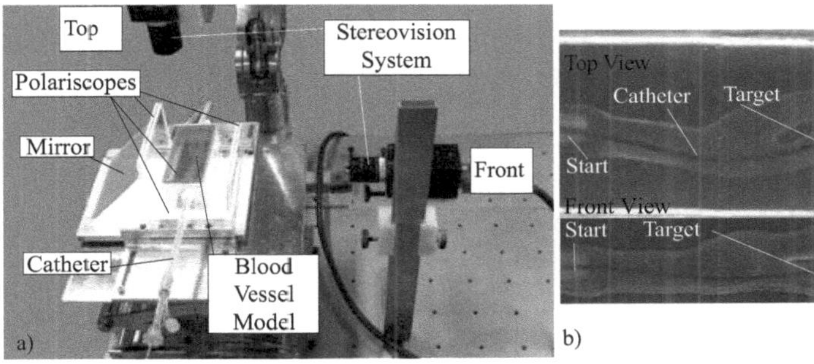

Figure 4.25 (a) Bi-planar vision system for real-time catheter trajectory analysis adapted for a carotid artery vasculature model. (b) Example of the captured images with the bi-planar vision system, showing the catheter inserted from the start point to the target branch.

4.11.2 Filtering

To distinguish between these three regions we use two static images for each plane. For the XY-plane, $L_{XY,i,k}$ is the frame number k of a trajectory video $L_{XY,i}$ (i, $k \in N$), the optical path length image D_{XY} and the first frame of the studied trajectory $L_{XY,i,0}$. The first one is used to calculate the stress value, D_{XY} is deduced from the blue light transmittance. For preventing noise in this image we calculate the noise indicator $S_{x,y}$ for each pixel of D_{XY}:

$$S_{x,y} = \sum_{x-5}^{x+5} (D_{n',y-5} + D_{n',y+5}) + \sum_{y-5}^{y+5} (D_{x-5,n'} + D_{x+5,n'}) \quad (4.40)$$

where $D_{x,y}$ is the optical path length at the pixel located in the row x of the column y (n' is the mute index for each sum). If $S_{x,y}$ is below 4000 and $D_{x,y}$ is above 200 μm, then the $D_{x,y}$ is considered noise and equalled to zero. The same considerations are applicable for $S_{x,z}$ and $D_{x,z}$ in XZ plane.

The second static image is the first frame of the source trajectory, $L_{XY,i,0}$ is used for comparing the shape and luminace of the studied frame with the original conditions. At $L_{XY,i,0}$ the catheter in not present and the model has no deformation in all cases, from it the luminance value of the pixels belonging to the model is registered. For $L_{XY,i,k}$, a pixel belongs to the model if its luminance value is above 56 and identified as deformation if its corresponding pixel in $L_{XY,i,0}$ belongs to the background. A reduction on the luminance value respect to $L_{XY,i,0}$ and with values below 68 in the red channel and 55 in the blue were considered as part of the catheter. Respective threshold values of 56, 70 and 65 for $L_{XZ,i,0}$ and $L_{XZ,i,k}$ are used in XZ-plane.

4.11.3 Stress and Deformation Measurement

Stress is calculated from measurements of green light intensity. We use the following equation to represent the relation between stress and retardation for XY plane:

$$(\sigma_1 - \sigma_2)_{x,y} = \frac{\mathrm{Re}_{x,y}}{CD_{x,y}} \quad (4.41)$$

where $\mathrm{Re}_{x,y}$ is the retardation in the pixel located at column x of the row y and is measured in nanometers and $(\sigma_1 - \sigma_2)_{x,y}$ the difference between principal stresses on each side of a biaxial stress field at that pixel. Considering the simplification of equation (4.39), equation (4.24) is applied at pixel level:

$$(I_{GN})_{x,y} = \sin^2 \frac{\pi \mathrm{Re}_{x,y}}{\lambda_G} \quad (4.42)$$

For the pixels identified as part of the blood vessel model in both planes, the difference between principal components of stress was calculated after measuring the green light intensity. With I_{bmax} = 181.92, I_{Gmin} = 51, I_{Gmax} = 225.5 calibration values for the XY-plane.

Similar equations and respective calibration values of 184.78, 29.5, 225 can be applied to calculate the stress for XZ-plane. $\sigma_{XY,\ i,\ k}$ and $\sigma_{XZ,\ i,\ k}$ are the pixels with stress above 640 mmHg, and are counted during the measurement done for each frame.

The blood vessel membrane deformation is obtained from comparing the area that belongs to the vasculature model in pixels between the two consecutive frames using:

$$G_{XY,i,k} = \left|F_{XY,i,k} - F_{XY,i,k-1}\right| \quad (4.43)$$

where $G_{XY,\ i,\ k}$ is the deformation value and $F_{XY,\ i,\ k}$ is the number of pixels belonging to the phantom for frame $L_{XY,\ i,\ k}$. The initial conditions are defined as follows:

$F_{XY,\ i,\ 0} = F_{XY,\ i,1}$ and $G_{Li,0} = 0$. In the same way, $G_{XZ,\ i,\ k}$ values for the XZ-plane are obtained.

4.11.4 Catheter Tip Search

Let us suppose that for frame $L_{XY,\ i,k\text{-}1}$ the tip was found pixel $(c_{x,k\text{-}1};\ c_{y,k\text{-}1})$. For frame $L_{XY,\ i,k}$ the search starts then at the pixel $(c_{x,k\text{-}1}$-40; 0$)$. Every even row in column $c_{x,k\text{-}1}$ will be inspected for a pixel belonging to the catheter; if it is found, the inspection direction changes following the priorities described in Fig. 4.26 for a radius of 2 pixels. If the pixel located in the direction with highest priority does not belong to the catheter, the following priority is inspected. If the next pixel belonging to the catheter is found, then search moves to that pixel and repeats the inspection in radius 2. If not, the search radius increases to 10 pixels in the same directions and priorities. If there are no pixels belonging to the catheter in both ranges, it means that the catheter tip location $(c_{x,\ k};\ c_{y,\ k})$ was found for frame k. For speeding up the search and enable motion capture during the catheter extraction, if $c_{x,\ k} > 60$ the search for $L_{\ XY,i,k-1}$ starts at column $c_{x,\ k}$-40. For other cases, the search starts in column 20. This way, the catheter tip coordinates are determined for both planes. The tip search algorithm relies on the proximity of the centreline of the blood vessel with X axis, where source images do not have this characteristic curved multi-planar reformation can be used to correct it.

While applying the threshold levels, the pixels belonging to the guide wire may be counted. As the diameter of guide wire is known, the guide wire is considered bent if there are more pixels per column than the guide wire diameter size. This can be used to flip the search direction when the bent condition is detected [10].

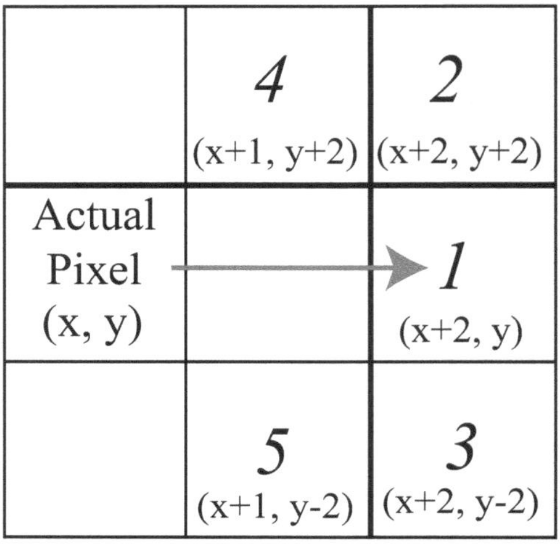

Figure 4.26 Search priorities for catheter tip location inside the image.

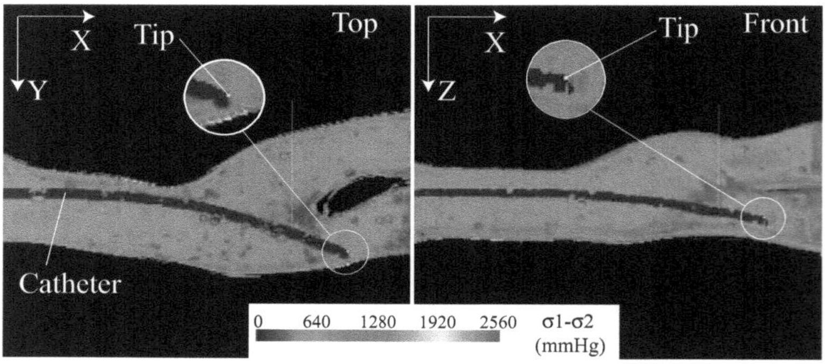

Figure 4.27 An example of result images after applying the filtering, measurement and search algorithm described in this section to the photoelastic images. The frame rate of the real-time analysis using the bi-planar vision system is above 3 fps.

4.11.5 Reference Trajectory Construction

The image-processing software described in the previous section was used for and stress analysis, morphology extraction, catheter recognition, and catheter tip detection in the XY-plane and the XZ-plane cameras. A point belonging to a source trajectory is defined as

$$L_{i,k}(c_x, c_y, c_z, \sigma_{XY}, G_{XY}, \sigma_{XZ}, G_{XZ}) \tag{4.44}$$

where c_x, c_y, and c_z are the catheter tip locations, σ_{XY} and σ_{XZ} are the stress levels, and G_{XY} and G_{XZ} are the deformation levels at frame k for the source trajectory i. These parameters are calculated from the video frames $L_{XY,i,k}$ and $L_{XZ,i,k}$. In this study, five source trajectories (L_1–L_5) between the start point and the target were registered; they have different duration and include the rotation of the catheter to select the target branch. They were built using a standard 5F catheter.

Let us name T_{Li} the amount of frames captured between the catheter tip detection at the start point until it reaches the target for trajectory L_i. To be compared, these trajectories must be normalized in time; the normalized frame rate for each trajectory is obtained from the inverse of T_{Li} and named SR_{Li}. And for all trajectories to be represented at least with a point during a reference trajectory point construction, the reference trajectory normalized frame rate SR_R is obtained from

$$SR_R = \frac{2\sum_{i=1}^{I} SR_{Li}}{I} \tag{4.44}$$

where I is the number of studied trajectories. The number of points composing the reference trajectory T_R is given by rounding to the next natural number the inverse of SR_R. The reference trajectory is constructed using the following equation:

$$R_n(c_x, c_y, c_z, \sigma_{XY}, G_{XY}, \sigma_{XZ}, G_{XZ}) = \frac{\sum_{t=t_{n-1}}^{t_n} \sum_{i=1}^{5} L_{i,t}(c_x, c_y, c_z, \sigma_{XY}, G_{XY}, \sigma_{XY}, G_{XY})}{p_n} \tag{4.45}$$

where n is reference trajectory frame number (between zero and the number of frames of the reference trajectory T_R) that occurs at normalized time t_n, and p_n is the number of points belonging to L_i occurring after t_{n-1} and before t_n. Figures 4.28–4.30 show the variation in normalized time for the components of the trajectories L_1–L_5 and the resulting reference trajectory R obtained by applying equation (4.45) to these components.

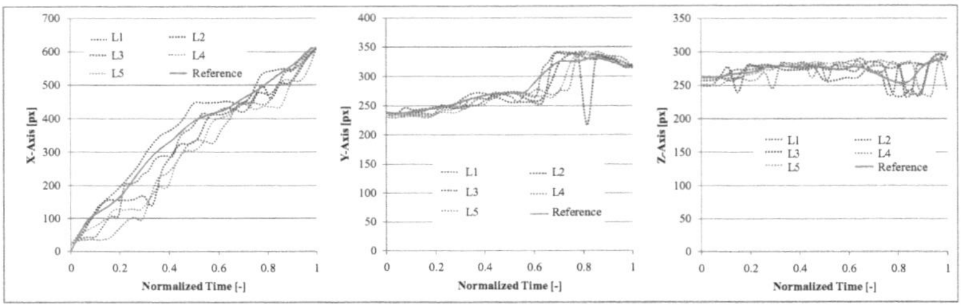

Figure 4.28 Catheter tip position in the three axes in normalized time for trajectories L_1–L_5 and the obtained reference trajectory.

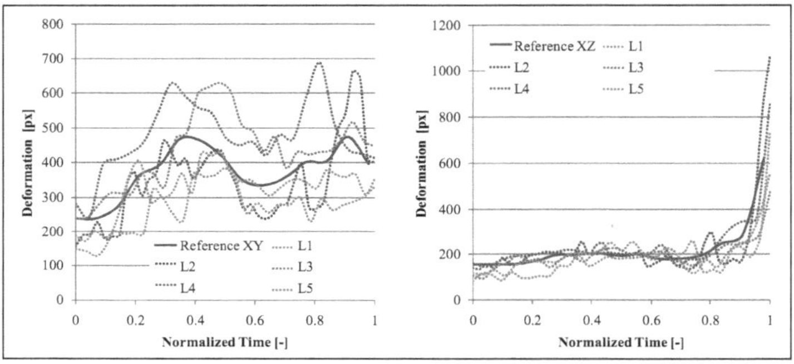

Figure 4.29 Blood vessel wall deformation in normalized time for trajectories L_1–L_5 and the obtained reference trajectory.

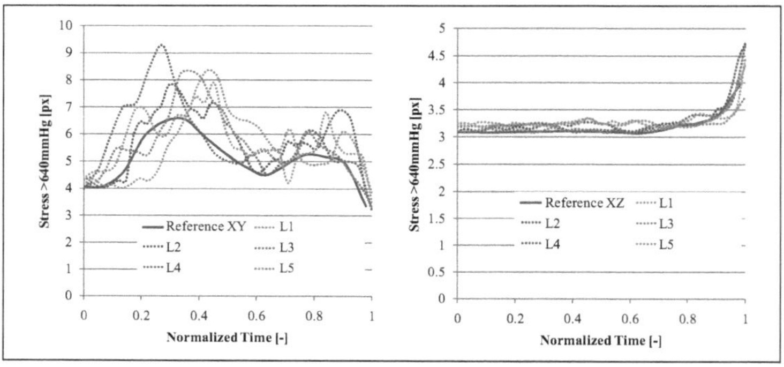

Figure 4.30 Stress level in normalized time for trajectories L_1–L_5 and the obtained reference trajectory.

The reference trajectory construction method was applied in a mono-vision configuration to compare the catheter motion and stress levels variation of reference trajectory with trajectories while the catheter was driven by inexperienced human and robot.

A vasculature model with similar morphology as shown in Fig. 4.27 was used. The reference trajectory was constructed from a set of five trajectories and is shown in Fig. 4.31. The robot used for catheter manipulation was presented in [11].

The inexpert user succeeded in entering the correct branch after six trials; the back and forth motion is shown in Fig. 4.32, where the value of c_z has local maxima at t = 0.26, t = 0.34, t = 0.52, t = 0.63 and t = 0.77 before reaching the target. At t = 0.70, the inexpert user inserted the catheter into a branch not leading to the target and then pulled back the catheter between t = 0.70 and t = 0.84 from pixel (530; 157) to (339; 188), during that motion, the stress level reached its global maximum at 12.43% of pixels with stress above 1280 mmHg. The average stress value along the trajectory is 7.31%. During the robotic manipulation, Fig. 4.33 shows a variation of c_z corresponding to the stepping motion of the catheter insertion robot and shows an insertion to the correct branch in the first trial. The average stress value along the trajectory is 6.61% and its global maximum is 7.44%.

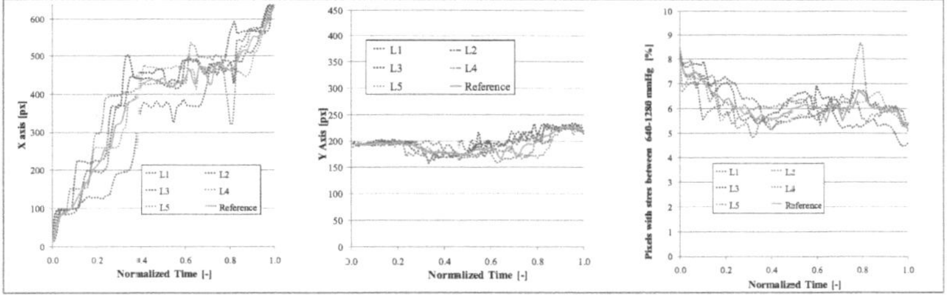

Figure 4.31 Reference trajectory construction for motion capture and stress variables.

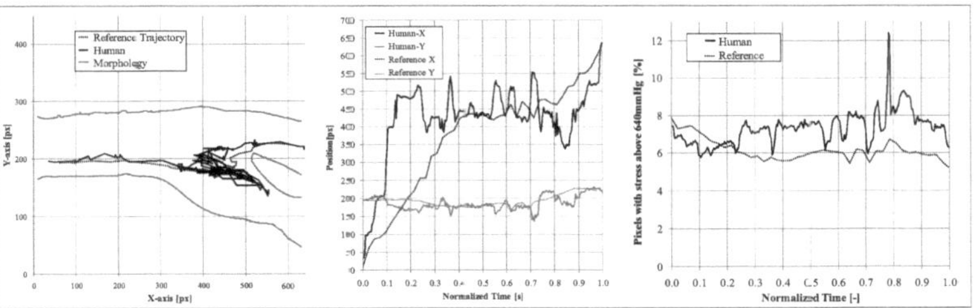

Figure 4.32 Catheter driven by an inexperienced person: Insertion path projection in the XY-plane compared to blood vessel morphology and reference trajectory, catheter tip tracking in normalized time for both axes, stress level variation in normalized time.

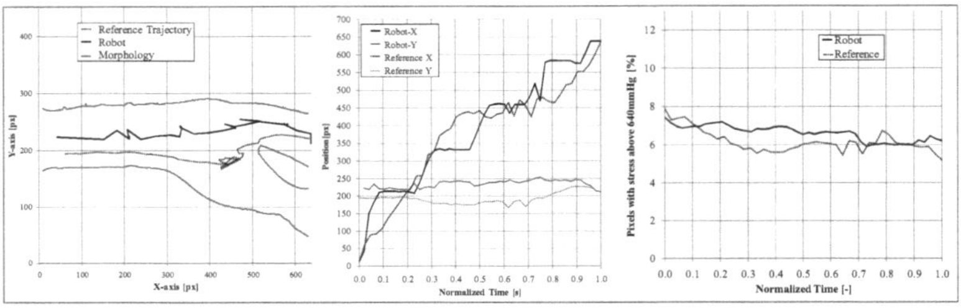

Figure 4.33 Catheter driven by robot: Insertion path projection in the XY-plane compared with blood vessel morphology and reference trajectory, catheter tip tracking in normalized time for both axes, stress level variation in normalized time.

4.12 Summary

From our studies in photoelastic stress analysis, we can say that vasculature modeling with photoelastic materials enabled to visualize the stress produced by standard intravascular tools such as guide wires, guide catheters, micro-coils and micro catheters. This enabled different kinds of stress analysis to represent numerically respect for tissue integrity during endovascular surgery simulator-based training. The presented 3D scanner was able to register the stress produced by the guide wire to the blood vessel wall; however, several seconds are necessary to collect the data and calculate the reconstruction of the three-dimensional model of the stress field and guide wire. This enables the study of the stress produced by intravascular tools after being deployed, as a stent delivered in a model of stenosis, but is a limitation for the real-time analysis of the catheter progress. On the other hand, the bi-planar vision system enabled to register the catheter trajectory in terms of the catheter tip motion capture and tissue respect parameters at 3 fps. This processing speed can be increased by reducing the resolution of the analysis, by focusing the analysis to a region of interest or increasing the computing power. This analysis area, time and resolution should be configured for specific applications.

Photoelastic stress analysis made possible to visualize and identify behaviors that may compromise tissue integrity by misuse and repeated trials to achieve a task. Actually there are at least nine catheters and guide wires driving devices [12], and an autonomous catheter insertion system to reconstruct catheter trajectories through vasculature phantoms relying on magnetic trackers [11]. The measurement of tissue respect while these robotic systems are manipulating the catheter is useful for designing and for demonstration of the reliability of those robotic systems.

The presented reference trajectory construction method is useful not only for evaluation purposes but also for storing the abilities of driving a catheter of one or a group of experts. And since tailor-made models of vasculature are used, this reference trajectory could be used for path planning during an endovascular intervention in terms of tissue respect.

To apply the presented methodology for simulator-based training, other evaluation criteria of Global Rating Index of Technical Skills must be included [13–14], such as bimanual dexterity, instrument knowledge, knowledge of the procedure and flow of operation. For that, the vision system must be adapted to enable tracking simultaneously a guide catheter and a guide wire, as well as consider the motion capture of the catheter within the catheter insertion port and motion capture of the operator hands. Vasculature modeling with photoelastic materials should be developed for enabling the modeling of large vasculature section with photoelastic materials with high photoelastic coefficient.

References

1. Ikeda S, Fukuda T, Arai F, *et al.* Patient-specific neurovascular simulator for evaluating the performance of medical robots and instruments. In *Proceedings of the IEEE-ICRA*, 2006, 625–630.
2. Tanimoto M, Tanimoto F, Arai T, *et al.* Study on micro force sensor for minimum invasive surgery. *Trans Japan Soc Mech Eng*, 1998, C 64–620, JSME, 150–155.
3. Tuji J, *et al.* Koudansei Jikkenhou. Tokyo: Nikkan Kougyou Shinbun-sya, 1965.
4. Kuske A, Robertson G. *Photoelastic Stress Analysis*, Wiley-Interscience Publication, New York, 1974, 87–109, 263–274.
5. Goldstein D. *Polarized Light.* New York: Marcel Dekker, 2003, Chapter 11.
6. Tercero C, Ikeda S, Tijerino E, *et al.* Human blood pressure simulation for stress analysis in model of vasculature using photoelastic effect. *Int J Automation Technol*, 2009, 3:5, 533–540.

7. Tercero C, Ikeda S, Matsushima M, Fukuda T, Negoro M. Photoelastic stress analysis error quantification in vasculature models for robot feedback control. *IEEE/ASME Trans Mechatron*, 2010, 15:4, 520–526.
8. Matsushima M, Matsushima C, Tercero S, Ikeda T, Fukuda F, Arai M, Negoro I. Takahashi, photoelastic stress analysis in blood vessel phantoms: three-dimensional visualization and saccular aneurysm with bleb. *Int J Med Robotics Comput Assist Surg*, 2011, l.7:1, 33–41.
9. Li H, Pugh D. *Mechanical Behavior Materials under Pressure*, Elsevier Publishing Co. Ltd., Chapter 2.
10. Tercero C, Ikeda S, Fukuda T, *et al.* Catheter insertion reference trajectory construction method using photoelastic stress analysis for quantification of respect for tissue during endovascular surgery simulation. *Int J Optomechatron*, 2011, 5:4, 322–339.
11. Tercero C, Tercero S, Uchiyama T, *et al.* Autonomous catheter insertion system using magnetic motion capture sensor for endovascular surgery. *Intl J Med Robotics Comput Assist Surg*, 2007, 3:1, 52–58.
12. Govindarajan S, Thenkurussi K, Xinyan L. Design and fabrication of a robotic mechanism for remote steering and positioning of interventional devices. *Intl J Med Robotics Comput Assist Surg*, 2010, 6, 160–170.
13. Moorthy K, Munz Y, Sarker S, Darzi A. Objective assessment of technical skills in surgery. *Br Med J*, 2003, 327, 1032–1037.
14. Doyle JD, Webber EM, Sidhu RS. A universal global rating scale for the evaluation of technical skills in the operating room. *Am J Surg*, 2007, 193, 551–555.

Chapter 5

Numerical Simulation for Blood Flow

Table 5.1 Definition of variables used in this chapter

Variables	Description	Unit
ρ	Density	kg/m^3
u	Velocity component of x direction	m / s
v	Velocity component of y direction	m / s
w	Velocity component of z direction	m / s
F	Force acting on unit mass	N/kg
M	Mass	kg
α	Acceleration	m/s^2
$O(\delta^2)$	High order terms more than second	
τ	Shear stress	Pa
σ	Normal stress	Pa
p	Pressure	Pa
μ	Viscosity coefficient	Pa·s
$\nu = \mu/\rho$	Kinematic viscosity coefficient	m^2/s
ϕ	Vector containing variable values at the grid nodes	
U	Characteristic velocity	m/s
L	Characteristic length	m
$Re = U{\cdot}L/\nu$	Reynolds number	

Microsurgery: Advances, Simulations, and Applications
Edited by Toshio Fukuda and Carlos Tercero

ISBN 978-981-4364-69-0 (Hardcover), 978-981-4364-70-6 (eBook)
www.panstanford.com

5.1 Basic Equations of Flow Analysis

5.1.1 Introduction

Owing to rapid development in computer hardware in recent years, numerical simulation has been highlighted as an approach to solve problems in science and engineering. The numerical calculation is widely accepted in the field of engineering, because it has several advantages in swiftness, safeness and cost compared with experimental trial. As it has the potential to simulate complex phenomena that are difficult to measure and analyze mathematically, it is also the principal method in fundamental research. The solution of phenomena in flow by numerical calculation and the academic field of numerical calculation method itself are called numerical fluid dynamics and also computational fluid dynamics. The numerical calculation has a wide field application such as flow in space, weather and ocean, and also airplanes, ships, trains and cars. Moreover, biological flows in human vessel and organ are included. As these flow fields contain complex elements in boundary shape, external force and fluid property, it is difficult to express in analytical formulation. In the area of numerical fluid dynamics, basic equation of flow is numerically calculated and flow field is reconstructed in calculator to predict and clarify various phenomena in flow. Basic equations for continuum fluid have been established from conservation law. They are law of conservation of mass, momentum and energy, respectively. As necessary, gas equation and chemical reaction equation is solved incrementally. The numerical simulation of flow means numerical calculation in the adequate initial and boundary condition and simulation of actual flow. To be more precise, it is also said to be reproduction of flow phenomena such as momentarily, velocity, density and temperature in some points of flow field.

Depending on the nature of flow, we decide whether we solve the Navier–Stokes equation or Euler's equation which is simplified by ignoring fluid viscosity. We chose physics model such as turbulence model and non-Newtonian constitutive equation if needed. As a result, partial differential equation (PDE) for numerical calculation is decided.

We discretize PDE with finite-difference method (FDM), finite-element method (FEM), or finite-volume method (FVM) and then set the grid for calculation depending on the method. As a consequence, the algebraic equation for numerical calculation is derived Subsequently we decide numerical solution of algebraic equations and create programs to bring out the full potential of the calculator. Although the result of numerical calculation is exported as big amounts of data, it is difficult to understand flow phenomena only with these data. Therefore, it is important to visualize with graph, computer graphics and animation to understand the phenomena. Figure 5.1 shows a flowchart of numerical simulation.

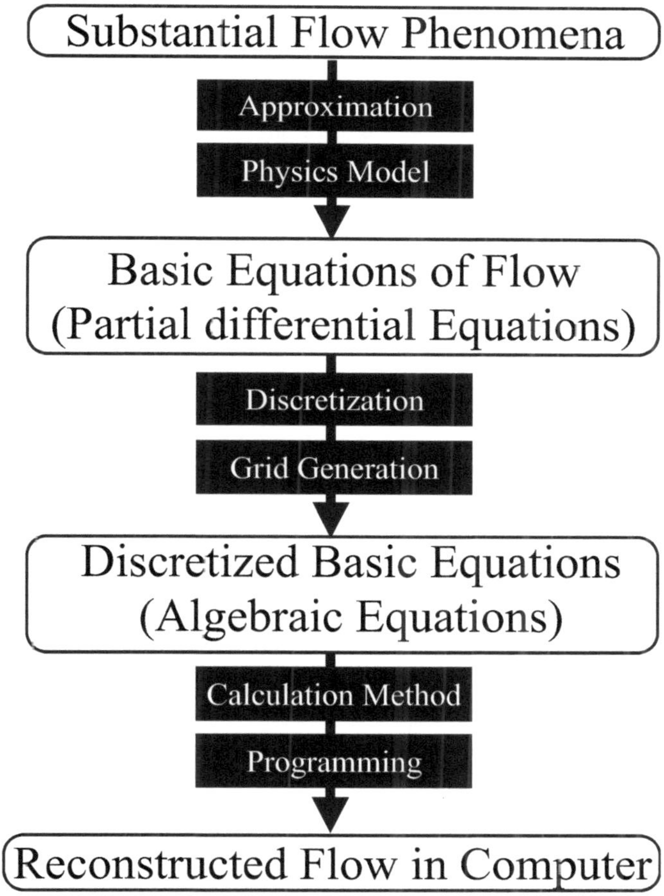

Figure 5.1 Flowchart of numerical simulation.

5.1.2 Law of Conservation of Mass

We suppose an imaginary minimal rectangular area centered around the point (x, y, z). We consider ρu as a parameter and Taylor series of above equation around the point (x, y, z) is obtained. Mass of fluid going through two faces perpendicular to x axis and entering and leaving inspection area between the time t and $t + \delta t$ is

$$\begin{aligned}\rho\left(x-\frac{\delta x}{2}, y, z, t\right)\cdot u\left(x-\frac{\delta x}{2}, y, z, t\right)\delta y\delta z\delta t \\ \simeq (\rho u)+\frac{\partial(\rho u)}{\partial x}\cdot\left(-\frac{\delta x}{2}\right)_{x,y,z,t}\delta y\delta z\delta t\end{aligned} \tag{5.1}$$

and then, as the fluid flow out from the face on the right-hand perpendicular to x axis, we reverse the direction of the Eq. (5.1) as follows:

$$\begin{aligned}-\rho\left(x+\frac{\delta x}{2}, y, z, t\right)\cdot u\left(x+\frac{\delta x}{2}, y, z, t\right)\delta y\delta z\delta t \\ \simeq -(\rho u)+\frac{\partial(\rho u)}{\partial x}\cdot\left(\frac{\delta x}{2}\right)_{x,y,z,t}\delta y\delta z\delta t\end{aligned} \tag{5.2}$$

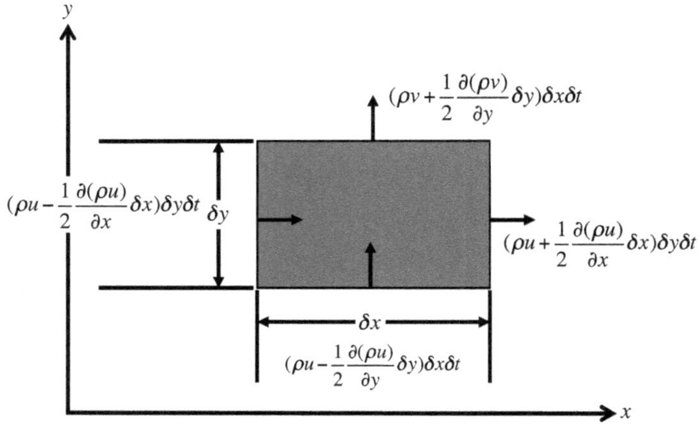

Figure 5.2 Fluid flow of rectangular area.

From the sum of Eqs. (5.1) and (5.2), the increase of mass from the face perpendicular to x axis is $-\partial(\rho u)/\partial x \delta x \delta y \delta z \delta t$. The increase of mass through the face of cuboid perpendicular y axis and z axis is

$-\partial(\rho v)/\partial y\delta x\delta y\delta z\delta t$ and $-\partial(\rho w)/\partial z\delta x\delta y\delta z\delta t$, respectively. The sum of these three terms must be equal to the increase of mass (increase of density × volume) within rectangle, $(\partial\rho/\partial t) \times \delta x\delta y\delta z\delta t$. Therefore, we obtain an equation of continuity as a law of conservation of mass.

$$\frac{\partial\rho}{\partial t}+\frac{\partial(\rho u)}{\partial x}+\frac{\partial(\rho v)}{\partial y}+\frac{\partial(\rho w)}{\partial z}=0 \tag{5.3}$$

The symbol "∇" means $\nabla = (\partial/\partial x, \partial/\partial y, \partial/\partial z)$ and $\nabla\cdot\mathbf{v}$ is inner product of vector.

$$\nabla\cdot\mathbf{v}=\left(\frac{\partial u}{\partial x}+\frac{\partial v}{\partial y}+\frac{\partial w}{\partial z}\right) \tag{5.4}$$

We apply substantial derivative symbol of operation to above expression and we have

$$\frac{D\rho}{Dt}-\rho\nabla\cdot\mathbf{v}=0 \tag{5.5}$$

When the first term on the left-hand of Eq. (5.4) is equal to zero,

$$\frac{D\rho}{Dt}\equiv u\frac{\partial\rho}{\partial x}+v\frac{\partial\rho}{\partial y}+w\frac{\partial\rho}{\partial z}=0 \tag{5.6}$$

The working fluid is considered an incompressible fluid. If the fluid is not compressible, substantial derivative of ρ is zero.

$$\frac{\partial\rho}{\partial x}+\frac{\partial\rho}{\partial y}+\frac{\partial\rho}{\partial z}=0 \tag{5.7}$$

Therefore, we obtain

$$\nabla\cdot\mathbf{v}=0(or\ div\ \mathbf{v}=0) \tag{5.8}$$

5.1.3 Law of Conservation of Momentum

Dynamics of mass point and rigid body is based on Newton's second law.

$$F = M\alpha \tag{5.9}$$

where F is force, M mass and α acceleration. Although fluid dynamics is always discussed on the basis of this law, unlike mass point and rigid body, fluid is sort of infinite form. Therefore, we need to change this law in the way of fluid dynamics. Let us suppose a rectangular area with plane perpendicular to every coordinate axis (center point: x, y, z, length of side: δx, δy, δz). We apply Eq. (5.9) to this area.

At first, mass M of this minute fluid body is

$$M = \rho\delta x\delta y\delta z \tag{5.10}$$

and acceleration is obtained from substantial derivative of velocity. That is

$$\alpha_x = \frac{\partial u}{\partial t} + u\frac{\partial u}{\partial x} + v\frac{\partial u}{\partial y} + w\frac{\partial u}{\partial z} = \frac{Du}{Dt} \tag{5.11}$$

$$\alpha_y = \frac{\partial v}{\partial t} + u\frac{\partial v}{\partial x} + v\frac{\partial v}{\partial y} + w\frac{\partial v}{\partial z} = \frac{Dv}{Dt} \tag{5.12}$$

$$\alpha_z = \frac{\partial w}{\partial t} + u\frac{\partial w}{\partial x} + v\frac{\partial w}{\partial y} + w\frac{\partial w}{\partial z} = \frac{Dw}{Dt} \tag{5.13}$$

Force acting on fluid without viscosity (ideal fluid) is just only mass force and pressure. Pressure p works in the direction of internally normal stress toward a plane. As a difference of pressure acting on two faces with a distance δx perpendicular to x axis, the external force in the x axis is

$$\{p(x - \frac{\delta x}{2}, y, z, t) - p\{p(x + \frac{\delta x}{2}, y, z, t)\}\delta y\delta z \tag{5.14}$$

$$= -\frac{\partial p}{\partial x}\delta x\delta y\delta z + O((\delta x)^2)\delta y\delta z$$

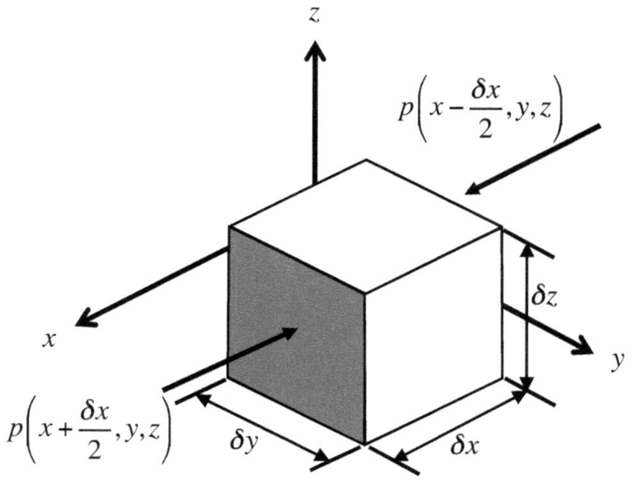

Figure 5.3 Pressure acting on a surface of fluid body.

Here, $O(\delta^2)$ means higher order terms more than second such as, δx^2, δt^2, δx^3, δt^3.... Pressure acting on a plane perpendicular to the y axis and the z axis is similarly derived as $(-\partial p/\partial y)\delta x\delta y\delta z$ and $(-\partial p/\partial z)\delta x\delta y\delta z$, respectively

$$F = \rho F \delta x \delta y \delta z = \rho \begin{pmatrix} F_x \\ F_y \\ F_z \end{pmatrix} \delta x \delta y \delta z \tag{5.15}$$

Therefore, these relations are substituted into Eq. (5.9) and then mass equations of fluid are obtained.

$$\frac{\partial u}{\partial t} + u\frac{\partial u}{\partial x} + v\frac{\partial u}{\partial y} + w\frac{\partial u}{\partial z} = F_x - \frac{1}{\rho}\frac{\partial p}{\partial x} \tag{5.16}$$

$$\frac{\partial v}{\partial t} + u\frac{\partial v}{\partial x} + v\frac{\partial v}{\partial y} + w\frac{\partial v}{\partial z} = F_y - \frac{1}{\rho}\frac{\partial p}{\partial y} \tag{5.17}$$

$$\frac{\partial w}{\partial t} + u\frac{\partial w}{\partial x} + v\frac{\partial w}{\partial y} + w\frac{\partial w}{\partial z} = F_z - \frac{1}{\rho}\frac{\partial p}{\partial z} \tag{5.18}$$

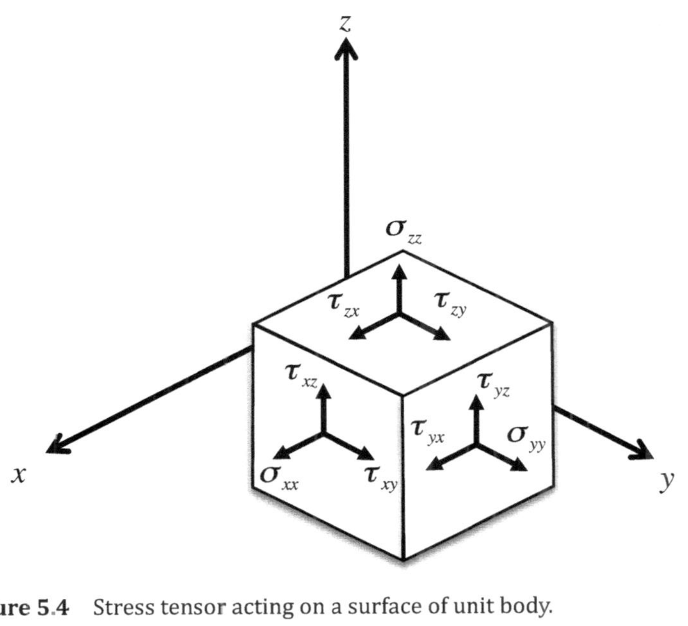

Figure 5.4 Stress tensor acting on a surface of unit body.

These are Euler's equations of motion.

$$\frac{Du}{Dt} = F_x - \frac{1}{\rho}\frac{\partial p}{\partial x} \tag{5.19}$$

$$\frac{Dv}{Dt} = F_y - \frac{1}{\rho}\frac{\partial p}{\partial y} \tag{5.20}$$

$$\frac{Dw}{Dt} = F_z - \frac{1}{\rho}\frac{\partial p}{\partial z} \tag{5.21}$$

We define the symbol and the direction of stress. First, consider unit surface perpendicular to the x axis. We set the normal line on the face where force is applied and consider its direction of a face. This stress applied to unit surface is divided into two forces. One is normal stress at right angle to the face and another is shear stress parallel to the face. Moreover, the latter is divided into the component toward the y axis and the z axis and consists of three components in the end. These three components are expressed as $\tau_{\alpha\beta}$. The first additional character α indicates the face what we are focusing on and the second additional character indicates the force direction of component. For example, three components of forces on the face perpendicular to the x axis are $\tau_{xx}\tau_{xy}\tau_{xz}$; τ_{xx} is usually written as σ_{xx}. Force direction at a center point of faces must have three directions of axis, such as x, y, z. Therefore, stress at the points in fluid has nine components:

$$\mathrm{P} = \begin{bmatrix} \sigma_{xx} & \tau_{yx} & \tau_{zx} \\ \tau_{xy} & \sigma_{yy} & \tau_{zy} \\ \tau_{xz} & \tau_{yz} & \sigma_{zz} \end{bmatrix} \tag{5.22}$$

We call this a stress tensor. The coordinate axis whose off-diagonal component of stress tensor has zero is called principal axis. Temperature and density at a point in fluid are stated as quantities without directional property. Velocity at a point that has amplitude and direction is called vector. Stress tensor is diagonally tensor whose diagonal component is identical. This means,

$$\tau_{xy} = \tau_{yx,} \quad \tau_{yz} = \tau_{zy,} \quad \tau_{zx} = \tau_{xz} \tag{5.23}$$

Stress in fluid comes from the result of relative movement of fluid. Stress tensor defined by Eq. (5.22) must be associated with fluid motion. Although the motion of the minimal element of the fluid can be divided into displacement (translation·rotation) and deformation (extension shear deformation), inner stress arises during fluid's deformation process.

The simplest deformation occurs when a viscous fluid is placed between parallel plates, and one of the plates moves at constant rate (Fig. 5.5).

$$\tau_{yx} = \mu \frac{du}{dy} \tag{5.24}$$

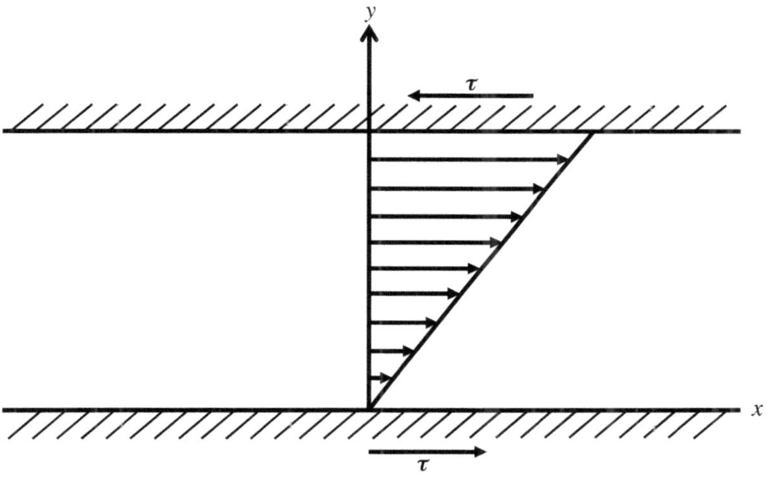

Figure 5.5 Wall shear stress on a flat plate.

Here, because shear stress acts in a direction of x axis in a plane vertical to y axis. This resistance law can be generalized as "shear stress τ proportional to shear deformation γ acts in viscous fluid."

$$\tau_{xy} = \tau_{yx} = \mu \gamma_{xy} = \mu \left(\frac{\partial v}{\partial x} + \frac{\partial u}{\partial y} \right) \tag{5.25}$$

$$\tau_{yz} = \tau_{zy} = \mu \gamma_{yz} = \mu \left(\frac{\partial w}{\partial y} + \frac{\partial v}{\partial z} \right) \tag{5.26}$$

$$\tau_{zx} = \tau_{xz} = \mu \gamma_{zx} = \mu \left(\frac{\partial u}{\partial z} + \frac{\partial w}{\partial x} \right) \tag{5.27}$$

In the case of ideal fluid, only pressure "p" is considered normal stress. As for linkage between normal stress and deformation in viscous fluid, distortion effect $\gamma_{xx} = \partial u/\partial x + \partial u/\partial x$ is on the pattern of $\gamma_{xy} = \partial v/\partial x + \partial u/\partial y$ and regarded as $2\mu\partial u/\partial x$. Also, an effect of distortion (term of volume change) by pressure is added.

$$\sigma_{xx} = -p + 2\mu\frac{\partial u}{\partial x} + \lambda\Theta \tag{5.28}$$

$$\sigma_{yy} = -p + 2\mu\frac{\partial v}{\partial y} + \lambda\Theta \tag{5.29}$$

$$\sigma_{zz} = -p + 2\mu\frac{\partial w}{\partial z} + \lambda\Theta \tag{5.30}$$

Here, volume change Θ is

$$\Theta = div\mathbf{v} = \left(\frac{\partial u}{\partial x} + \frac{\partial v}{\partial y} + \frac{\partial w}{\partial z}\right) \tag{5.31}$$

In an incompressible fluid $\Theta = 0$. Although positive direction of pressure p is set to the direction of the surface, positive direction of stress is generally coincident with the direction of coordinate axis and negative sign is added on right hand of Eq. (5.31). We sum equations above and consider $p = -(\sigma xx + \sigma yy + \sigma zz)/3$.

$$\lambda = -\frac{2}{3}\cdot\mu \tag{5.32}$$

Here, μ and λ are first viscosity coefficient and second viscosity coefficient, respectively.

Now, we could make a correlation between stress deriving from viscosity and motion of fluid (deformation rate), we build this relationship into motion equation. Here, we consider a force in x direction. From Fig. (5.4), normal stress $-\sigma_{xx}dydz$ works on x-plane and also $\sigma_{xx} + ((\partial\sigma_{xx}/\partial x)\partial x)dydz$ on $(x + dx)$-plane and the balance of them comes out $((\partial\sigma_{xx}/\partial x)dx)dydz$. Next, stress in the direction of x axis on y-plane is $-\tau_{yx}dxdz$ and x component on $(y + dy)$-plane is $\tau_{yx} + ((\partial\tau_{yx}/\partial y)dy)dxdz$. Their difference is $((\partial\tau_{yx}/\partial y)dx)dydz$ in the end. The x component of stress on two faces perpendicular to the z axis is similarly $((\partial\tau_{zx}/\partial z)dx)dydz$. Stress per unit mass in x axis on this cube will be

$$\frac{1}{\rho}\left(\frac{\partial\sigma_{xx}}{\partial x} + \frac{\partial\tau_{yx}}{\partial y} + \frac{\partial\tau_{zx}}{\partial z}\right) \tag{5.33}$$

A similar relation is obtained in other components. Equation 5.33 is added to Euler's equation, and motion equation for viscosity fluid is obtained. For example, as for the x component of the coordinate axis,

$$\frac{Du}{Dt}=F_x+\frac{1}{\rho}\left(\frac{\partial\sigma_{xx}}{\partial x}+\frac{\partial\tau_{yx}}{\partial y}+\frac{\partial\tau_{zx}}{\partial z}\right) \tag{5.34}$$

However, pressure term is included in σ_{xx}. Substituting relational expression between stress tensor (5.25), (5.26), (5.27) and deformation rate (5.28), (5.29), (5.30) into the equation above, we obtain the following. These are Navier–Stokes equations.

$$\frac{Du}{Dt}=F_x-\frac{1}{\rho}\frac{\partial p}{\partial x}+\frac{1}{3}\frac{\mu}{\rho}\frac{\partial\Theta}{\partial x}+\frac{\mu}{\rho}\left(\frac{\partial^2 u}{\partial x^2}+\frac{\partial^2 u}{\partial y^2}+\frac{\partial^2 u}{\partial z^2}\right) \tag{5.35}$$

$$\frac{Dv}{Dt}=F_y-\frac{1}{\rho}\frac{\partial p}{\partial y}+\frac{1}{3}\frac{\mu}{\rho}\frac{\partial\Theta}{\partial y}+\frac{\mu}{\rho}\left(\frac{\partial^2 v}{\partial x^2}+\frac{\partial^2 v}{\partial y^2}+\frac{\partial^2 v}{\partial z^2}\right) \tag{5.36}$$

$$\frac{Dw}{Dt}=F_z-\frac{1}{\rho}\frac{\partial p}{\partial z}+\frac{1}{3}\frac{\mu}{\rho}\frac{\partial\Theta}{\partial z}+\frac{\mu}{\rho}\left(\frac{\partial^2 w}{\partial x^2}+\frac{\partial^2 w}{\partial y^2}+\frac{\partial^2 w}{\partial z^2}\right) \tag{5.37}$$

Or else,

$$\frac{D\mathbf{v}}{Dt}=\mathbf{F}-\frac{1}{\rho}grad\,p+\frac{1}{3}\frac{\mu}{\rho}grad\,\Theta+\frac{\mu}{\rho}\nabla^2\mathbf{v} \tag{5.38}$$

Here,

$$\frac{D}{Dt}=\frac{\partial}{\partial t}+u\frac{\partial}{\partial x}+v\frac{\partial}{\partial y}+w\frac{\partial}{\partial z}=\frac{\partial}{\partial t}+\mathbf{v}\cdot grad \tag{5.39}$$

$$\nabla^2=\left(\frac{\partial^2}{\partial x^2}+\frac{\partial^2}{\partial y^2}+\frac{\partial^2}{\partial z^2}\right) \tag{5.40}$$

$$grad=\left(\frac{\partial}{\partial x'}\frac{\partial}{\partial y'}\frac{\partial}{\partial z}\right) \tag{5.41}$$

$$\Theta=\frac{\partial u}{\partial x}+\frac{\partial v}{\partial y}+\frac{\partial w}{\partial z} \tag{5.42}$$

$v=\mu/\rho$ is kinematic viscosity. In the case of incompressible fluid, Navier–Stokes equations are rewritten as follows:

$$\frac{Du}{Dt}=F_x-\frac{1}{\rho}\frac{\partial p}{\partial x}+\frac{\mu}{\rho}\left(\frac{\partial^2 u}{\partial x^2}+\frac{\partial^2 u}{\partial y^2}+\frac{\partial^2 u}{\partial z^2}\right) \tag{5.43}$$

$$\frac{Dv}{Dt}=F_y-\frac{1}{\rho}\frac{\partial p}{\partial y}+\frac{\mu}{\rho}\left(\frac{\partial^2 v}{\partial x^2}+\frac{\partial^2 v}{\partial y^2}+\frac{\partial^2 v}{\partial z^2}\right) \tag{5.44}$$

$$\frac{Dw}{Dt}=F_z-\frac{1}{\rho}\frac{\partial p}{\partial z}+\frac{\mu}{\rho}\left(\frac{\partial^2 w}{\partial x^2}+\frac{\partial^2 w}{\partial y^2}+\frac{\partial^2 w}{\partial z^2}\right) \quad (5.45)$$

In another expression,

$$\frac{Dv}{Dt}=F-\frac{1}{\rho}grad\,p+v\nabla^2 v \quad (5.46)$$

5.2 Discretization Algorithm

Before solving a differential equation such as Navier–Stokes equations, it is necessary to transform the equation into a machine-readable equation. For it is impossible for computer to solve a differential equation analytically, the procedure of writing an algebra equation and relational expression for unknown value on a point of grid within calculated region is needed. This operation is called discretization. Based on the discrete equation, unknown value is evaluated by solving algebra equation and operating four fundamental rules of arithmetics. In the field of numerical simulation of flow, there are three typical methods for discretization.[1]

Finite-difference method: Finite-difference method, which is widely used in numerical analysis, expresses derivative term in differential with Taylor series. The derivative term is approximated by combination of relational equation expressed with Taylor series on the point of grid and neighboring grids depending on degree of precision for discretization. It is based on differential form of the equation of flow and the solution exists on the point of grid. Basically, the space between dots is not considered. Structured grid is used in FDM.

Finite-element method: The basic equation is not a differential form but a weak form derived from the integration of multiplied basic function. By substituting this formula into the governing equation and integrating after weighting, we obtain relational equation of variable arranged at the vertex of cell. In many cases, calculating area is composed of minim triangular elements. Therefore, it does not matter whether it is structured grid or unstructured grid.

Finite-volume method: Finite-volume method discretizes differential equation after integration to satisfy the conservation law of mass and momentum. It is based on the integral form of the equation of flow. Variable is not arranged on the points of grid but at the center of cells as representative value. Conservation law is expressed with inflow and outflow from the boundary of cells and also occurrence and extinction from within cells. The calculation of this relation in conservation law is the concept of the FVM. Both structured grid and unstructured grid are able to be adapted in this method.

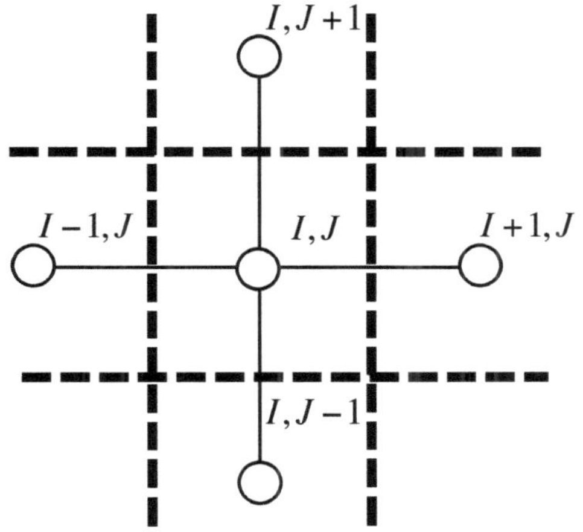

Figure 5.6 Finite-difference method.

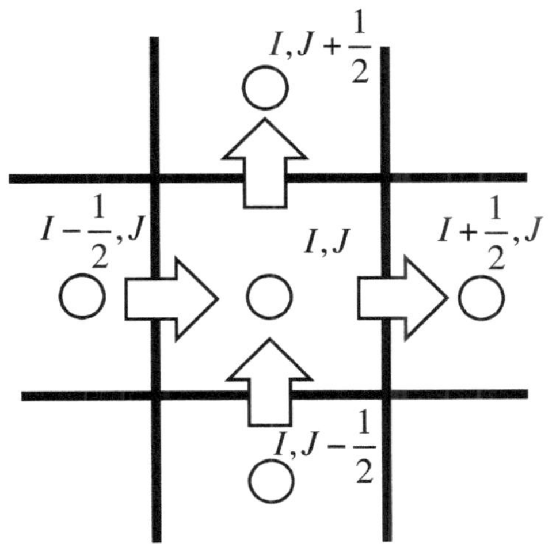

Figure 5.7 Finite-volume method.

As for the election of the grid system and the computation method, it is important to consider the purpose of analysis and the feature of flow. Table 5.2 shows the general properties of a structure grid and an unstructured grid. The structure grid, consisting of a regular grid, has an advantage in geometry accuracy and efficiency per grid and is suitable for high-grade calculations for fundamental research. However, the calculation with a structured grid is limited to relatively simple shape. The unstructured grid is at the other end, and it is practicable calculation method for complicated shapes of flow field.

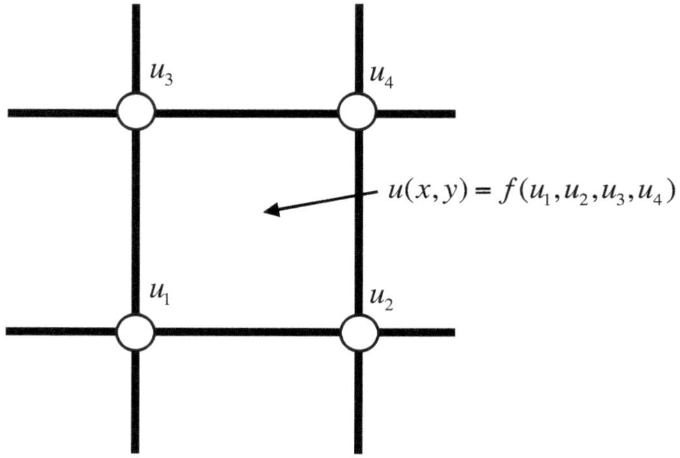

Figure 5.8 Finite-element method.

Table 5.2 Feature comparison between three discretization algorithms

	Structured grid	Unstructured grid
Calculation method	FDM-FVM	FVM-FEM
Geometric adaptability	Low	High
Total number of grid number	Inefficient	Efficient
Accuracy per a grid	High	Low
Calculation cost	Low	High

5.2.1 Finite-Difference Method

The first step obtaining a numerical solution is to discretize the geometric domain, i.e., a numerical grid must be defined. In finite-difference (FD) discretization methods, the grid is usually locally structured, i.e., each grid node may be considered the origin of a local coordinate system, whose axes coincide with grid lines. This also implies that two grid lines belonging to same families, say ξ_1, do not intersect, and that any part of grid lines belonging to different families, say ξ_1 = const. and ξ_2 = const., intersect only once. In three dimensions, three grid lines intersect at each node; the nodes of three lines intersect each other at any other point. Each node is uniquely identified by a set of indices, which are the indices of the grid lines that intersect at it, (i, j) in 2D and (i, j, k) in 3D. The neighbor nodes are defined by increasing or reducing one of the indices by unity. The generic scalar conservation equation is differential form. As it is linear in ϕ, it will be approximated by a system of linear algebraic equations, in which the variable values at the grid nodes are the unknown values. The solution to this system approximates the solution to the PDE. Each node thus has one unknown variable value associated with it and must provide one algebraic equation. The latter is a relation between the variable value at that node and those at some of the neighboring nodes. It is obtained by replacing each term of the PDE at the particular node by a finite-difference approximation. Of course, the numbers of equation and unknown must be equal. At boundary nodes where variable value are given (Dirchlet conditions), no equation is needed.

When the boundary conditions involve derivatives (as in Neumann conditions), the boundary condition must be discretized to contribute an equation to the set that must be solved. The idea behind finite-difference approximations is borrowed directly from the definition of a derivative:

A geometrical interpolation is shown in Fig. 5.9 to which we shall refer frequently.[2] The first derivative $\partial\phi/\partial x$ at a point is the slope of the tangent to the curve $f(x)$ at that point, the line marked "exact" in the figure. Its slope can be approximated by the slope of a line passing through two nearby points on the curve. The dotted line shows approximation by the slope of a line passing through two nearby points on the curve. The dotted line shows approximation by a forward difference; the derivative at x_i is approximated by the slope of a line passing through the point x_i and another point at $x_i + \delta x$. The dashed line illustrates approximation by backward difference: for which the second point is $x_i - \delta x$. The line labeled "central" represents approximation by a central difference: It uses the slope of a line passing through two points lying on opposite sides of the point at which the derivative approximated.

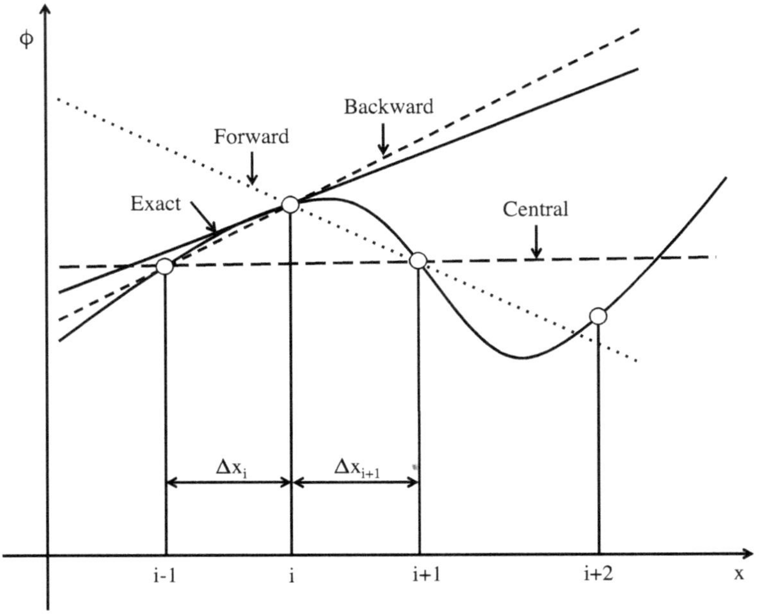

Figure 5.9 Derivative.

It is obvious from Fig. 5.9 that some approximations are better than others. The line for the central difference approximation has a slope very close to the slope of the exact line; if the function ϕ (x) were a second-order polynomial and the points were equality spaced in x direction, the slopes would match exactly. It is also obvious from Fig. 5.9 that the quality of the approximation improves when the additional points are close to x_i, i.e., as the grid is refined, the approximation improves. The approximations shown in Fig. 5.9 are few of many possibilities; the following sections outline the principal approaches to deriving approximations for the first and second derivatives. In the following two sections, only the one-dimensional case is considered. The coordinate may be either Cartesian or curvilinear; the difference is of little importance here. In multidimensional finite differences, each coordinate is usually treated separately so the methods developed here are readily adapted to higher dimensionality.

5.2.2 First Derivative

Discretization of the convective term in Navier–Stokes equation requires the approximation of the first derivative $\partial(\rho u\phi)/\partial x$. We shall now describe some approaches to the approximation of the first derivative of any quality. In the previous section, one means of deriving approximations to the first derivative was presented. There are more systematic approaches that are better suited to the derivation of more accurate approximations; some of these will be described later.

5.2.3 Taylor Series Expansion

Any continuous differentiable function $\phi(x)$ can, in the vicinity of x_1, be expressed as a Taylor series:

$$\phi(x)=\phi(x_i)+(x-x_i)\left(\frac{\partial\phi}{\partial x}\right)i+\frac{(x-x_i)^2}{2!}\left(\frac{\partial^2\phi}{\partial x^2}\right)i$$

$$+\frac{(x-x_i)^3}{3!}\left(\frac{\partial^3\phi}{\partial x^3}\right)i+\cdots+\frac{(x-x_i)^n}{n!}\left(\frac{\partial^n\phi}{\partial x^n}\right)i+H.O.T \quad (5.47)$$

where H.O.T means "higher order terms." By replacing x with x_{i+1} or x_{i-1} in this equation, one obtains expressions for the variable values at these points in terms of the variable and its derivatives at x_i. This can be extended to any other point near x_i in terms of the function values at neighboring points. For example, using Eq. (5.47) for ϕ at x_{i+1}, we can show that

$$\left(\frac{\partial\phi}{\partial x}\right)_i=\frac{\phi_{i+1}-\phi_i}{x_{i+1}-x_i}-\frac{x_{i+1}-x_i}{2}\left(\frac{\partial^2\phi}{\partial x^2}\right)_i$$

$$-\frac{(x_{i+1}-x_i)^2}{6}\left(\frac{\partial^3\phi}{\partial x^3}\right)_i+H.O.T \quad (5.48)$$

Another expression may be derived using the series expression (5.47) at x_{i-1}:

$$\left(\frac{\partial\varphi}{\partial x}\right)_i=\frac{\phi_i-\phi_{i-1}}{x_i-x_{i-1}}-\frac{x_i-x_{i-1}}{2}\left(\frac{\partial^2\phi}{\partial x^2}\right)_i$$

$$-\frac{(x_i-x_{i-1})^2}{6}\left(\frac{\partial^3\phi}{\partial x^3}\right)_i+H.O.T \quad (5.49)$$

Still another expression may be obtained by using Eq. (5. 47) at both x_{i-1} and x_{i+1}:

$$\left(\frac{\partial\phi}{\partial x}\right)_i = \frac{\phi_{i+1}-\phi_{i-1}}{x_i - x_{i-1}} - \frac{(x_{i+1}-x_i)^2-(x_i-x_{i-1})^2}{2(x_i-x_{i-1})}\left(\frac{\partial^2\phi}{\partial x^2}\right)_i$$

$$-\frac{(x_{i+1}-x_i)^3+(x_i-x_{i-1})^3}{6(x_{i+1}-x_{i-1})}\left(\frac{\partial^3\phi}{\partial x^3}\right)_i + H.O.T \tag{5.50}$$

Equations (5.48–5.50) are exact if all terms on the right-hand side are retained. Because the higher-order derivatives are unknown, these expressions are not of great value as they stand. However, if the distance between the grid points i.e., $x_i - x_{i-1}$ and $x_{i+1} - x_i$ are small, the higher-order terms are locally very large. Ignoring the latter possibility, approximations to the first derivative result from truncating each of the series after the first terms on the right-hand sides:

$$\left(\frac{\partial\phi}{\partial x}\right)_i \approx \frac{\phi_{i+1}-\phi_i}{x_{i+1}-x_i} \tag{5.51}$$

$$\left(\frac{\partial\phi}{\partial x}\right)_i \approx \frac{\phi_i-\phi_{i-1}}{x_i-x_{i-1}} \tag{5.52}$$

$$\left(\frac{\partial\phi}{\partial x}\right)_i \approx \frac{\phi_{i+1}-\phi_{i-1}}{x_{i+1}-x_{i-1}} \tag{5.53}$$

These are the forward- (FDS), backward- (BDS), and central-difference (CDS) schemes, respectively. The terms that were deleted from the right-hand sides are called the truncation errors; they measure the accuracy of the approximation and determine the rate at which the error truncated term is usually the principal source of error. The truncation error is the sum of products of a power of the spacing between the points and higher order derivatives at the point $x = x_i$:

$$\varepsilon\gamma = (\delta x)^m \alpha_{m+1} + (\delta x)^{m+1}\alpha_{m+2} + \cdots + (\delta x)^n \alpha_{n+1} \tag{5.54}$$

where δx is the spacing between the points (assumed all equal for the present) and the α's are higher-order derivatives multiplied by constant factors. From Eq. (5.54), we see that the terms containing higher powers of δx are smaller for small spacing so that the leading term (the one with the smaller exponent) is the dominant one. As δx is reduced, the above approximations converge to the exact derivatives with an error proportional to $(\delta x)^m$, where m is the exponent of the leading truncation error term. The order of an approximation indicates how fast the error is reduced when the grid is refined; it does not indicate the absolute magnitude of the error. The error is thus reduced by a factor of 2, 4, 8, or 16 in the first-, second-, third- or fourth-order approximations, respectively. It should be remembered that this rule is valid only for sufficiently small spacing; the definition of small enough depends on the profile of the function $\phi(x)$.

5.2.4 Approximation of the Second Derivative

Second derivatives appear in the diffusive terms in Navier–Stokes equation. To estimate the second derivative at a point, one may use the approximation for the first derivative twice. This is the only possible approach when the fluid properties are variable, since we need the derivative of the product of diffusion coefficient and the first derivative. Next, we consider approximation to the second derivative; application to the diffusive term in the conservation equation will be discussed later. Geometrically, second derivative is the slope of the line tangent to the curve representing the first derivative (see Fig. 5.9. By inserting approximations for the first derivatives at locations x_{i+1} and x_i, an approximation for the second derivative is obtained:

$$\left(\frac{\partial^2\varphi}{\partial x^2}\right)_i \approx \frac{\left(\frac{\partial\phi}{\partial x}\right)_{i+1} - \left(\frac{\partial\phi}{\partial x}\right)_i}{x_{i+1} - x_i} \qquad (5.55)$$

All such approximations involve data from at least three points. In Eq. (5.55), the outer derivative was estimated by FDS. For inner derivatives, one may use a different approximation, e.g., BDS; this results in the following expression:

$$\left(\frac{\partial^2\varphi}{\partial x^2}\right)_i = \frac{\phi_{i+1}(x_i - x_{i-1}) + \phi_{i-1}(x_{i+1} - x_i) - \phi_i(x_{i+1} - x_{i-1})}{(x_{i+1} - x_i)^2(x_i - x_{i-1})} \qquad (5.56)$$

One could also use the CDS approach which requires the first derivative at x_{i-1} and x_{i+1}. A better choice is to evaluate $\partial\phi/\partial x$ at points halfway between x_i and x_{i+1} and x_i and x_{i-1}. The CDS approximations for these first derivatives are

$$\left(\frac{\partial\phi}{\partial x}\right)_{i+\frac{1}{2}} \approx \frac{\phi_{i+1} - \phi_i}{x_{i+1} - x_i} \text{ and } \left(\frac{\partial\phi}{\partial x}\right)_{i-\frac{1}{2}} \approx \frac{\phi_i \phi_{i-1}}{x_i - x_{i-1}} \qquad (5.57)$$

respectively. The resulting expression for the second derivatives is

$$\left(\frac{\partial^2 \phi}{\partial x^2}\right)_i \approx \frac{\left(\frac{\partial \phi}{\partial x}\right)_{i+\frac{1}{2}} - \left(\frac{\partial \phi}{\partial x}\right)_{i-\frac{1}{2}}}{\frac{1}{2}(x_{i+1} - x_{i-1})}$$

$$\approx \frac{\phi_{i+1}(x_i - x_{i-1}) + \phi_{i-1}(x_{i+1} - x_i) - \phi_i(x_{i+1} - x_{i-1}}{\frac{1}{2}(x_{i+1} - x_{i-1})(x_{i+1} - x_i)(x_i - x_{i-1})} \tag{5.58}$$

For equidistant spacing of the points, expression (5.56) becomes simple:

$$\left(\frac{\partial^2 \phi}{\partial x^2}\right)_i \approx \frac{\phi_{i+1} + \phi_{i-1} - 2\phi}{(\delta x)^2} \tag{5.59}$$

Higher-order approximations for the second derivative can be obtained by including more data points, say x_{i-2} or x_{i+2}.

Finally, one can use interpolation to fit a polynomial degree n through n + 1 data points. From that interpolation, approximations to all derivatives up to the nth can be obtained by differentiation.

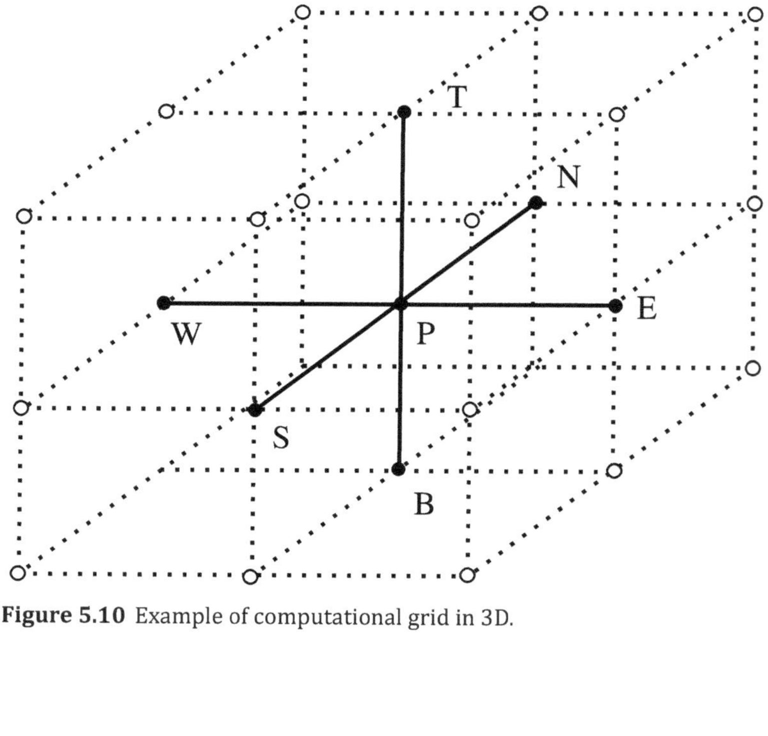

Figure 5.10 Example of computational grid in 3D.

5.2.5 The Algebraic Equation system

Table 5.3 Correspondence table of indices on a grid

Grid location	Notation	Location
i, j, k	P	$I = (k - 1)N_jN_i + (i - 1)N_j + j$
$i - 1, j, k$	W	$I - N_J$
$i, j - 1, k$	S	$l - 1$
$i, j + 1, k$	N	$l + 1$
$i + 1, j, k$	E	$l + N_j$
$i, j, k - 1$	B	$l - N_iN_j$
$i, j, k + 1$	T	$l + N_iN_j$

A finite-difference approximation provides an algebraic equation at each grid node; it contains the variable value at that nodes as well as values at neighboring nodes. If the differential equation is non-linear, the approximation will contain some non-linear terms. The numerical solution process will then require linearization. We only consider the linear case. The methods described are applicable in the non-linear case as well. For this case, the result of discretization is a system of linear algebraic equations of the form:

$$A_P\phi_P + \sum_l A_l\phi_l = Q_P \tag{5.60}$$

where P denotes the node at which the PDE is approximated and index l runs over the neighbor nodes involved in finite-difference approximations. The node P and its neighbors from the so-called computational molecule; two examples, which result from second- and third-order approximations, are shown in Figs. 5.12 and 5.13. The coefficients A_l depend on geometrical quantities fluid properties and, for non-linear equations, the variable values themselves. Q_P contains all the terms that do not contain unknown variable values; it is presumed known.

The numbers of equations and unknowns must be equal, i.e., there has to be one equation for each grid node. Thus, we have a large set of linear algebraic equations, which must be solved numerically. This system is sparse meaning that each equation contains only a few unknowns. The system can be written in the matrix notation as follows:

$$A\phi = Q \tag{5.61}$$

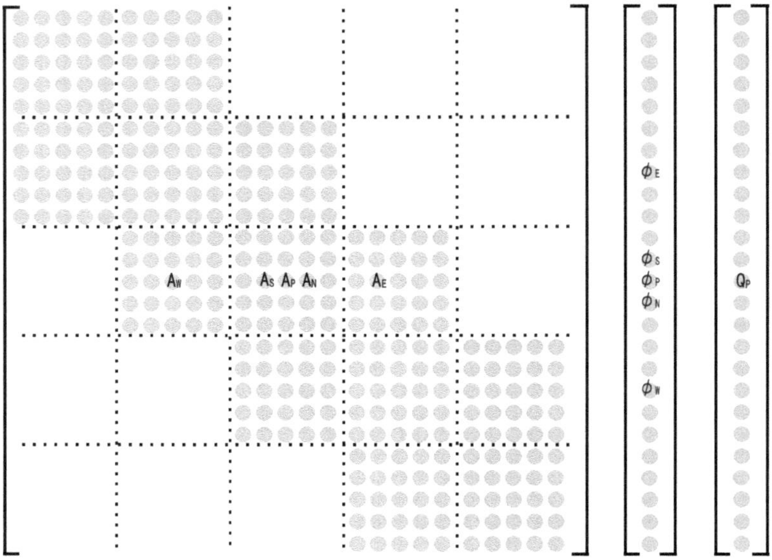

Figure 5.11 Structure of matrix.

where A is the square sparse coefficient matrix, ϕ is a vector (or column matrix) containing the variable values at the grid nodes, and Q is the vector containing the terms on the right-hand side of Eq. (5.60).

The structure of matrix A depends on the ordering of variable in the vector ϕ. For structured grids, if the variables are labeled starting at a corner and traversing line after line in a regular manner (lexicographic ordering), the matrix has a poly-diagonal structure. For the case of a five-point computational molecule, all the non-zero coefficients lie on the main diagonal, the two neighboring diagonals, and two other diagonals removed by N positions from the main diagonal, where N is the number of nodes in one direction. All other coefficients are zero. This structure allows the use of efficient iterative solvers.

Throughout this book we shall, for the sake of definiteness, order the entries in vector ϕ starting at the southwest corner of the domain, proceeding northwards along each grid line and then eastward across the domain (in three-dimensional cases we shall start at the bottom computational surface and proceed on each horizontal plane in the manner just described, and then go from bottom to top). The variables are normally stored in computers in one-dimensional arrays. The convention between the grid locations, compass notation, and storage locations is indicated in Table 5.1.

Because the matrix A is sparse, it does not make sense to store it as a two-dimensional array in computer memory (this is standard practice for full matrices). Storing the elements of each non-zero diagonal in a separate array of dimension $1 \times N_iN_j$, where N_i and N_j are the numbers of grid points in the two coordinate directions, requires only $5N_i^2N_j^2$ words of storage; full array storage would require $N_i^2N_j^2$ words of storage. In three dimensions, the numbers are $7N_iN_jN_k$ and $N_i^2N_j^2N_k^2$, respectively. The difference is sufficiently large that the diagonal-storage scheme may allow the problem to be kept in main memory when the full-array scheme does not.

If the nodal values are referenced using the grid indices, say ϕ_{ij} in 2D, they look like matrix elements or components of a tensor. Since they are actually components of a vector ϕ, they should have only single index indicated in Table 3.2. The linearized algebraic equations in two dimensions can now be written in the following form:

$$A_{l,l-N_j}\phi_{l-N} + A_{l,l-1}\phi_{l-1} + A_{l,l}\phi_l + A_{l,l+1}\phi_{l+1} + A_{l,l+N_j}\phi_{l+N_j} = Q_l \quad (5.62)$$

As noted above, it makes little sense to store the matrix as an array. If, instead, the diagonals are kept in separate arrays, it is better to give each diagonal a separate name. Since each diagonal represents the connection to the variable at a node that lies in a particular direction with respect to the central node, we shall call them A_W, A_S, A_p, A_N, and A_E; their locations in the matrix for a grid with 5×5 internal nodes are shown in Fig. 5.18. With this ordering of points, each node is identified with an index l, which is also the relative storage location. In this notation, Eq. (5.62) can be written as follows:

$$A_W\,\phi_W + A_W + A_S\,\phi_S + A_P\,\phi_P + A_N\,\phi_N + A_E\,\phi_E = Q_P \quad (5.63)$$

where the index l, which indicated rows in Eq. (5.62) is understood, and the index indicating column or location in the vector has been replaced by the corresponding letter. We shall use this shorthand notation from now on. When necessary for clarify, the index will be inserted. A similar treatment applies to three-dimensional problems.

For block-structured and composite grids, this structure is preserved within each block, and the solvers for regular structured grids may be used. For block-structured grids, the coefficient matrix remains sparse, but it no longer has banded structure. For a 2D grid of quadrilaterals and approximations that use only the four nearest neighbor nodes, there are only five non-zero coefficients that lie within a certain range of the main diagonal but not necessarily on definite diagonals. A different type of iterative solver must be used for such matrices.

5.2.6 Finite-Volume Method

Finite-volume method basically consists of the integral form of the conversation equation as follows:

$$\int_S \rho\phi\mathbf{v}\cdot\mathbf{n}dS = \int_S \Gamma grad\phi\cdot\mathbf{n}dS + \int_\Omega qd\Omega \tag{5.64}$$

The solution area is subdivided into a finite number of small control volume (CVs) by a grid, compared with FDM. Control volume has volume boundaries, not the computational node. The integral conservation equation (5.76) is applied to not only each CVs but also whole solution area. If whole CVs equations are summed up, the global conservation equation is obtained, since surface integrals over inner CV faces cancel out. The global conservation is built into the method automatically because of its integral form and this is the advantage of FVM.

Figures 5.12 and 5.13 are typical 2D and 3D control volumes, respectively. The CV surface consists of four (in 2D) or six (3D) plane faces, denoted by lower case index corresponding to their direction (e, w, n, s, b, and t) with respect to the central point *P*. The net flux thorough CV boundary is the sum of integrals over the four (2D) or six (3D) CV faces.

$$\int_S fdS = \sum_k \int_{S_k} fds \tag{5.65}$$

Here f is the component of the convective ($\rho\phi\,\mathbf{v}\cdot\mathbf{n}$) or the diffusive ($\Gamma\text{grad}\phi\cdot\mathbf{n}$) flux vector in the direction normal to CV face. For maintaining conservation, it is important that CVs do not overlap. Each CV face is independent on the two CVs, which lie on either side of it. To calculate the surface integral in Eq. 5.77, the integral f needs to be known on the surface S_e This information is not available, as only the nodal (CV center) values of ϕ are calculated, so an approximation must be introduced.

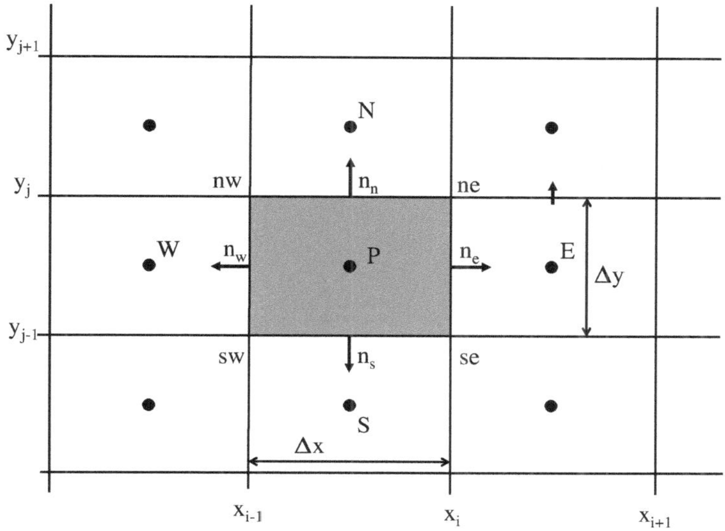

Figure 5.12 Two-dimensional grid for finite-volume method.

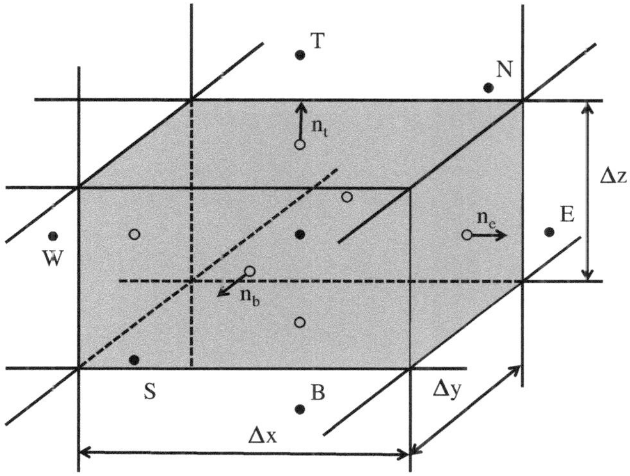

Figure 5.13 Three-dimensional grid for finite-volume method.

5.3 Numerical Solution

We dealt with the FDM and FVM for discretization of a generic conservation. The discretion principles described there apply to the momentum and continuity equations. In this section, to calculate discretization equations on the computer specific example of numerical solution is described. As an example of discretization of steady incompressible viscous flow, Navier–Stokes equation is discretized by using FVM.

$$\frac{\partial}{\partial x}(\rho u \cdot u)+\frac{\partial}{\partial y}(\rho v \cdot u)=\frac{\partial}{\partial x}\left(\mu\frac{\partial u}{\partial x}\right)+\frac{\partial}{\partial y}\left(\mu\frac{\partial u}{\partial y}\right)+\frac{\partial p}{\partial x} \tag{5.66}$$

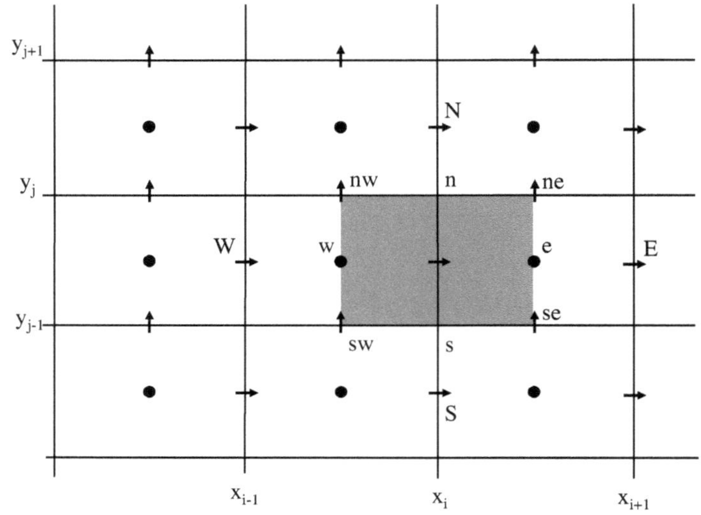

Figure 5.14 Diagram of discrete Navier–Stokes equation in *x* direction.

Volume integration in *x* direction of first convection term

$$\iint \frac{\partial}{\partial x}(\rho u \cdot u)dxdy = \Delta y \int \frac{\partial}{\partial x}(puu)dx$$
$$= \Delta y \times \int \{(\rho uu)_e - (\rho uu)_w\}$$
$$= \Delta y \times \{0.5 \times (\rho \times u_{i.j} + \rho \times u_{i+1.j}) \times u_e$$
$$-0.5 \times (\rho \times u_{i-1.j} + \rho \times u_{i.j}) \times u_w\}$$
$$= 0.5 \times \Delta y \times \rho\{(u_{i.j} + u_{i+1.j})(u_{i+1.j} + u_{i.j})$$
$$-(u_{i-1.j} + u_{i.j})(u_{i.j} + u_{i-1.j})\} \tag{5.67}$$

second convection term

$$\iint \frac{\partial}{\partial y}(\rho v \cdot u)dxdy = \Delta x \int \frac{\partial}{\partial y}(\rho vu)dy$$

$$= \Delta x \times \{(\rho \times v_{i.j+1} + \rho v_{i-1.j+1}) \times u_n$$
$$-0.5 \times (\rho \times v_{i.j} + \rho \times v_{i-1.j}) \times u_s\}$$
$$= 0.5 \times \Delta x \times \rho\{(v_{i.j+1} + v_{i-1.j+1})(u_{i.j+1} + u_{i.j})$$
$$-(v_{i.j} + v_{i-1.j})(u_{i.j} + u_{i.j-1})\} \tag{5.68}$$

First viscous term

$$\iint \frac{\partial}{\partial x}\left(\mu \frac{\partial u}{\partial x}\right) dxdy = \Delta y \int \frac{\partial}{\partial x}\left(\mu \frac{\partial u}{\partial x}\right) dx$$
$$= \Delta y \left\{\left(\mu \frac{\partial u}{\partial x}\right)_e - \left(\mu \frac{\partial u}{\partial x}\right)_w\right.$$
$$= \frac{y_j - y_{j-1}}{2} \times \mu \left\{\frac{2}{x_{i+1} - x_{i-1}} \times (\partial u)_e - \frac{2}{x_i - x_{i-2}} \times (\partial u)_w\right\}$$
$$= \mu \frac{y_j - y_{j-1}}{x_{i+1} - x_{i-1}} (u_{i-1.j} - u_{i.j}) - \mu \frac{y_j - y_{j-1}}{x_i - x_{i-2}} (u_{i.j} - u_{i-1.j}) \tag{5.69}$$

Second viscous term

$$\iint \frac{\partial}{\partial y}\left(\mu \frac{\partial u}{\partial y}\right) dxdy = \Delta x \int \frac{\partial}{\partial y}\left(\mu \frac{\partial u}{\partial y}\right) dy$$
$$= \Delta x \left\{\left(\mu \frac{\partial u}{\partial y}\right)_n - \left(\mu \frac{\partial u}{\partial y}\right)_s\right\}$$
$$= \frac{x_i - x_{i-1}}{2} \times \mu \left\{\frac{2}{y_{j+1} - y_{j-1}} \times (\partial u)_n - \frac{2}{y_i - y_{j-2}} \times (\partial u)_s\right\}$$
$$= \mu \frac{x_i - x_{i-1}}{y_{j+1} - y_{j-1}} (u_{i.j+1} - u_{i.j}) - \mu \frac{x_i - x_{i-1}}{y_j - y_{j-2}} (u_{i.j} - u_{i.j-1}) \tag{5.70}$$

Pressure gradient form

$$\iint \left(-\frac{\partial p}{\partial x}\right) dxdy = -\Delta y \int \frac{\partial p}{\partial x}$$
$$= -\Delta y(P_e - P_w)$$
$$= \frac{y_j}{y_{j-1}} \times (P_{i-1.j} - P_{i.j}) \tag{5.71}$$

Combined discrete equations of Navier–Stokes equations in x direction are shown as follows:

$$0.5\times\Delta y\times\rho\{(u_{i.j}+u_{i+1.j})(u_{i+1.j}+u_{i.j})-u_{i-1.j}+u_{i.j})(u_{i.j}+u_{i-1.j})\}$$

$$+0.5\times\Delta x\times\rho\{(v_{i.j+1}+v_{i-1.j+1})(v_{i.j+1}+v_{i.j})-(v_{i.j}+v_{i-1.j})(v_{i.j}+v_{i.j-1})\}$$

$$=\mu\frac{y_j-y_{j-1}}{x_{i+1}-x_{i-1}}(v_{i+1.j}-v_{i.j})-\mu\frac{y_j-y_{j-1}}{x_i-x_{i-2}}(v_{i.j}-v_{i-1.j})$$

$$+\mu\frac{x_i-x_{i-1}}{y_{j+1}-y_{j-1}}(v_{i.j+1}-v_{i.j})-\mu\frac{x_i-x_{i-1}}{y_j-y_{j-2}}(v_{i.j}-v_{i.j-1})$$

$$+\frac{y_j}{y_{j-1}}\times(P_{i-1.j}-P_{i.j}) \tag{5.72}$$

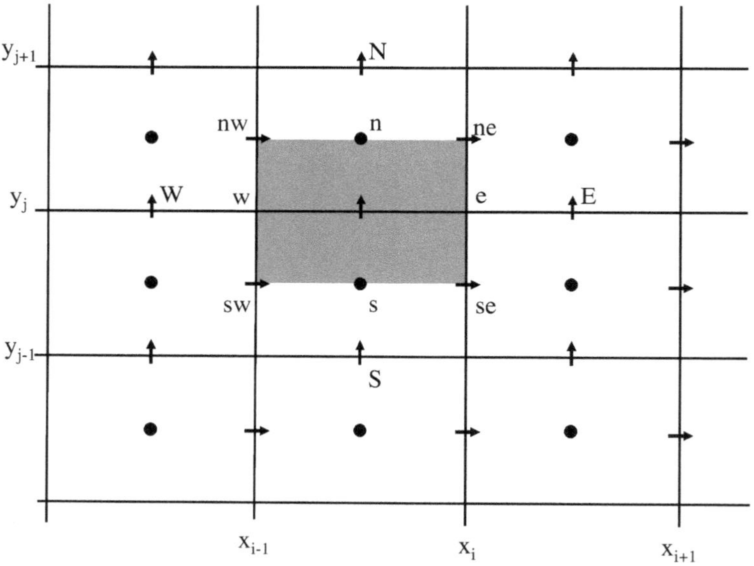

Figure 5.15 Diagram of discrete Navier–Stokes equation in y direction.

Discretization of Navier–Stokes equations in y direction is shown as the same way in x direction. Thus, Navier–Stokes equations are transformed into a machine-readable equation, we can solve the equations by programmed calculation.

$$\frac{\partial}{\partial x}(\rho u\cdot v)+\frac{\partial}{\partial y}(\rho v\cdot v)=\mu\left(\frac{\partial v}{\partial x}\right)+\frac{\partial}{\partial y}\left(\mu\frac{\partial v}{\partial y}\right)-\frac{\partial p}{\partial y} \tag{5.73}$$

$$0.5\times\Delta y\times\rho\{(u_{i.j}+u_{i+1.j})(v_{i+1.j}+v_{i.j})$$
$$-(u_{i-1.j}+u_{i.j})(v_{i.j}+v_{i-1.j})\}$$
$$+0.5\times\Delta x\times\rho\{v_{i.j+1}+v_{i-1.j+1})(u_{i.j+1}+u_{i.j})$$
$$-(v_{i.j}+v_{i-1.j})(u_{i.j}+u_{i.j-1})\}$$
$$=\mu\frac{y_j-y_{j-1}}{x_{i+1}-x_{i-1}}(u_{i+1.j}-u_{i.j})-\mu\frac{y_j-y_{j-1}}{x_i-x_{i-2}}(u_{i.j}-u_{i-1.j})$$
$$+\mu\frac{x_i-x_{i-1}}{y_{j+1}-y_{j-1}}(u_{i.j+1}-u_{i.j})$$
$$-\mu\frac{x_i-x_{i-1}}{y_j-y_{j-2}}(u_{i.j}-u_{i.j-1})+\frac{y_j}{y_{j-1}}\times(P_{i-1.j}-P_{i.j}) \qquad (5.74)$$

5.3.1 The Choice of Grid

When geometry is regular (e.g., rectangular or circular), choosing the grid is simple: The grid lines usually follow the coordinate directions. In complicated geometries, the choice is not at all trivial. The grid is subject to constraints imposed by the discretization method. If the algorithm is designed for curvilinear orthogonal, non-orthogonal grids cannot be used; if the CVs are required to be quadrilaterals or hexahedra, grids consisting of triangles and tetrahedra cannot be used, etc. When the geometry is complex and the constraints cannot be fulfilled, compromises have to be made.

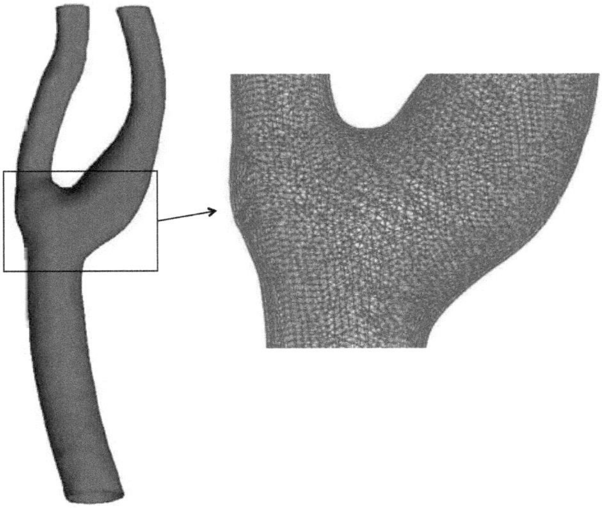

Figure 5.16 Geometry data and surface mesh of bifurcation area of ICA.

5.3.2 Approximation Using Regular Grids

The simplest approach uses orthogonal grids (Cartesian or polar-cylindrical). To apply such a grid to solution domains with inclined or curved boundaries, the boundaries have to be approximated by staircase-like steps. This approach has been used, but it raises two kinds of problems:

1. The number of grids points (or CVs) per grid line is not constant, as it is in a fully regular grid. This requires either indirect addressing, or special arrays have to be created that limit the index range on each line. The computer code may need to be changed for each new problem.
2. The steps at the boundary introduce errors into the solution, especially when the grid is coarse. The treatment of the boundary conditions at stepwise walls also requires special attention. This approach is a last resort, to be used when an existing solution method cannot be quickly adapted to a grid that fits boundary better. It is not recommended, except when the solution algorithm allows local grid refinement near the wall.

Figure 5.25 shows example of trimmer mesh applied to vessel 3D data of internal carotid artery (ICA). This method is the derived from stepwise approximation.

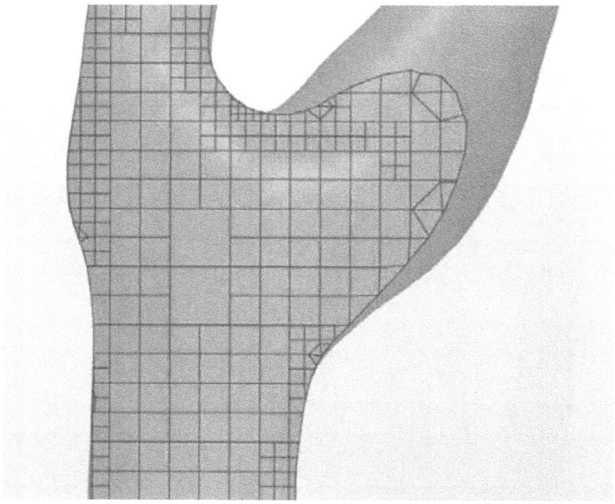

Figure 5.17 Trimmer mesh.

5.3.3 Grid Generation

The generation of grids for complex geometries is an issue that requires too much space to be dealt with in great detail here. We shall present only some basic ideas and the properties that a grid should have. More details about various methods of grid generation can be found in books and conference proceedings devoted to this topic (e.g., Thompson *et al.*, 1985; Arcilla *et al.*, 1991). Even though necessity demands that the grid be non-orthogonal, it is important to make it as nearly orthogonal as possible. In FV methods orthogonality of grid lines at CV vertices is unimportant — It is the angle between the cell face surface normal vector and the line connecting the CV centers on either side of it that matters. Thus, a D grid made of equilateral triangles is equivalent to an orthogonal grid, since lines connecting cell centers are orthogonal to cell faces. Cell topology is also important. If the midpoint rule integral approximation, linear interpolation, and central difference are used to discretize the equations, then the accuracy will be higher if the CVs are quadrilaterals in 2D and hexahedra in 3D, than if we use triangles and tetrahedra, respectively. The reason is that parts of the errors made at opposite cell faces when discretizing diffusion terms cancel partially (if cell faces are parallel and of equal area, they cancel completely) on quadrilateral and hexahedral CVs. To obtain the same accuracy on triangles and tetrahedra, more sophisticated interpolation and difference approximations must be used. Especially near solid boundaries it is desirable to have quadrilaterals or hexahedra, since all quantities vary substantially there and accuracy is especially important in this region.

Accuracy is also improved if one set of grid lines closely follows the streamlines of the flow, especially for the convective terms. This cannot be achieved if triangles and/or tetrahedra are used, but is possible with quadrilaterals and hexahedra. Non-uniform grids are the rule rather than exception when complex geometries are treated. The ratio of the sizes of adjacent cells should be kept under control, as accuracy is adversely affected if it is large. Especially when block-structured grids are used, one should take care that the cells are of nearly equal size near block interfaces; affector of two variation should be the maximum. An experienced user may know where strong variation of velocity, pressure, temperature, etc., can be expected; the grid should be fine in these regions since the errors are most likely to be large there. However, even an experienced user will encounter occasional surprises and more sophisticated methods are useful in any event. It is possible, however, to start with a coarse grid and later refine it locally according to an estimate of the discretization error; methods for doing this are called solution-adaptive grid methods.

Finally, there is the issue of grid generation. When the geometry is complex, this task usually consumes the largest amount of user time by far; it is not unusual for a designer to spend several weeks generating a single grid. Since the accuracy of the solution depends as much (if not more) on the grid quality as on the approximations used for discretization of the equations, grid optimization is a worthwhile investment of time.

Many commercial codes for grid generation exist. The automation of the grid generation process, aimed at reducing the user time and speeding up the process, is the major goal in this area. Overlapping grids are easier to generate, but there are geometries in which the application of this approach is difficult due to the existence of too many irregular places. The generation of triangular and tetrahedral meshes is easier to automate, which is one of the reasons for their popularity. One usually specifies mesh points on the bounding surface and process from there toward the center of the domain. When a surface grid has been created, tetrahedra is continued toward the center of the volume along a marching front; the entire process is something like solving an equation by a marching procedure and, indeed, some methods are based on the solution of elliptic or hyperbolic PDEs.

Tetrahedral cells are not desirable near walls if the boundary layer needs to be resolved because the first grid point must be very close to the wall while relatively large grid sizes can be used in the directions parallel to the wall. These requirements lead to long thin tetrahedra, creating problems in the approximation of diffusive fluxes. For this reason, some grid generation methods generate first layer of prisms or hexahedra near solid boundaries, starting with a triangular or quadrilateral discretization of the surface, on top of this layer, a tetrahedral mesh is generated automatically in the remaining part of the domain. An example of such a grid is shown in Fig. 5.18.

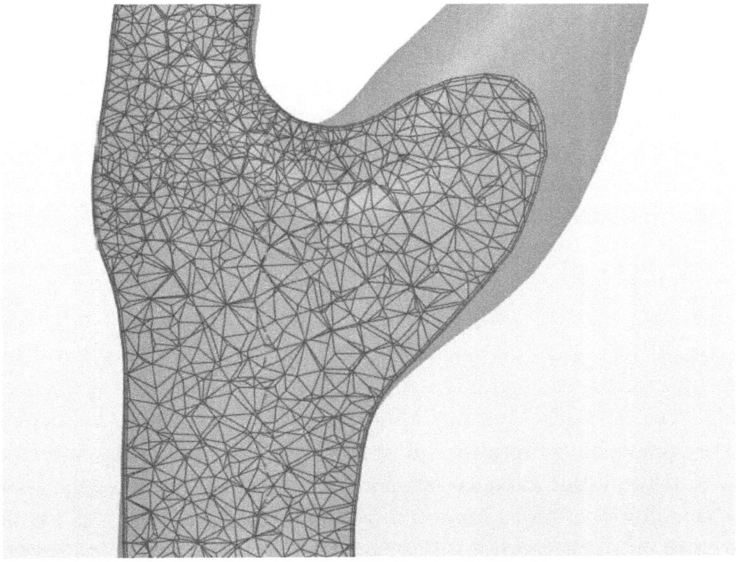

Figure 5.18 Tetrahedral volume mesh with prism layered mesh.

This approach enhances grid quality near the walls and leads to both more accurate solutions and better convergence of numerical solution methods; however, it can be only used if the solution method allows for mixed control volume types. In principle, any type of method (FD, FV, FE) can be adapted to this kind of grid.

Another approach to automatic grid generation is to cover the solution domain with a Cartesian grid and adjust the cells cut by domain boundaries to fit the boundary. The problem with this approach is that the cells near boundary are irregular and may require special treatment. However, if this is done on a very coarse level and the grid is then refined several times, the irregularity is limited to a few locations and will not affect the accuracy much, but the degree of boundary irregularity is limited.

To move the irregular cells further away from walls, one can first create a layer of regular prisms or hexahedra near walls; the outer regular grid is then cut by the surface of the near-wall cell layer. This approach allows fast grid generation but requires a solver that can deal with the polyhedral cells crated by cutting regular cells with arbitrary surface. Again, all types of methods can be adapted to this type of grid.

If the solution method can be applied on an unstructured grid with cells of varying topology, the grid generation program is subject to few constraints. For example, local grid refinement by subdivision of cells into smaller ones is possible. A non-refined neighbor cell, although it retains its original shape (e.g., hexahedron), becomes a logical polyhedral since a face is replaced by a set of sub-faces. The solution domain can first be divided into blocks that can be subdivided into grids with good properties; one has the freedom to choose the best grid topology (structured H-, O-, or C-grid, or unstructured tetrahedral or hexahedral grid) for each block. The cells on the block interfaces then faces of regular shape and have to be treated as polyhedra. An example of such a grid is shown in Fig. 5.19. The grid contains a block interface on which the faces are irregular. The generation of grids with non-matching interfaces is much simpler than creation of a single-block grid fitted to the whole domain. We again note that the solution method has to allow treatment of polyhedral CVs with an arbitrary numbers of faces.

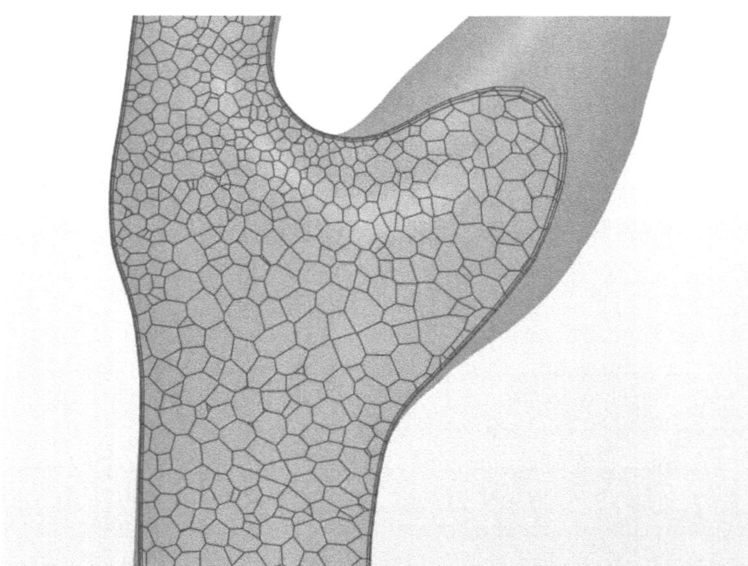

Figure 5.19 Polyhedral volume mesh with prism layered mesh.

5.4 Blood Flow Simulations for Internal Carotid Artery

5.4.1 Introduction

According to the statistics published by the Ministry of Health in Japan, one of three major causes of death is cerebral vascular disturbance and more than 130,000 people die from this disease every year. Most of subarachnoid hemorrhage (SAH) is said to arise from the rupture of a brain aneurysm at bifurcation of cerebral vessel.[3] Even if patients survive the rupture of a brain aneurysm, their rehabilitation is impeded by the aftereffect. Therefore, SAH is a very serious disease. However, the exact cause of SAH is still unclear. This is because the interaction between endothelial cell functions and hemodynamics stimulus is not well understood. Moreover, parameters of artery, such as wall shear stress, blood pressure and blood velocity are difficult to measure *in vivo* because cerebral vessels are narrow and complicated. However, with the development of computing capability, various sorts of parameters in cerebral vessels are obtained in more details by numerical simulations. In recent years, computational fluid dynamics studies have actively attempted to clarify the mechanism of cerebral vascular disease with the data obtained from computed tomography (CT) and magnetic resonance image (MRI). Flow dynamics of cerebral vessels were studied in numerous experimental models and clinical studies.[4–9] As a result, new findings from these researches help us make a guideline for diagnosis of the cerebral vascular disease. In this research, fluid structure interaction (FSI) analysis in ICA model was carried out.

5.4.2 Studied Morphology

Arterial model was extracted from CT data and used as computational model. This model is part of the ICA bifurcation. The shape of this model is shown in Fig. 5.20. The length of the whole model is 65.8 mm and the diameter of the inlet is 6.0 mm, and the diameters of two outlets are 3.35 and 3.09 mm, respectively. The thickness of this arterial wall is assumed to be uniform and 0.3 mm. The mesh was generated with unstructured grid as polyhedral cells to fit shape of the model. The number of vertices and cells were 4465184 and 840414 for the fluid region and 1727357 and 562478 for the wall region, respectively.

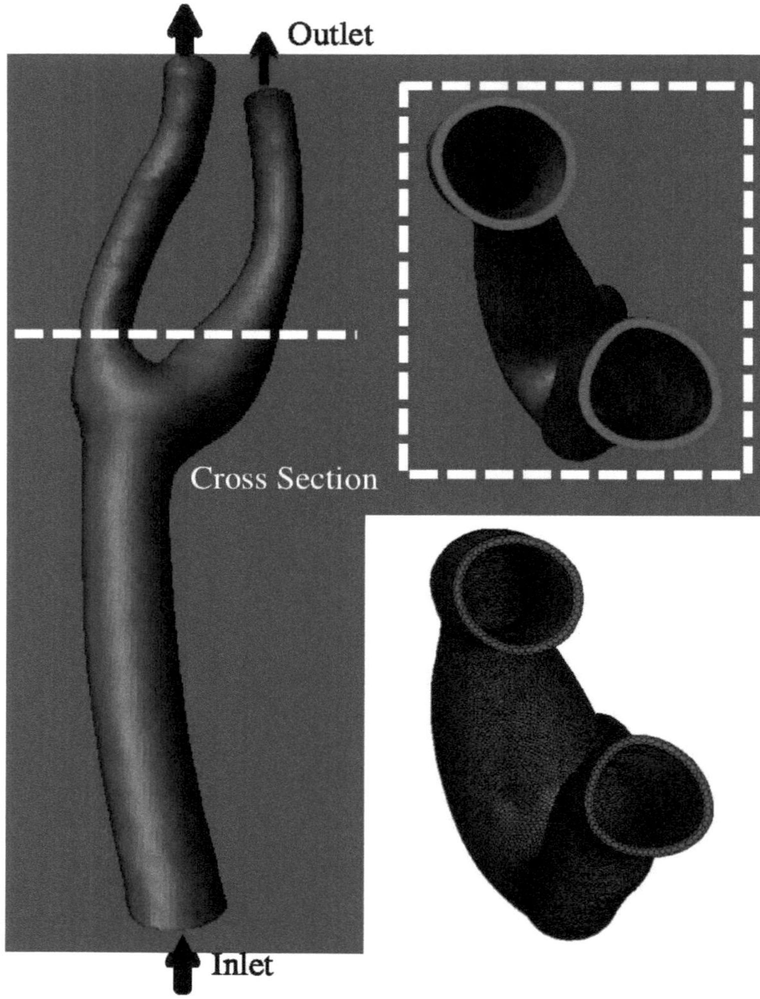

Figure 5.20 Geometry data of internal carotid artery.

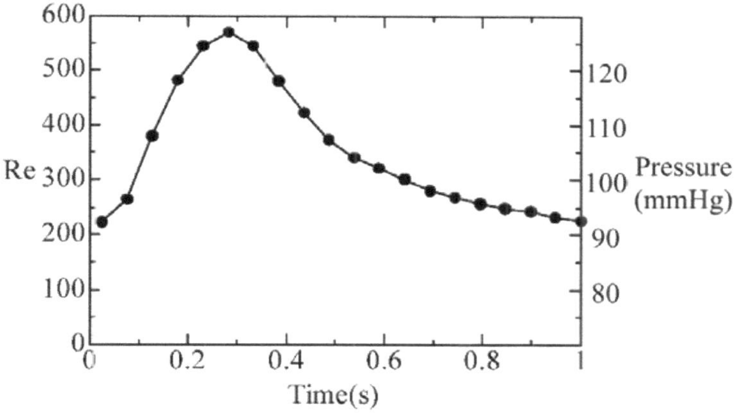

Figure 5.21 Pulsed wave form.

5.4.3 Mechanical Properties

In this study, we deal with relatively large diameters ($d > 3$ mm), the fluid is assumed to be Newtonian.[10] The density and viscosity of the blood were set to 1050 kg/m^3 and 4.0×10^{-3} Pa s. We set the elastic modulus to 3.0 MPa and Poisson's ratio is set to 0.45.

5.4.4 Fluid Calculation Method

The governing equations for fluid are the Navier–Stokes equations of incompressible flows and equation of continuity.

$$\nabla \cdot \mathbf{u} = 0 \tag{5.75}$$

$$\frac{\partial \mathbf{u}}{\partial t} + \mathbf{u} \cdot \nabla \mathbf{u} = -\frac{1}{\rho} \nabla p + \nu \nabla^2 \mathbf{u} \tag{5.76}$$

where ρ is the density and u is the velocity. In this calculation, we adopted large eddy simulation (LES) as turbulence model, regarding that pulsatile unsteady and vortex, separated flow within vessels will occur. To obtain accurate data, subgrid scale of LES is wall-adapting local eddy viscosity (WALE) model. The WALE subgrid scale model provides the following mixing-length type formula for the subgrid scale viscosity:

$$\mu_t = \rho \delta^2 S_w \tag{5.77}$$

where δ is the length scale or grid filter width defined in terms of the cell volume V as

$$\delta = C_W V^{1/3} \tag{5.78}$$

$$\delta = \min\,(k_d \cdot C_W V^{1/3}) \tag{5.79}$$

where k_d (= 0.41) is the von Karman constant. If length scale limit is not applied, δ is defined as Eq. (5.78), and if length scale limit is applied, δ is defined as Eq. (5.79). The deformation parameter S_w is the strain rate tensor defined as

$$S_w = \frac{S_d \cdot s_d^2 / 3}{S_d \cdot s_d^{5/4} + S_d \cdot s_d^{5/2}} \tag{5.80}$$

where S is the strain rate tensor defined as

$$S = \frac{1}{2}(\nabla u + \nabla u^T) \tag{5.81}$$

and the tensor S_d is defined as

$$S_d = [\nabla u \cdot u + (\nabla u \cdot \nabla u^T)] \tag{5.82}$$

The coefficient C_w (= 0.544) is experimentally obtained and works well for homogeneous isotropic decaying turbulence and for channel flow. This calculation employs the FVM, appropriate for complex geometry of vessels. As for the calculation scheme, the three-dimensional segregated solver and second-order implicit formulation in time were selected. Spatial discretization used the second-order upwind scheme for viscous and momentum term of Navier–Stokes equation and algebraic multigrid (AMG) method for pressure–velocity coupling. The minimum time step was 0.001 s (for a 1 s cycle) and the maximum number of iteration is 30 in every time step.

5.4.5 Boundary Conditions

Pulsed blood flow at blood vessel inlet is constructed from seven patients' data. The data were obtained as flow volume at same cross section of ICA using MRI with 20 Hz resolution. Because a cycle length of pulsating time differs from one patient to another, the cycle length of flow pulsation is normalized as 1 Hz and averaged with seven patients' data and expressed as Re number. The highest Re number is 571 at cardiac systole and the lowest at cardiac diastole. Outlet pressure also fluctuates along with waveform of pulsation from 80 to 120 mmHg. The range of blood pressure is determined by referring to the healthy human condition.

5.4.6 Wall Shear Stress

Wall shear stress is most interesting hemodynamic parameter, because it has relation to degeneration of vessel wall. Figure 5.22 shows the result of wall shear stress distribution on ICA at three different states. The first state is at 0.10 s of the pulsation cycle and is shown in Fig. 5.22(a). The second is systole phase at 0.25 s and is shown in Fig. 5.22(b). The third is diastole phase of 0.50 s, shown in Fig. 5.22(c). The maximum wall shear stress is approximately 15 Pa at the bifurcation of ICA model, and its phase is diastole phase at 0.25 s. The lower limit of physiological wall shear stress ranges 10 Pa.[11] Locally high wall shear stress at the bifurcation is more than 15 Pa. Furthermore, higher wall shear stress occurs as the diameter of blood vessel becomes narrow. The wall shear stress on the outside of larger bifurcation is lower than 4 Pa and is nearly equal to 0 at all times.

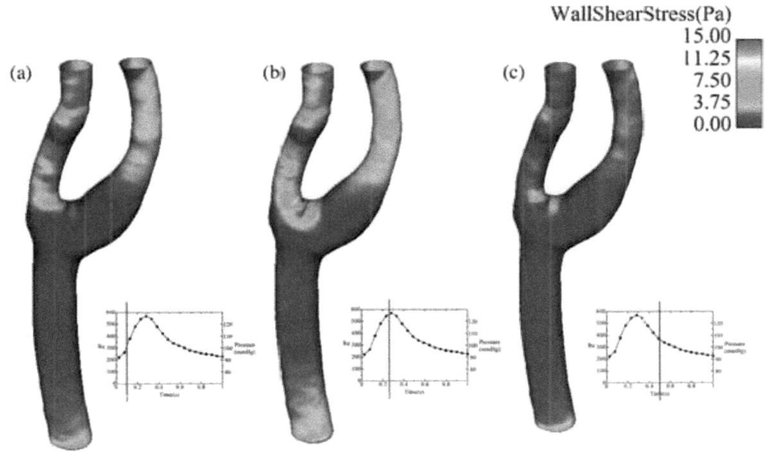

Figure 5.22 Wall shear stress. See also Color Insert.

5.4.7 Velocity Magnitude

Dashed line in Fig. 5.20 represents the cutting plane at separated branch of ICA model to be a cross section. Also the velocity magnitude at the cross section is shown in Figure (5.23) and the maximum velocity magnitude reaches 0.30 m/s. As is the case with wall shear stress distribution, three phases of pulsed wave are shown in Figs. 5.23(a–c). The common characteristic in three phases, the velocity magnitude of the internal vessel, is higher than that of the outer vessel. Additionally, at the peak systole, the difference of velocity between the inside and outside of the vessel becomes clear. At the diastole phase (t = 0.50 s), vortex structure appears outside of the vessel and the velocity decreases nearly to 0 m/s.

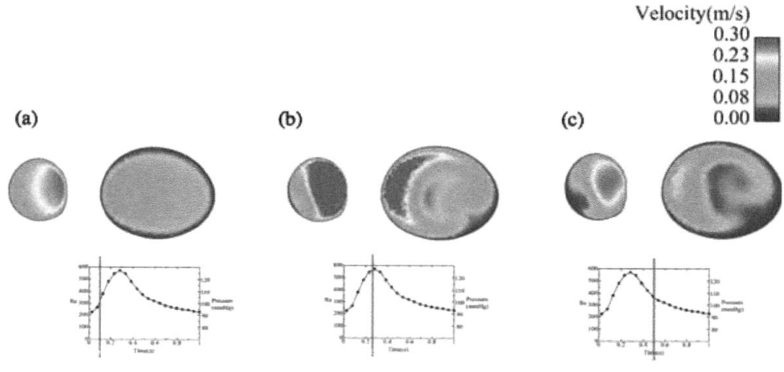

Figure 5.23 Velocity. See also Color Insert.

5.4.8 Streamline

Figure 5.24 shows the streamline between three calculation states. Blood flow from inlet impact at the bifurcation and divides into two branches, and the velocity of the narrow branch is faster than that of the larger branch. In the first state shown in Fig. 5.24(a) (t = 0.1 s), the streamline describes the lines along with the shape of the vessel wall and the velocity of blood flow decreases at the larger branch. Vortex flow appears at systole phase shown in Fig. 5.24(b) (t = 0.25 s) after inlet flow impacts the bifurcation. The velocity of blood flow is the maximum in this phase and more than 3.0 m/s. In the next phase shown in Fig. 5.24(c), vortex flow maintains its structure and the velocity is slowed down.

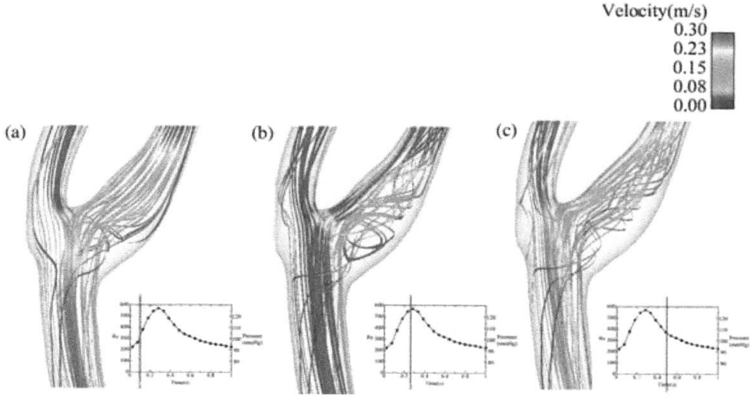

Figure 5.24 Streamline. See also Color Insert.

5.4.9 Mises Stress

Mises stress is defined as follows:

$$\sigma = \sqrt{\frac{(\sigma_1 - \sigma_2)^2 + (\sigma_2 - \sigma_3)^2 + (\sigma_3 - \sigma_1)^2}{2}} \tag{5.83}$$

where σ_1, σ_2, and σ_3 are the maximum principal stress, the intermediate principal stress, and the minimum principal stress, respectively; σ is scalar and used as index value to identify the yield condition of materials under multiaxial stress field. Figure 5.25

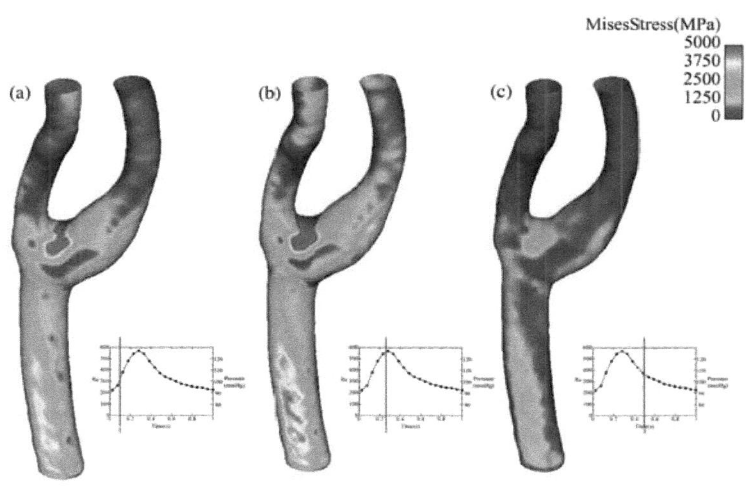

Figure 5.25 Mises stress. See also Color Insert.

shows the Mises stress for the ICA model. The maximum Mises stress occurs at the bifurcation and near the base of bifurcation at systole phase. In addition, Fig. 5.25 also shows that the stress on the front face is higher than that on the lateral face.

5.4.10 Discussion

Aneurysm initiation is believed to be caused by the interaction of abnormally high-flow hemodynamic forces and the arterial wall.[12,13] From Fig. 5.22, oscillated locally high wall shear stress more than 15 Pa at the bifurcation can cause the fragmentation of the internal elastic of blood vessel and can be a trigger of aneurysm initiation. Mises stress is also focused on the bifurcation. Therefore, stress on a vessel wall also can be an index value of aneurysm initiation as the same way of wall shear stress. A wall shear stress value lower than 4 Pa is reported to cause atherosclerotic change on the vessel wall [10], and this low wall shear stress causes the degeneration of the vessel, such as thinning and weakness of vessel wall, also vascular narrowing by the aggregation of plaque. From the results shown in Figs. 5.23(a–c), vortex flows in blood vessel invoke stagnation. As shown in Fig. 5.22, the wall shear stress outside larger bifurcation where vortex flow exists is nearly equal to 0. Therefore, the region of low wall shear stress can develop stenosis. As a future work, the validation between numerical simulation and experimental results needs to be confirmed. Furthermore, we apply FSI analysis for complex cases, such as atherosclerosis and aneurysm models.

5.4.11 Conclusion

FSI analyses were carried out for one ICA model reconstructed based on CT images. Moreover, we constructed the geometry data of the vasculature model with thickness and its material elasticity, such as Young's modulus and Poisson ratio is considered. The results of wall shear stress and Mises stress show that oscillatory locally high stress can cause aneurysm initiation and low wall shear stress leads to atherosclerotic change on the vessel wall. From the result of velocity magnitude after bifurcation, blood flow speed of the internal vessel is faster than that of the outer. At the diastole phase ($t = 0.50$ s), vortex structure outside of vessel decreases the velocity of blood flow to nearly 0 m/s and stagnation of flow. The streamline at systole phase ($t = 0.25$ s) shows vortex flow at larger branch after bifurcation. Vortex flow decreases blood flow speed and causes the stagnation of flow.

5.5 Outlook for the Future of Computer Fluid Dynamics

Computer fluid dynamics (CFD) utilizing a supercomputer will play an important role in the field of medicine and have possibilities of clinical applications. With the advancement of medical technique, improvement of CT technology and high-resolution MRI make it possible to detect a smaller unruptured aneurysm that was not detected by previous medical devices. As for the frequency of aneurysm rupture, in 1998, it was reported that the annual rupture rate of an aneurysm smaller than 10 mm was approximately 0.05% and for an aneurysm larger than 25 mm was around 6%.[1] Moreover, according to ISUIA (International Study of Unruptured Intracranial Aneurysm),[14] the rupture rate of an anterior aneurysm smaller than 7 mm is less than 0.5% and for the aneurysm larger than 7 mm is from 0.5% to 8% depending on the size; the rupture rate of posterior aneurysm smaller than 7 mm is from 0.5% to 0.7%, and for aneurysm larger than 7 mm is around 3% . Also, the annual rupture rate of giant aneurysm — defined as the aneurysm of size over 25 mm — is said to be around 10% independent of its location. It has been clear that all aneurysms do not have possibilities of rupture. There are some cases in which clinical follow-up ensures sufficient safety of aneurysm. Therefore, in some cases, an aggressive medical treatment for aneurysm is more likely to cause rupture compared with spontaneous rupture within a year. There is an urgent need of guidelines for aneurysm treatment that help doctors make decisions of therapeutic strategy by identifying the risk factor of rupture based on the result of CFD. In the near future, with the advancement of processing power and high-speed networks, it will be possible to evaluate rupture risk of an aneurysm in a short time with high degree of accuracy through a remotely connected server that has large amount of patients' historical data of ruptured and unruptured aneurysms. Therefore, "medical image data (CT, MRI, DSA) + CFD" will be a medical standard for the diagnosis and management of aneurysm. Moreover, with the advancement of technology for CFD analysis, surgery rehearsal by "virtual surgery" on a computer could be realized. Although it is desirable to have a rehearsal before an actual operation by using actual stent and embolization coil, it is unrealistic considering the cost of endovascular devices and time for preparing a patient's specific phantom vascular model. Hence, in advance of actual survey, virtual surgery by CFD technology makes it possible to provide further insights into surgery, such as size and placement of stent and material and volume of the embolization coil. As a result, virtual surgery technology achieves the optimization of clinical surgery and contributes to minimally invasive therapies.

References

1. Arakawa C. *Computational Fluid Dynamics for Engineering*, 7th ed., Tokyo University Press, Inc., 1994.
2. Ferzinger JH, Peric M. *Computational Methods for fluid Dynamics*, 3rd ed., Springer, 2002.
3. The International Study of Unruptured Intracranial Ameurysms Investigation. Unruptured intracranial aneurysms — risk of rupture and risks of surgical intervention, *N Engl J Med.* 1998, 339: 1725–1733.
4. Nakatani H, Hashimoto N, Kang Y. Cerebral blood flow patterns at major vessel bifurcations and aneurysm in rats. *J Neurosurg.* 1991, 74: 258–262.
5. Gonzalez CF, Choi YI, Ortega V. Intracranial aneurysms: flow analysis of their origin and progression. *AJNR Am J Neuroradiol.* 1992, 13: 181–188.
6. Tenjin H, Asakura F, Nakahara Y. Evaluation of intraaneurysmal blood velocity by time-density curve analysis and digital subtraction angiography. *AJNR Am J Neuroradiol.* 1998, 19: 1303–1307.
7. Gobin YP, Counard JL, Flaud P. *In vitro* study of haemodynamics in a giant saccular aneurysm model: influence of flow dynamics in the parent vessel and effects of coil embolization. *Neuroradiology.* 1994, 36: 530–536.
8. Ujiie H, Yoshimoto T, Hori T, Inagaki S, Ishibashi Y, Takeda S. Flow visualization studies of bifurcation aneurysms. *Surg Cereb Stroke.* 1999, 27: 439–445.
9. Tateshima S, Murayama Y, Villablanca JP. Intraaneurysmal flow dynamics study featuring an acrylic aneurysm model manufactured

using computerized tomography angiogram as a mold. *J Neurosurg.* 2001, 95: 1020–1027.

10. McDonald DA. *Blood Flow in Arteries*, 2nd ed., Edward Arnold, 1974.
11. Malek AM, Alper SL, Izumo S, Hemodynamic shear stress and its role in atherosclerosis. *J Am Med Assoc.* 1999, 282: 2035–2042.
12. Matsuda M, Handa J, Saito A, Matsuda I, Kamijyo Y, Ruptured cerebral aneurysms associated with arterial occlusion. *Surg Neurol.* 1983, 20: 4–12.
13. Kayembe KNT, Sasahara M, Hazama F. Cerebral aneurysms and variations of the circle of Willis. *Stroke.* 1984, 15: 846–850.
14. Do W. Unruptured intracranial aneurysms: natural history, clinical outcome, and risks of surgical and endovascular treatment. *Lancet.* 2003, 362: 103–110.

Chapter 6

Pumps for Human Blood Pressure Simulation

Panza [1] defines normal human systolic pressure as less than 120 mmHg and normal diastolic pressure less than 70 mmHg. De Bruyne *et al.* [2] measured coronary diastolic pressure at 87.5 mmHg, and a systolic pressure of 152.5 mmHg at 1.81 Hz. Ebenal *et al.* [3] measured aortic systolic pressure at 120 mmHg and diastolic pressure at 50 mmHg at 1.3 Hz. Regularly a local minimum above 50 mmHg must be recreated and pressure variation adjusted to realistically replicate human blood pressure waves. The simulation of human blood pressure is desirable to recreate a realistic motion of the silicone vasculature in the endovascular evaluator (EVE) and also for interacting with catheters, guide wires and other intravascular tools inserted within the silicone vasculature during the simulation. Even more the visualization of the stress produced by human blood pressure based on photoelastic effect is desirable to evaluate the natural conditions of stress in the vasculature membrane models. For that purpose homogeneous urethane elastomer models cannot be used, as they have a plastic deformation for the pressure ranges below 200 mmHg.

A piston pump was initially used to circulate fluids trough the vasculature of EVE. The piston pump creates a variable flow but cannot sustain the pressure and flow above zero during a complete

Microsurgery: Advances, Simulations, and Applications
Edited by Toshio Fukuda and Carlos Tercero

ISBN 978-981-4364-69-0 (Hardcover), 978-981-4364-70-6 (eBook)
www.panstanford.com

cycle. Several kinds of pumps were compared for this purpose: Peristaltic pumps introduce undesired vibrations on the flow at low flow rates. Diaphragm pumps not only introduce noise on the flow but also are too loud to be used for this kind simulation. Gear pumps and turbine pumps have big inertia to control precisely the waveform.

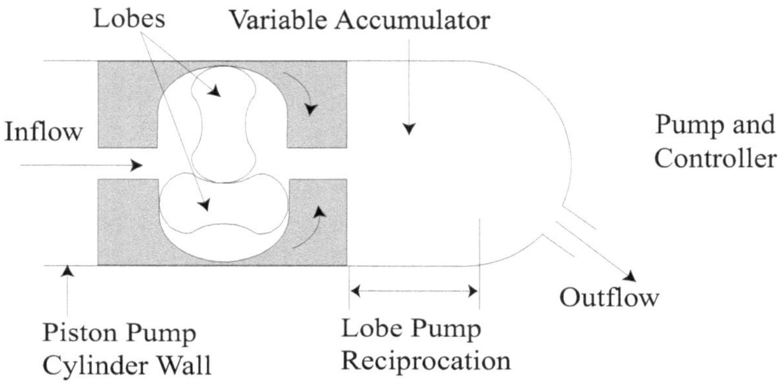

Figure 6.1 Schematic of the hybrid pump for human blood pressure simulation.

In this chapter, we will present two pumps for human blood pressure simulation and vasculature models enabling fluid circulation and photoelastic stress analysis. These vasculature models have a urethane elastomer layer shielded with silicone layers to prevent plastic deformation and to isolate urethane from the environment and fluid circulation. The first pump proposed is a hybrid pump that combines the features of lobe and piston pumps; the piston provides a carrier waveform at the heart beat frequency, and the lobe pump provides the minimum pressure and fine waveform tuning. It uses pressure and frequency feedback control to reconstruct a pressure waveform defined by polynomial functions. The second system uses a single magnetic drive pump with pressure feedback control for maximum and minimum pressures, the reproduced waveforms a result of open loop control of predefined functions and frequencies. A schematic of the hybrid pump is shown in Fig. 6.1 and the pressure variation in the outflow for the piston pump is shown in Fig. 6.2.

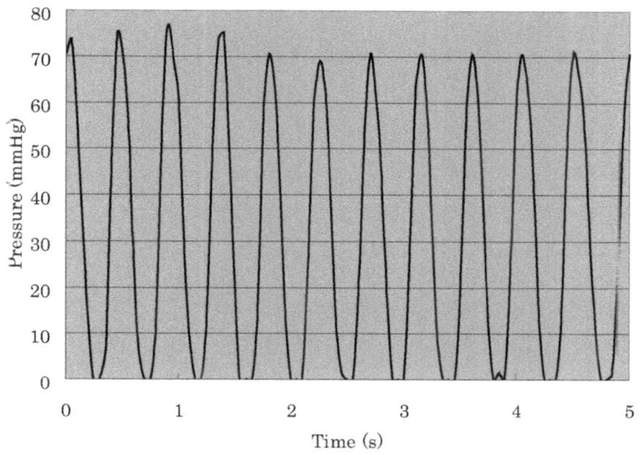

Figure 6.2 Pressure variation in the outflow port of the piston pump.

6.1 Multilayer Urethane Model Elaboration

The urethane elastomer was shielded by deep-coating wax vasculature fabricated as detailed in [4]. The first coating was silicone elastomer at 0.1 mm/s, then urethane elastomer coating twice at 1 mm/s after the silicone dried. The silicone coating was repeated after the urethane elastomer dried, and then wax was melted to complete shielding.

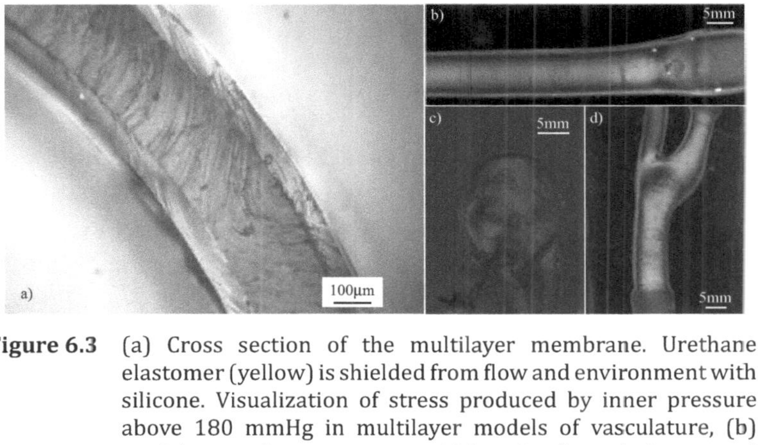

Figure 6.3 (a) Cross section of the multilayer membrane. Urethane elastomer (yellow) is shielded from flow and environment with silicone. Visualization of stress produced by inner pressure above 180 mmHg in multilayer models of vasculature, (b) straight vasculature segment within a circular polariscope and green filter, (c) giant aneurysm within elliptical polariscope, and (d) carotid artery within a circular polariscope. See also Color Insert.

Deep coating yielded a model wall with a 250 μm-thick urethane elastomer layer and two 60 μm-thick silicone layers shielding the urethane, as shown in Fig. 6.3(a). Water was circulated inside the model and no plastic deformation occurred at pressure below 200 mmHg. Different models built with this technique and showing the photoelastic properties of the material are shown in Fig. 6.3(b).

6.2 Lobe Pump Design

Simulating blood pressure requires a capacity pump to provide a constant flow and a variable volume accumulator to introduce flow pulsation. When the piston travels along the pump cylinder, water volume inside the cylinder changes; the inlet valve opens when the volume increases and the exhaust valve opens when this volume decreases. The piston pump cylinder has sufficient volume to install inside a piston-lobe pump. Removing both piston pump check valves and sealing the inlet results in a hybrid pump suitable for our simulation purpose. Lobe pumps are rotary positive-displacement pumps, which force fluid between two rotary elements and a static casing. The pump flow is a function of rotational lobes speed, and is independent from the pressure differential.

6.2.1 Lobe Profiles and Mechanism

Lobe profiles were modeled with parametric equations following the lobe profile synthesis proposed by Tong *et al.* [5]. We considered a required flow of 1 lpm, and the cross-sectional diameter is limited to 42 mm by the piston pump cylinder diameter, and the maximum desired motor speed is 400 rpm. The flow function is sinusoidal, the flow range is between 0.847 and 1.059 lpm, the frequency is 3.2 kHz, and the two-rotor lobe configuration has a rotation speed of 800 rpm.

Using nomenclature of [5] for the parameters selected above, corresponding non-circularity h is 0.75, cavity volume V is 2352 mm^3 and w = 20 mm, yielding the following lobe profile generation parametric equations are shown in Fig. 6.4.

Inner profile:

$$G_{1x} = 6\cos(\theta) + 3\cos(2\theta)\cos(\pi - \theta) \qquad (6.1)$$

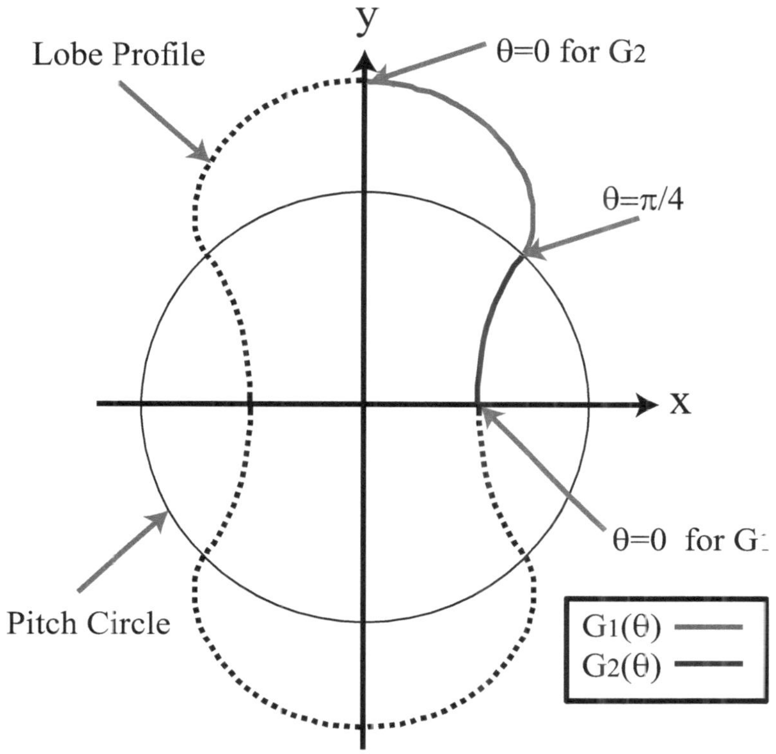

Figure 6.4 Lobe profile design.

$$G_{1y} = 6\sin(\theta) + 3\cos(2\theta)\sin(\pi - \theta) \quad (6.2)$$

Outer profile:

$$G_{2x} = 6\cos(\pi/2 - \theta) - 3\cos(2\theta)\cos(3\pi/2 - 3\theta) \quad (6.3)$$

$$G_{2y} = 6\sin(\pi/2 - \theta) - 3\cos(2\theta)\sin(3\pi/2 - 3\theta) \quad (6.4)$$

For both equations, θ varies from 0 to $\pi/4$. For the first quadrant of the plane, Eqs. 6.1 and 6.2 define the profile inside the *pitch* circle and Eqs. 6.3 and 6.4 outside it. The profile in the other three quadrants is generated symmetrically. Figure 6.5 shows the hollow shaft and transmission used for coupling reciprocating and rotational actuators to the piston-lobe pump, the gearbox, and the lobe pump static casing.

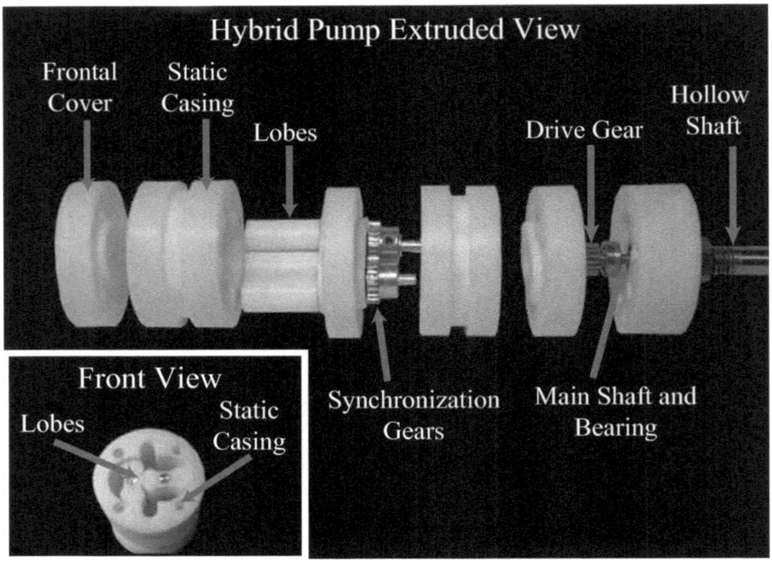

Figure 6.5 Mechanism inside the piston-lobe pump. Hollow shaft couples the lobe pump to the reciprocation and rotational motions.

6.1.2 Feedback Control

Feedback control was applied in two closed loops for controlling reciprocation frequency and pressure in the pump output, the diagram of the system is shown in Fig. 6.6. Command signal $F(z)$ is a step function with the desired frequency, and $P(z)$ is a polynomial approximation of the blood pressure reported in [10]. *A*, *L*, and *Pi* are transfer functions of the accumulator; lobe pump and piston pump, C_1 and C_2 are PI type controllers. Function $P(z)$ is periodical and synchronized with the reciprocation cycle repeated every 2500 steps. When accumulator volume reaches its minimum, an optical switch state change marks the start of the reciprocation cycle, $P(z)$ is defined as follows:

If $z < 1250$

$$P(z) = -7.25\times10^{-18}z^6 + 6.59\times10^{-14}z^5 - 2.61\times10^{-10}z^4 + 4.25\times10^{-7}z^3 - 3.1\times10^{-4}z^2 + 4.3\times10^{-2}z + 152.5 \quad (6.5)$$

If $z > 1250$

$$P(z) = -3.3\times10^{-15}z^5 + 2.6\times10^{-11}z^4 - 8.5\times10^{-8}z^3$$
$$+ 1.3\times10^{-4}z^2 - 4\times10^{-2}z + 87.5 \quad (6.6)$$

The system was tested to simulate human blood pressure at 1 Hz, and then used to circulate flow while measuring stress on the multilayered vasculature wall.

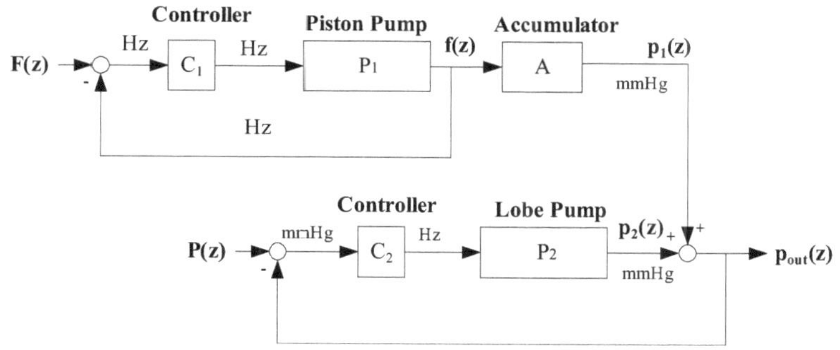

Figure 6.6 Control diagram of the hybrid pump.

6.3 Image Processing Software for Stress Measurement

The system shown in Fig. 6.7 was constructed for stress measurements. Analyzed and result images are 640 × 480 pixels, with x and y used to define pixel locations in rows and columns. N_0 is the image of the model when filled with and surrounded by water and placed between the light source and camera. $D_{(x, y)}$ is an optical path length pixel belonging to N_1 and is obtained by applying Eq. 6.7 to the blue intensity of pixel $I_{B(x, y)}$ of N_0 in a similar way as shown in Chapter 4. To eliminate the noise of N_1 produced by the background, all pixels with optical path length equal to 0.2 mm or less were set to zero (black). Using the polariscope detailed in [6–7], we captured images in which the photoelastic effect is visible. N_2 and N_3 are images of the vasculature model inside the polariscope, N_2 when there is no flow circulation inside the model and N_3 when the pressure inside the model is 182 mmHg. Based on previous results [6, 8], we modeled the relation between retardation $Re_{(x, y)}$ and green light intensity $I_{G(x, y)}$ in urethane elastomer model image as follows:

$$I_{G(x,y)} = (I_{G\max} - I_{G\min})\sin^2 \frac{\pi}{\lambda_G} \mathrm{Re}_{(x,y)} + I_{G\min} \tag{6.7}$$

I_{Gmin} is the minimum green light intensity between all pixels of N_2, $I_{G\max}$ the maximum green light intensity between all pixels of N_3, and λG = 538 nm the green light wavelength. We consider F a collection of images (F_i with $i \in \mathbb{N}$ and with same dimensions than N_1) where photoelastic effect is visible. To calculate retardation $Re_{(x,y)}$ in pixel (x, y) of F_i we used Eq. (6.7) where $I_{G(x,y)}$ is the intensity of green registered at pixel (x, y) of F_i. We obtain optical path length $D_{(x,y)}$ from pixel (x, y) of N_1 (for photoelastic coefficient calibration, refer to Chapter 4) and using Eq. 6.8, we calculate the magnitude of principal stress component $(\sigma_1 - \sigma_2)_{(x,y)}$ in pixel (x, y) of F_i.

$$(\sigma_1 - \sigma_2)_{(x,y)} = \frac{\mathrm{Re}_{(x,y)}}{D_{(x,y)}C} \tag{6.8}$$

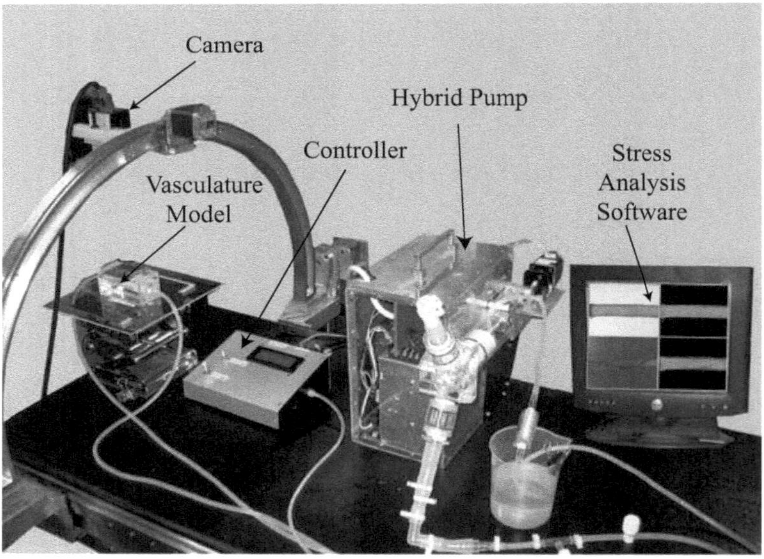

Figure 6.7 Experimental setup for photoelastic stress analysis on the vasculature model wall membrane while human blood pressure is simulated within the model.

Stress was measured for eight images: N_2 for reference, F a collection of five video frames captured when the hybrid pump was circulating fluid through the model ($N_3 = F_3$), and N_4, N_5 are pressure maxima of flow circulation with a guide wire applying stress to model wall. In N_5 the guide wire deforms the model wall. System integration is shown in Fig. 6.7.

Human blood pressure was simulated with an average error of 5.6% at a frequency of 1.01 Hz, as shown in Fig. 6.8. When the pump was tested with the urethane elastomer model, the reciprocation stroke was increased for testing the shielding endurance, achieving a pressure range between 98 and 182 mmHg for registration of N_3, N_4, N_5 and F. Figure 6.9 shows that pressure produced by the fluid was within the same range whether the guide wire was inside or outside the model, so stress calculation differences are caused by the catheter and model deformation.

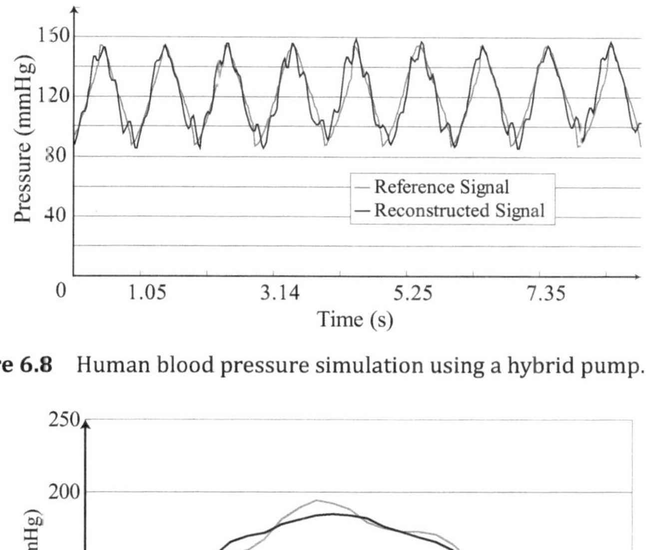

Figure 6.8 Human blood pressure simulation using a hybrid pump.

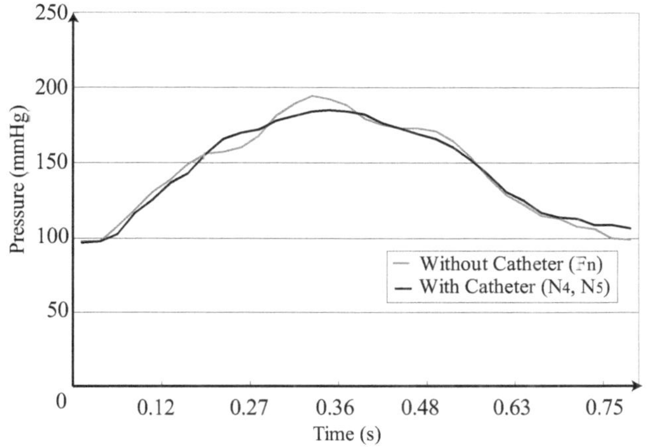

Figure 6.9 Estimated pressure ranges during registration of F, N_4 and N_5.

6.4 Stress Measurement

Images N_0 and N_1 are shown in Figs. 6.10(a) and 6.10(b), respectively. N_1 shows an optical path length between 640 and 840 μm for the 55% of pixels, and 13.4% of pixels of N_1 have a shorter optical path length. Images N_2 and N_3 are shown in Figs. 6.10(c) and 6.11, I_{Gmin} = 53 and I_{Gmax} = 221. The threshold of 20 μm was adequate and no black pixels appeared in the image area corresponding to blood vessels. Background noise has nearly disappeared. After analyzing N_2 we obtained N'_2, we measured that the 98.2% of pixels of N_2 had $(\sigma_1 - \sigma_2)_{(x,y)}$ below 32 kPa (Fig. 6.10(d)). Results for all the analyzed images are summarized in Table 6.1 and shown in Figs. 6.11 and 6.12. Estimated flow pressure variation during registration of F is shown in Fig. 6.9.

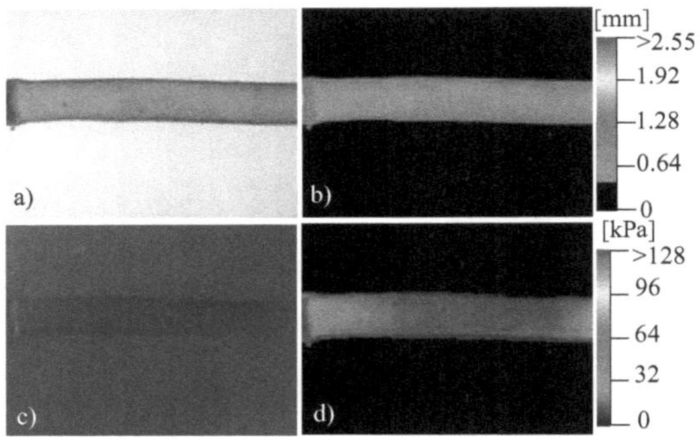

Figure 6.10 Images used and obtained during the calibration process (a) image N_0, used to calculate the optical path length. (b) N_1 optical path length quantification. (c) Image N_2 used to find I_{Gmin}, and (d) N'_2 stress quantification of N_2.

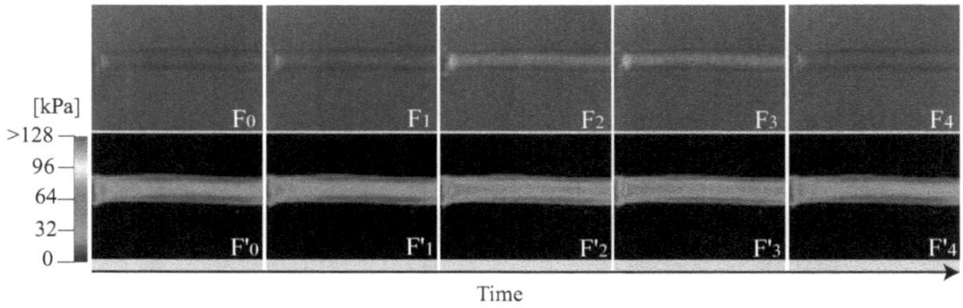

Figure 6.11 F image collection and their respective stress measurements F'. $N_3 = F_3$ is the image used to find I_{Gmax} during the calibration process.

Table 6.1 Percentage of pixels per stress range in all analyzed images

Stress range	N'_2 (%)	F'_0 (%)	F'_1 (%)	F'_2 (%)	F'_3 (%)	F'_4 (%)	N'_4 (%)	N'_5 (%)
R_0: σ_1–σ_2 < 32 kPa	98.20	85.12	74.58	50.96	50.42	81.55	57.46	50.37
R_1: 32< σ_1–σ_2 < 65 kPa	1.71	14.70	25.22	48.44	48.88	18.17	41.30	47.98
R_2: 64 < σ_1–σ_2 < 97 kPa	0.08	0.17	0.19	0.52	0.60	0.25	1.04	1.15
R_3: 96 < σ_1–σ_2 < 128 kPa	0.00	0.00	0.00	0.06	0.08	0.00	0.18	0.49

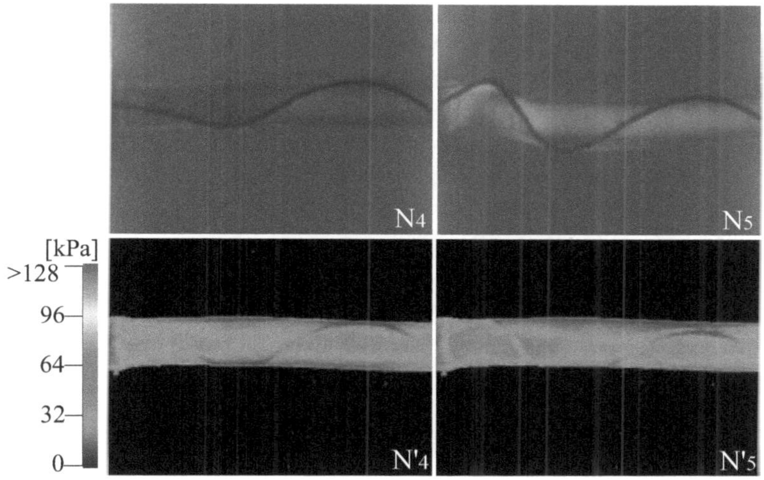

Figure 6.12 Measurement of stress produced by a guide wire of catheter and pressure on the membrane of the model of vasculature obtained from N_4 and N_5.

Human blood pressure simulation and urethane shielding in the vasculature model enabled us to analyze the principal stress component in the vasculature wall at different pressures. Calibration using vasculature images under normal stress enabled quantitative comparison of stress distribution as shown in Table 6.1 and Figs. 6.10–6.12. Normal stress parameters are defined by Table 6.1: F'_0, as normal low parameter, and F'_3 as normal high parameter. As expected, the values of F'_1 and F'_2 fall within the range defined by F'_0 F'_3 and N'_2 is below. N'_4 and N'_5 have values slightly above range

F'_0 F'_3 for stress ranges R_2 and R_3 and values that may be considered normal high for other parameters.

We must consider that we used a fixed image to calculate the value of the optical path length in all pixels. Image N_5 represents considerable deformation in the blood vessel model and high birefringence. In N'_5, the deformation of the blood vessel model by the guide wire is not visible and where this deformation occurs the stress measurements accuracy is compromised. For this reason, values of N'_5 are close to normal high values. This problem is solved by calculating the optical path length for all images. To implement real-time stress analysis solving this problem, a prismatic lens may be used enabling simultaneous use of two cameras, one with a polarizer and the other without.

For building a system that evaluates medical tools and human skills in real time, complementary stress measurement methods that use pressure sensors within the catheter [8] are not suitable, as introducing foreign objects into the catheter increases its stiffness and changes its original design and functions. Therefore, the number of possible evaluations with these sensors is limited. Human blood pressure levels are useful for calibrating the stress-analysis software, as they give a reference of normally high stress levels, as in this chapter stress due to pressure variation inside urethane vasculature models was compared quantitatively to stress produced with a guide wire. This methodology is applicable to catheter insertion robots safety evaluation [9–10], and comparison of medical tools performance and surgical skills. The simulation of human blood pressure is important also for the *in-vitro* environment as interaction between silicone vasculature with medical tools (catheters, stents, grafts, and coils) becomes more realistic, enabling better evaluation of tools prototypes.

6.5 Blood Pressure Simulation in Saccular Aneurysm Model with Bleb

During the aneurysm-growing process in cerebral arteries, a blister called bleb appears in the aneurysm membrane areas with low shear stress. The relation of the presence of the bleb with the rupture possibilities in aneurysm is being investigated [11]. Provided the presented techniques for stress measurements, in this section, we will present a study of stress analysis in saccular aneurysm model with bleb. Figure 6.13 shows the fluoroscopic image of an aneurysm with bleb.

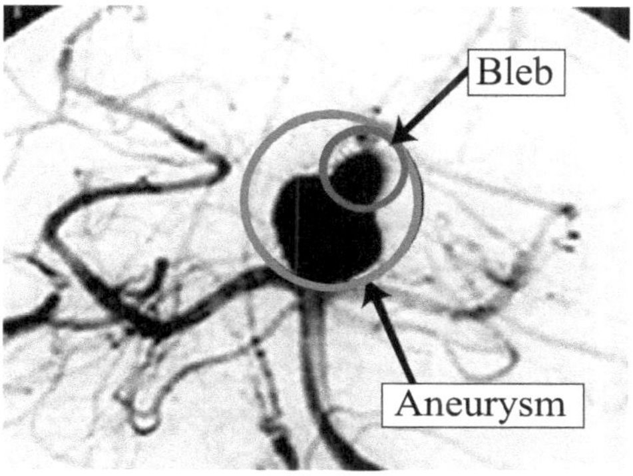

Figure 6.13 Fluoroscopic image of cerebral aneurysm with bleb.

6.5.1 Saccular Aneurysm with Bleb Model Design

As cerebral arteries are prone to the formation of saccular aneurysms, anterior cerebral artery was used for reference. This artery has an average inner diameter of 4 mm, as general guideline for aneurysms with 5 mm diameter in this blood vessel present a rupture risk of 1% within the first year and surgery is not recommended at this stage [11–12].

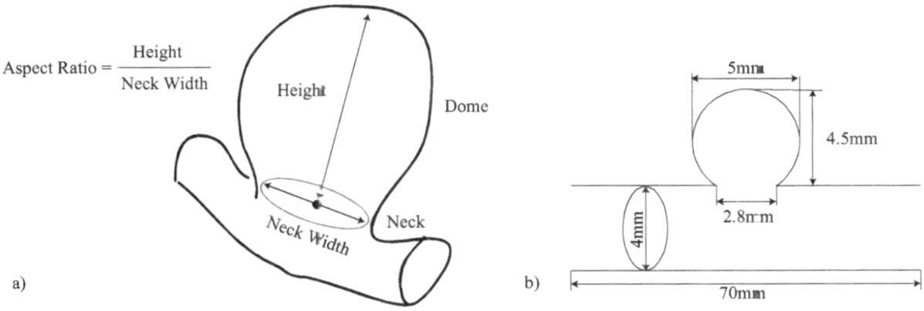

Figure 6.14 (a) Definition of aspect ratio for aneurysms. (b) Designed aneurysm.

Also when the aneurysm aspect ratio is above 1.6 the risk of rupture is high, this rupture risk increases with this aspect ratio [13–14]. Therefore, the aneurysm diameter was selected to be

5 mm and the aspect ratio 1.6 (Fig. 6.14). As the next step in the design process, computer fluids dynamics (CFD) was used for the simulation of the shear stress in the saccular designed aneurysm to find the minimum of shear stress in the aneurysm dome and locate the bleb model. The simulated fluid has a sine wave shape with frequency of 1 Hz, with Reynolds number in the range of 200–900 and a viscosity of 3.91 mPa·s^{-1}. Figure 6.15 shows the CFD result for wall shear stress variation during one period of the sine wave. Wall shear stress is higher in neck area and the dome side where the flow enters the aneurysm. From this image, the location of the bleb model was selected as depicted in Fig. 6.16(a). Following that design, a multilayer model of the aneurysm with bleb was constructed, for enabling plastic deformation of the model wall with inner pressure the silicone layers shielding was not present in a perimeter of about 1 mm over the aneurysm model surface. The multilayer model is shown in Fig. 6.16(b).

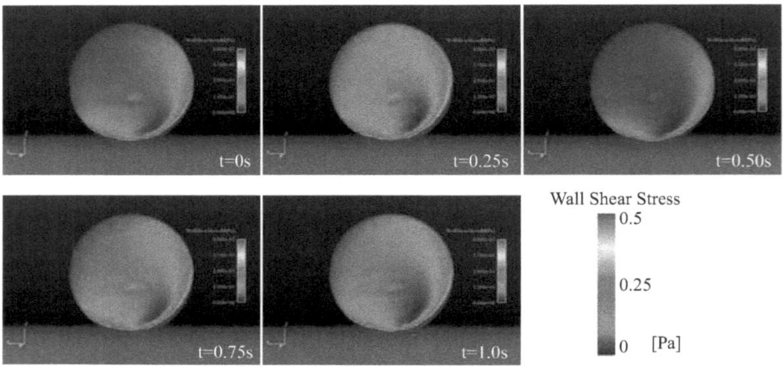

Figure 6.15 Wall shear stress analysis time line for the 1 Hz sine wave fluid inside the designed saccular aneurysm.

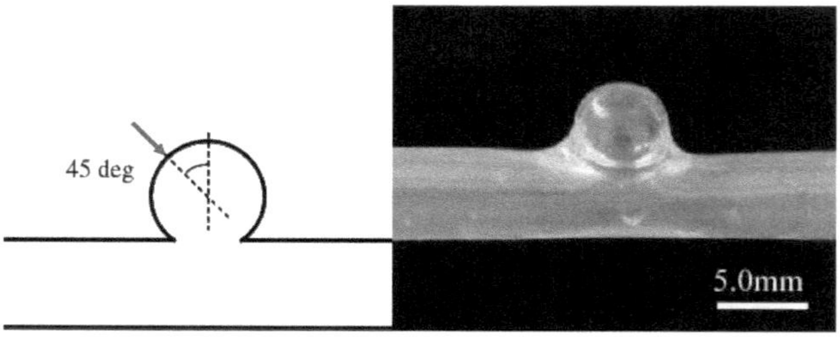

Figure 6.16 (a) Bleb location within the saccular aneurysm. (b) Multilayer model of saccular aneurysm with bleb.

6.5.2 Stress Analysis in the Bleb Model Using Static Pressure

The multilayer model of the saccular aneurysm with bleb was connected to closed loop pipe circuit filled with a glycerin solution at 60%. The pressure inside the circuit was increased from 60 mmHg to 200 mmHg. The deformation of the model with inner pressure is observed in Fig. 6.17. As we can see, the bleb shape appears at the designed location while pressure increases. Photoelastic stress analysis was applied for measuring the stress variation in the bleb area for pressure values near human blood pressure limits 60–140 mmHg showing a local maximum of stress in the bleb area (Fig. 6.18).

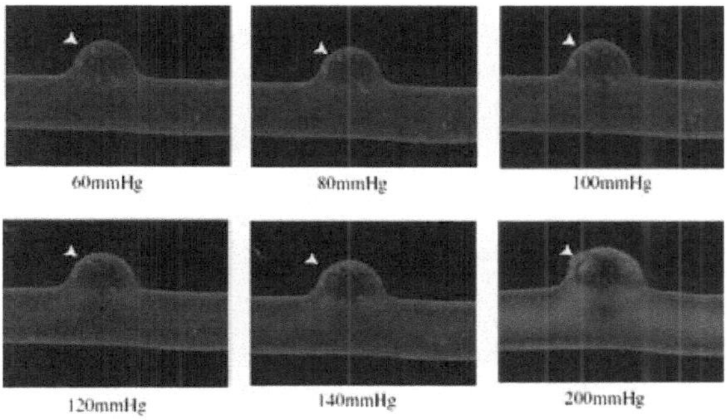

Figure 6.17 Saccular aneurysm model with bleb deformation with inner pressure, the bleb area is pointed with an arrowhead in all images.

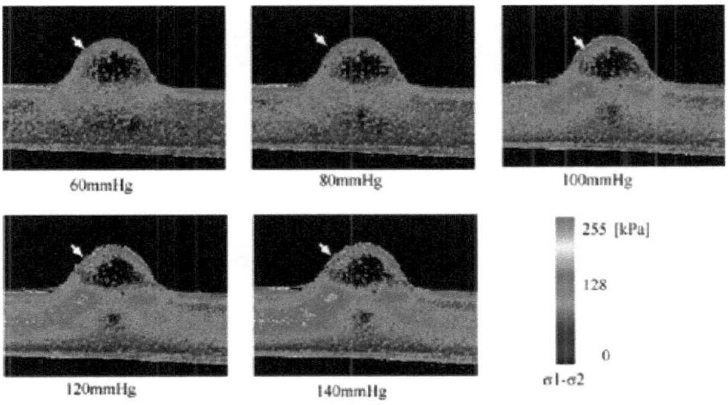

Figure 6.18 Photoelastic stress analysis in the saccular aneurysm model with bleb for pressures within the range of 60–140 mmHg. Arrows point at the bleb area.

6.5.3 Stress Analysis in the Bleb Model Using Blood Pressure Simulation

Photoelastic stress analysis was used for studying the stress in the multilayer model of saccular aneurysm with bleb while a pulsed flow was circulated through it. The flow waveform had a sine shape with a frequency of 1 Hz, a pressure range of 85–115 mmHg and flow rate of 250 mL/min. For an inner pressure of 85 mmHg, the number of pixels with stress above 960 mmHg is 0.2%. When pressure rose to 115 mmHg, that percentage increased to 7.1%. As consequence of plastic deformation and pressure change the bleb size increased while its membrane became thinner. The bleb membrane thickness reduction finally produced a rupture of the phantom, producing inner pressure loss and a reduction from 7.2% to 0.2% in percentage of pixels within the high stress range (Fig. 6.19) [16].

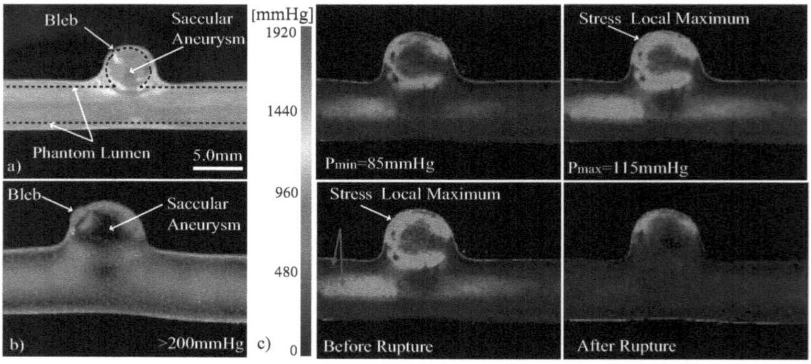

Figure 6.19 (a) Phantom of saccular aneurysm with bleb. (b) Source image example for stress measurement, green light intensity represents the stress in the phantom wall. Bleb shape is appreciated over 200 mmHg. (c) Stress analysis in Aneurysm Bleb phantom with pulsed flow circulation at: minimum pressure, maximum pressure, before rupture and after rupture. Red arrows point a measurement distortion appearing when the angle between the light beam and σ_3 increases. See also Color Insert.

6.6 Portable Simulator for Blood Pressure

A flow with temperature around 37°C and pressure variation around human normal ranges is desirable because endovascular tools physical properties change with temperature and dynamic flow. When the blood pressure is simulated within silicone vasculature, every flow pulse is visualized with the cross-section variation of silicone vasculature and when the *Y*-connectors used as catheter insertion ports are open, the circulating warm fluid will start leaking as blood does in a real intervention. For this purpose, we constructed a portable fluid control system that uses a magnetic drive pump with pressure feedback for bounding maximum and minimum pressure of pre-programmed waveforms profile. This simulator recreates a pulsed flow with frequency in the range of 50–150 bpm and with an amplitude range of 20–65 mmHg with minimum pressure range of 75–140 mmHg. Additionally temperature control was added to the circulating fluid to set the temperature in a range of 30–40°C (Figs. 6.20 and 6.21).

The system operation is controlled using a microcontroller PIC18F1320; the motion of the pump was controlled through a PWM interface and DC motor drive, the variation of the PWM signal is defined by the parameters set by the user and the pressure sensor feedback. The fluid is heated though a ceramic heater that is controlled in a binary state according to the desired temperature and the voltage level measured in a thermistor.

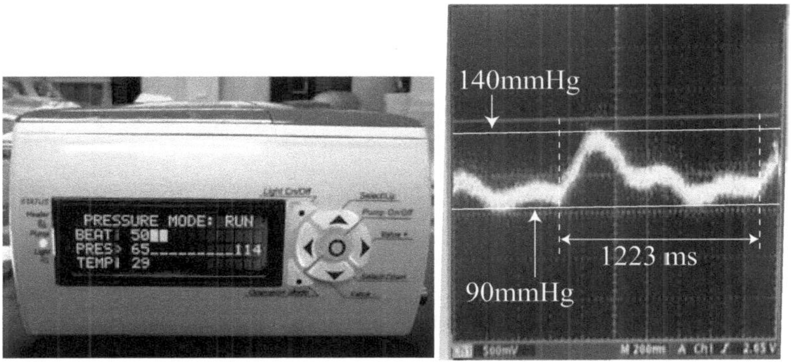

Figure 6.20 Portable system for human blood pressure simulation and flow temperature control (Left). Human pressure waveform at 50 bpm simulated using magnetic drive pump with feedback control of pressure limits and pre-programmed wave profile (right).

a) b) c) d)

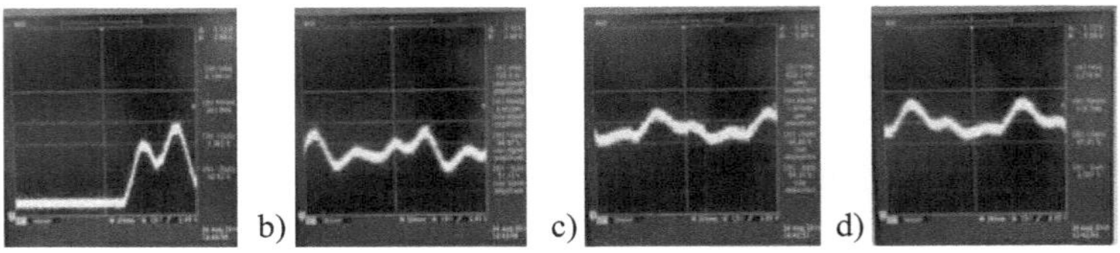

Figure 6.21 Minimum and maximum pressure settling for the preprogrammed waveform reconstruction at 50 bpm and 90–140 mmHg range. (a) $t = 0$ s, (b) $t = 14$ s, (c) $t = 20$ s, and (d) $t = 30$ s.

References

1. Panza J. High-normal blood pressure more high than normal. *N Engl J Med*, 2001, 345(18): 1337–1340.
2. De Bruyne B, Bartunek J, Sys SK, *et al.* Simultaneous coronary pressure and flow velocity measurements in humans. *Circulation*, 1996, 94: 1842–1849.
3. Ebenal J, Vasana S, Clinton C, Cox D, Shine T. Arterial blood pressure system modeling and signal analysis in *Proc IEEE-CIRA*, 2007, 386–391.
4. Ikeda S, Arai F, Fukuda T, Negoro M, Irie K. An *in vitro* patient specific biological model of the cerebral artery reproduced with a membranous configuration for simulating endovascular intervention. *J Robot Mechatron*, 2005, 17(3): 327–333.
5. Tong S, Yang D. Rotor profiles synthesis for lobe pumps with given flow rate functions. *J Mech Des*, 2005, 127: 287–294.
6. Ikeda S, Fukuda T, Arai F, *et al.* Patient-specific neurovascular simulator for evaluating the performance of medical robots and instruments in *Proceedings of the IEEE-ICRA*, 2006, 625–630.
7. Tercero C, Okada Y, Ikeda S, *et al.* Numerical evaluation method for catheter prototypes using photo-elastic stress analysis on patient-specific vascular model. *Int J Med Robot Comput Assist Surg*, 2007, 3(4): 349–354.
8. Tanimoto M, Arai F, Fukuda T, *et al.* Study on micro force sensor for minimum invasive surgery. *Trans. Jpn Soc Mech Eng, C 64-620, JSME*, 1998, 150–155.
9. Tercero C, Ikeda S, Fukuda T, *et al.* Robot manipulation and guidance using magnetic motion capture sensor and a rule-based controller. *J Robot Mechatron*, 2008, 20(1): 151–158.

10. Tercero C, Ikeda S, Uchiyama T, *et al.* Autonomous catheter insertion system using magnetic motion capture sensor for endovascular surgery. *Int J Med Robot Comp Assist Surg,* 2007, 3(1): 52–58.

11. Gaidcrain D, *Nihon Noudokku Gakkai Inou,* 3rd ed., 2008, p. 53.

12. International Study of Unruptured Intracranial Aneurysms Investigators, Unruptured intracranial aneurysms: risk of rupture and risks of surgical intervention. *N Engl J Med,* 1998, 339: 1725–1733.

13. Fukusaku K, Negoro M, The expectation for the intravenous treatment and fluid analysis of the cerebral aneurysm. *Computational Biomechanics Symposium*, 2001, 153–161.

14. Ujiie H, *et al.* Is the aspect ratio a reliable index for predicting the rupture of a saccular aneurysm? *Neurosurgery,* 2001, 48: 495–503.

15. Sakai H, Oshima M, Kobayashi T. Ketsuryu Shimyureshon ni okeru Ryunyu Jouken no ataeru eikyo no kentou. *Seisan Kenkyuu,* 2003, 55(2): 180–184.

16. Matsushima M, Tercero C, Ikeda S, *et al.* Photoelastic stress analysis in blood vessel phantoms: three-dimensional visualization and saccular aneurysm with bleb. *Int J Med Robot Comput Assist Surg,* 2011, 7(1): 33–41.

Chapter 7

Magnetic Trackers: Robot Control and Vasculature Imaging

This chapter presents the theory for robot control using magnetic trackers and sparse lookup tables. Silicone models of vasculature offer an environment to show two applications of this robot control theory. A robotic camera with three degrees of freedom is controlled to simulate the C-arm used in fluoroscopic systems; this camera is manipulated using the magnetic tracker to capture images of the catheter moving inside silicone vasculature. A second application is presented for catheter insertion robot feedback control; the guidance provided with magnetic tracker enabled to reconstruct catheter insertion paths in different silicone models. Finally this chapter presents a vasculature imaging technique based on magnetic trackers and intravascular ultrasounds.

7.1 Robot Control with Magnetic Trackers

The first step to achieve robot control with uncalibrated magnetic trackers in restricted environments, such as silicone models of vasculature, is to build a sparse LUT. This table is a collection of vectors representing all the reference point needed to control the robot behavior. Those vectors components are the coordinates of a point in space associated with a control parameter. Some examples of such parameters could be position, state or speed of the motors

Microsurgery: Advances, Simulations, and Applications
Edited by Toshio Fukuda and Carlos Tercero

ISBN 978-981-4364-69-0 (Hardcover), 978-981-4364-70-6 (eBook)
www.panstanford.com

of the robot to be controlled. The sparse LUT M is made of m vectors with $m \in \mathbb{N}$.

$$M = \begin{bmatrix} \vec{P}_0 \\ \vec{P}_1 \\ \dots \\ \vec{P}_m \end{bmatrix} \quad (7.1)$$

$$\overrightarrow{P_n} = (p_{xn}, p_{yn}, p_{zn}, C_{0n}, C_{1n}, \dots, C_{kn}) \quad (7.2)$$

p_{xn}, p_{yn}, and p_{zn} are the position coordinates of a reference point measured with the magnetic tracker. In those points, the robot should receive the commands defined by the variables C_{0n}, C_{1n}, ..., C_{kn} to reproduce a desired motion. $k - 1$ is the number of control parameters that can be modified in a robot to change its behavior.

Then a detection range, d_n, should be defined for each P_n. This range must ensure that the reference point is detected. $X^t{}_{mag}$, $Y^t{}_{mag}$, and $Z^t{}_{mag}$ are defined as the spatial coordinates given by the magnetic tracker in a time t after the construction of the LUT. The rule-based controller, shown in Fig. 7.1, compares M to the output of the magnetic tracker.

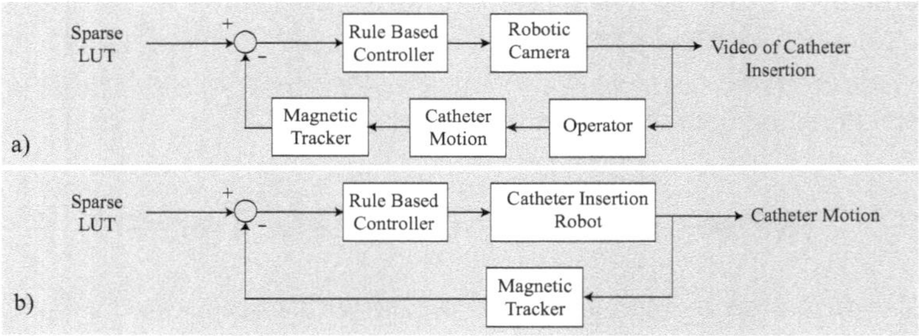

Figure 7.1 Diagrams for robot control based on magnetic trackers and sparse look up table for (a) robotic camera system and (b) catheter insertion robot.

The behavior of the robot changes only if for any vector P_n the following equation is true.

$$\sqrt{(p_{xn} - X_{mag}^t)^2 + (p_{yn} - Y_{mag}^t)^2 + (p_{zn} - Z_{mag}^t)^2} \leq d_n \quad (7.3)$$

Provided that the same magnetic tracker is used to control the robot and to build M, if the dispersion of samples of the magnetic tracker is smaller than the detection range, then the correction of the distortion of magnetic field of magnetic tracker is not necessary. It is necessary to study the dispersion of samples of the magnetic tracker when sensing a coordinate at fixed location and orientation in order to confirm that the sparse LUT is valid and that the controllers will react at the proper location. For that, the sensor is hold collecting samples for every reference point of M. If the MMCS remains at a static position $P \in M$ collecting samples, the average coordinate of all the collected samples $\overline{P} = (\overline{X}, \overline{Y}, \overline{Z})$, can be considered the correct coordinate. Then dispersion of the samples for the evaluated point P could be modeled with

$$E_P^+ = \sqrt{(X_{\max} - \overline{X})^2 + (Y_{\max} - \overline{Y})^2 + (Z_{\max} - \overline{Z})^2} \qquad (7.4)$$

$$E_P^- = \sqrt{(X_{\min} - \overline{X})^2 + (Y_{\min} - \overline{Y})^2 + (Z_{\min} - \overline{Z})^2} \qquad (7.5)$$

where E_P^+ and E_P^- are the dispersions above and below of the average. $X_{\max}$, $Y_{\max}$ and $Z_{\max}$ are the maximum coordinates reported in the three axes for all samples of P. In the same way $X_{\min}$, $Y_{\min}$, and $Z_{\min}$ are the minimum reported coordinates. If this dispersion is small compared to the detection range the control method can be applied.

7.2 Robotic Camera for Digital Subtraction Angiography Simulation

At surgical rooms equipped for Interventional Radiology (IVR), the fluoroscope is commonly attached to a C-arm; this robotic arm allows changing the orientation and position of the fluoroscope enabling to visualize the development of the endovascular intervention. When required an additional C-arm and fluoroscope is set in a bi-planar configuration, to register radiological images in perpendicular planes. In raw fluoroscopic images, blood vessels are visible only while contrast media is present in the blood stream; meanwhile catheters and guide wires are visible over the background. Therefore to visualize the progress of the catheter or guide wire through blood vessels image processing Digital Subtraction Angiography (DSA) [1] and *Road Map* technique are used (Fig. 7.2).

A *Road Map*, as described by Turski *et al.* [1], is an image processing technique used in IVR to create a video containing simultaneously the moving catheter and a static angiogram. First an angiogram is registered using X-rays when contrast media is injected through the vasculature; a clear and static image of the vasculature to be treated is obtained. When the catheter is inserted through the vasculature, only the catheter and bone is seen in the video obtained from the fluoroscope. If the position and orientation of the fluoroscope has not changed since the angiogram registration, then the static image of vasculature became a map of the desired trajectory to be followed by the catheter. When that image is subtracted to the video of the catheter inside the body, bone is eliminated as it appears in both images and the moving catheter appears over the static image of vasculature.

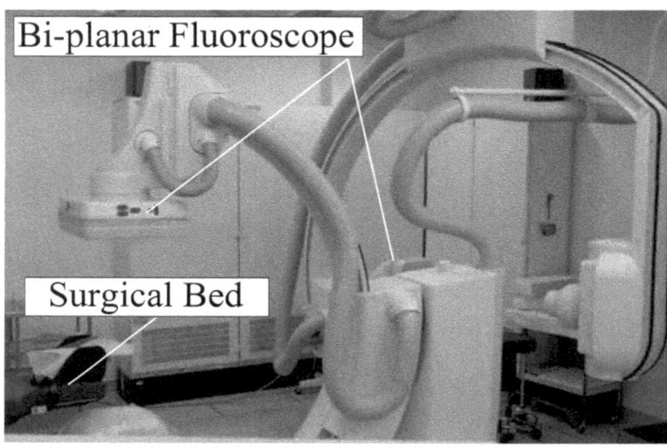

Figure 7.2 Surgical room equipped with Interventional Radiology (IVR) system.

The endovascular simulator EVE provides an excellent platform to recreate vascular lumen, simulating in a realistic way the interaction between the catheter and the vascular wall. But physicians and trainees using EVE, can see the catheter moving through the silicone vasculature directly. To have a more realistic simulator and a better training tool, it is needed that the person using EVE sees the catheter as he does during a real endovascular intervention. Therefore, it is needed to generate an image similar to used for IVR and to track catheter trajectory in order to have always the catheter inside the IVR simulated image.

7.2.1 Mechanical Design

A Surgical bed was adapted to hold on its sides a robotic system with three degrees of freedom. This covers a half cylinder shaped working space, similar to the X-rays device mounted on the C arm in IVR environment. The first DOF is linear, allowing reciprocation of the camera along the endovascular surgery simulator. The second DOF is an arch mounted over the DOF1, this allows changing the pitch angle of the camera. The third DOF allows rotating the camera along the arch, changing the yaw angle of the camera (Fig. 7.3).

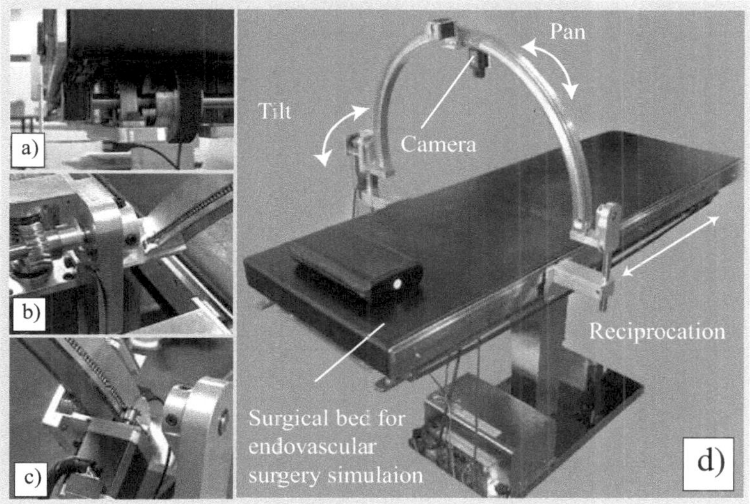

Figure 7.3 (a) Belt system of reciprocation motion of surgical bed. (b) Worm gear system to rotate arch of surgical bed. (c) Chain and sprocket for camera panning. (d) Robotic camera with three degrees of freedom for interventional radiology images simulation.

The selected mechanism for DOF1 is a belt system and a linear motion rail. A servo motor rotates one of the shafts of the belt system to reciprocate the arch at 0.012 mm/step over the bed surface. DOF2 is a worm gear mechanism to tilt the arch with a resolution of 0.002 deg/step. A damper was added to the shaft of the arch to reduce the oscillation produced by the intrinsic properties of the worm gear. Such gear system increases the torque and allows fixing the arch position while the system is turned off. DOF3 enables panning the camera along the arch. A roller cart is attached a motor and a camera, enabling the camera to move along the arch using chain and sprocket system. The chain was inserted on a canal opened in the upper surface of the arch.

Camera panning resolution is 0.036 deg/step. Table 7.1 summarizes the mechanical characteristics of the robotic camera.

Table 7.1 Mechanical specifications

Dimensions	2000 mm	Length		
	600 mm	Width		
	1041 mm	Height (min)		
	1351 mm	Height (max)		
Degrees of Freedom	1200 mm	Reciprocation		
	90 deg	Tilt		
	200 deg	Pan		
Motors	Speed (max)	Torque		
AIM 275AM-P5	600 rpm	1.1 Nm	Resolution	Steps/Rev
AIM 275AM-P25	120 rpm	5 Nm	0.012 mm/step	1000 or 500
PK246PA	400 rpm	0.93 Nm	0.002 deg/step	500

7.2.2 Electronics Design

A circuit was designed to control the stepping motors drivers using a microcontroller. The selected microcontroller is PIC16F8877A, its software was made on C language for CCS compiler.

Table 7.2 Circuit specifications

Item	Number	Description
Power Supply	1	12 VDC, 100 mA
	1	–12 VDC, 100 mA
	1	5 VDC, 2 A
Power Consumption	12.5	W (max)
Clock Speed	20	MHz
Digital Inputs	4	Bounce Free TTL
	4	TTL unprotected
Analog Inputs	6	10 bits resolution (Joystick Interface)
Digital Outputs	4	TTL
	12	High Current (100 mA max)
Communications	1	Serial RS-232
	1	I2C
Motor Drives	2	PWM
	6	Steppers with CW/CCW controls

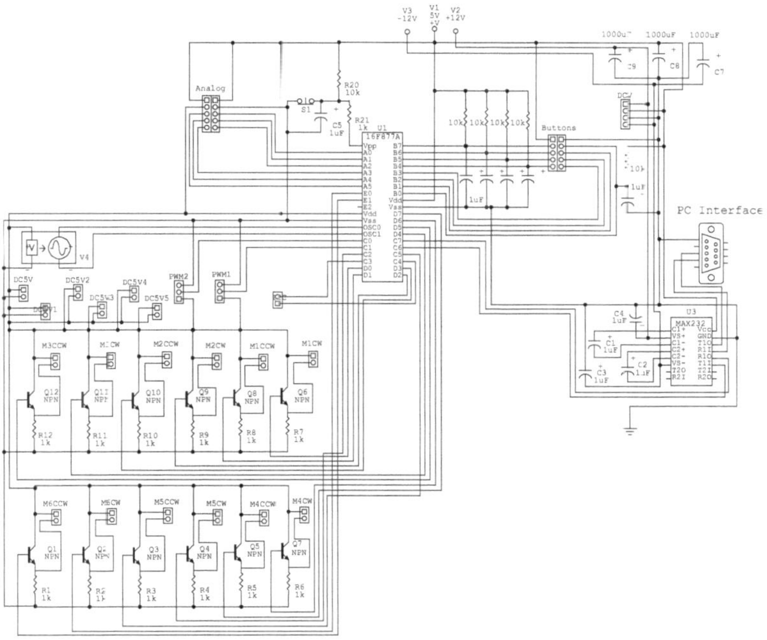

Figure 7.4 Circuit design for robotic camera control.

The three degrees of freedom are initialized using optical limit switches and then controlled using step count. The designed circuit offers an interface with a joystick and a serial port. The joystick allows a simple human interface to manipulate the three DOFs without the need of a computer. The serial port offers an option for controlling the Robotic Camera from the PC or by other electronic circuits. All those systems were integrated in printed circuit board. Additionally, an interface using foot pedals was built for controlling the image processing for the IVR environment simulation (Fig. 7.4, Table 7.2).

7.3 Robot Manipulation

7.3.1 Silicone Models of Vasculature

As the target of the robotic camera is to simulate endovascular surgery and catheter shaped coils are used, a silicone model of major vasculature offers a suitable environment for simulation. Sub-clavian and common carotid arteries were represented by a membranous silicone model mounted on an acrylic stand. This silicone model was built based on three-dimensional data of human arteries shown in Fig. 7.5(a). The lumen of the silicone model was filled with air and the surface of it was covered by surfactant (Figs. 7.5 and 7.6).

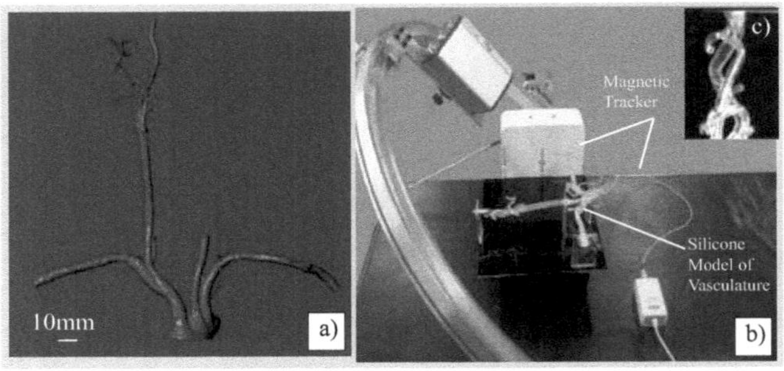

Figure 7.5 (a) Three-dimensional data use for modeling sub-clavian and carotid arteries. (b) Silicone model and MMCS placed over the DSAP for robot manipulation evaluation. (c) Image captured by the DSAP simulating endovascular surgery.

7.3.2 Magnetic Tracker

For controlling robots designed for endovascular surgery applications, Aurora (Northern Digital Inc.) with a catheter shaped micro coil with 5DOF (X, Y, Z, Yaw and Pitch) offered the most suitable tracking solution. The micro coil size is 0.8 mm diameter and 8 mm long and is embedded in to a 5F catheter. The physical properties of this magnetic tracker enable moving it inside silicone models of vasculature. The listed error by the manufacturer is 1.3 mm RMS in a working space of 500 mm $\times$ 500 mm $\times$ 500 mm. The magnetic field emitted by the transmitter of the sensing system is not oscillating; therefore, it minimizes Eddy currents allowing the use of it near stainless steel SUS300.

7.3.3 Controller Design

To evaluate robotic camera manipulation with magnetic trackers, a sparse LUT M_0 was designed with vectors P_n, using the method proposed in the previous section:

$$\overrightarrow{P_n} = (p_{xn}, p_{yn}, p_{zn}, C_{0n}, C_{1n}, C_{2n}) \quad (7.6)$$

C_{0n} corresponds to the reciprocation of the camera (DOF1), C_{1n} to the tilt angle of the camera (DOF2) and C_{2n} to the panning angle of the camera (DOF3) (Fig. 7.3). The detection range d_n must be equal or bigger than the arterial model cross section diameter to ensure that the reference points P_n are detected when the catheter shaped sensor passes through the arterial model. Therefore even if the maximum cross section diameter of the lumen of the arterial model is of 6 mm, the detection range was set to 10 mm. This gives time to the robotic system to place the camera at the correct location on time. With the silicone model placed inside the robotic camera work space, 7 sets of spatial coordinates were collected using the sensor inside of it (Fig. 7.6). Then, each set of coordinates was associated with a different configuration C of the robotic camera, building that way each vector of M_0. The configuration for each vector was designed in order to keep the catheter inside the field of view of the camera at all times.

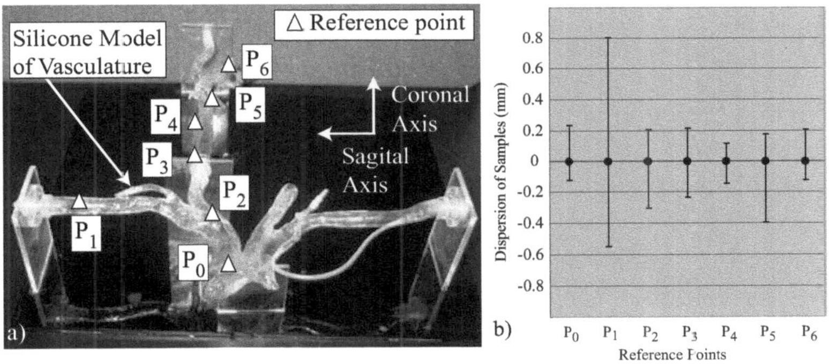

Figure 7.6 (a) Reference points inside silicone model of vasculature for robot manipulation experiment. (b) Dispersion of samples of the magnetic tracker at those locations.

In this experiment, the 3DOF of the camera were set to change automatically according to the software map M_0. The control diagram is shown in Fig. 7.7.

For the reference points P_n the maximum registered dispersion was 0.80 mm (Fig. 7.6). This allows the usage of those reference points to apply the control technique for M_0 as the detection range is of 10 mm. A PC was used to interface the robotic camera and the magnetic tracker, and then to store M_0. Robot manipulation is shown by controlling the camera motion using a micro-coil as human interface. The robotic camera was then brought to initial state and the controller M_0 was activated in the software of the computer. After initializing the magnetic tracker, the catheter was inserted manually inside the silicon model to trigger the motions of the robot described in M_0 to generate automatically the images of catheter insertion procedure.

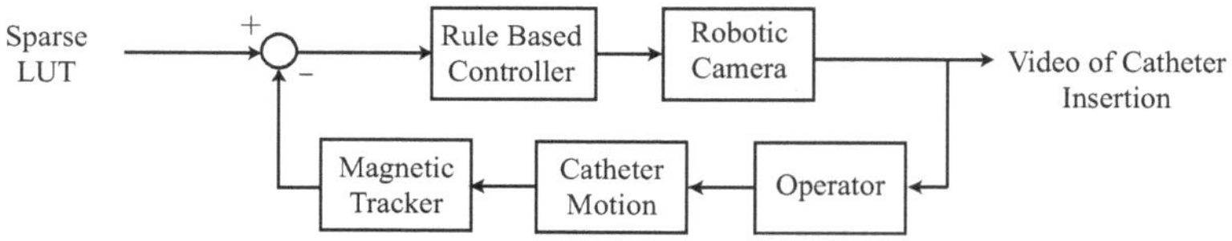

Figure 7.7 Controller for manipulation of the robotic camera.

7.3.4 Robot Manipulation

The seven reference points were detected inside predefined range of 10 mm, allowing a controlled motion on the robotic camera and capturing images of the catheter inside the silicone model automatically (Figs. 7.8 and 7.9. The registered motion of the motors and setting time of each reference point, experimental detection ranges are shown in Table 7.3. The robotic system reacted successfully to keep the catheter inside the vision range of the camera. For almost all reference points the system reacted just after entering the detection range. For point P_0, the system reacted at a range of 4.56 mm from the reference point.

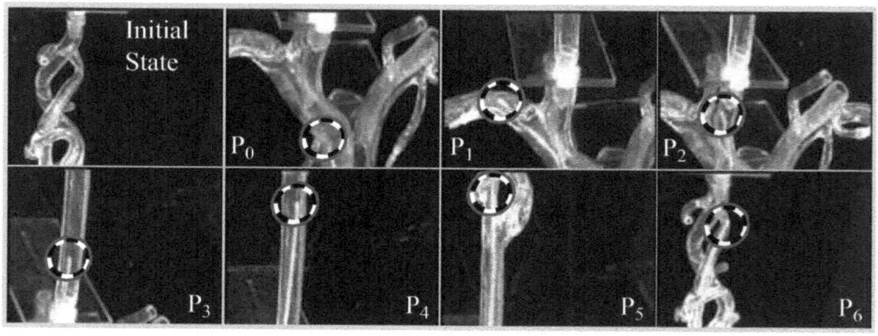

Figure 7.8 Image captured by the robotic camera at initial state and at every reference point.

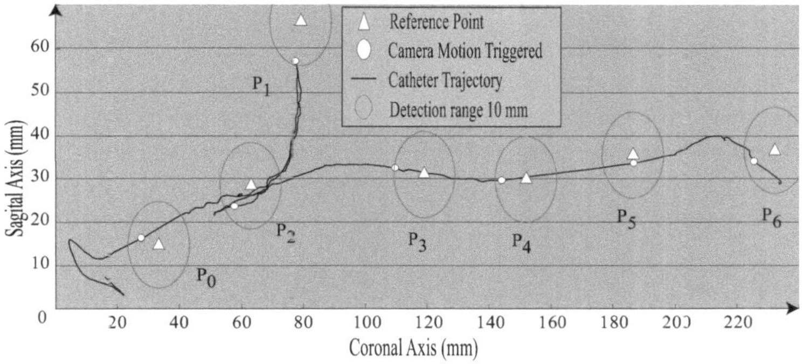

Figure 7.9 Magnetic sensor trajectory projection showing that the motion of the robotic camera was triggered inside the designed detection range [16].

This small reaction range may be explained by the location of P_0 near the silicone model entrance, but the small reaction range in P_5, 0.21 mm, can be explained by one of the following reasons:

(i) A local distortion in the magnetic field appeared between the LUT construction and the robot manipulation.
(ii) A change on silicone vasculature lumen shape between the LUT construction and the robot manipulation.
(iii) Excess of speed in the motion of the catheter.

Table 7.3 Comparison of defined and experimental detection ranges for all reference points in M_0 and system setting time

Reference point	Maximum defined detection range (mm)	Experimental reaction range (mm)	Setting time (s)	Step count		
				DOF1	DOF2	DOF3
P_0	10	4.56	9.6	27718	14812	1456
P_1	10	7.00	3.2	2836	8816	2342
P_2	10	6.68	0.4	0	0	1104
P_3	10	8.80	2.4	6334	1950	0
P_4	10	7.81	1.4	3740	0	218
P_5	10	0.21	2.2	5652	3396	0
P_6	10	7.03	2.6	9156	650	0

P_5 was successfully detected by the system to keep tracking the catheter. The most probable reason is the change in the vascular lumen shape, because P_5 was placed away from the acrylic frame holding the silicone model. For this application, the detection range could be increased up to 15 mm to give more time to the robotic camera to set. A larger detection range between P_3 and P_4 will produce ambiguities in M_0.

7.3.5 Integration of the Robotic Camera with EVE

Integrating EVE to the robotic camera enabled to perceive the silicone vasculature on the same way that human vasculature is perceived during an endovascular intervention. Improving the quality of the training system as the user will see the catheter through a video monitor as surgeons does in endovascular surgery. An optional control with joystick enabled to operate the camera as the fluoroscope is manipulated in IVR. But depending on the training purpose an automatic motion of the camera and DSA image generation is desired. As shown in Fig. 7.10, the vasculature is visible in all moments during the simulation procedure. To make the process more realistic is necessary that silicone vasculature is visible only when a contrast media is injected or after the application DSA process.

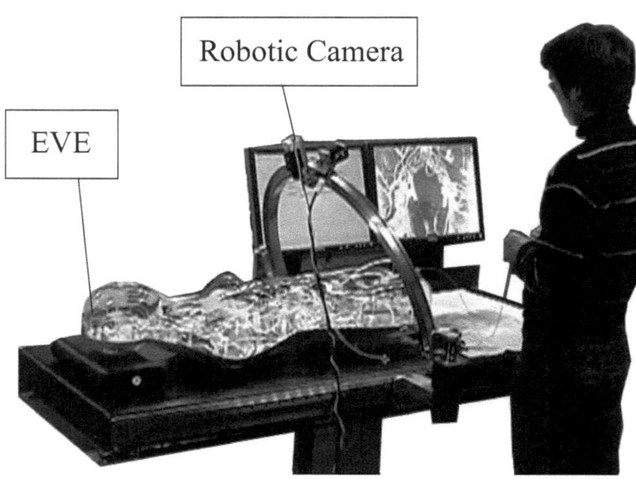

Figure 7.10 Endovascular intervention simulation with EVE and robotic camera.

7.4 Robot Guidance

In order to reduce the use of the fluoroscope in endovascular surgery, there is a need to develop autonomous catheter insertion systems. In this section, we show the catheter motion capture with a magnetic tracker for providing feedback control to a catheter insertion robot. With that feedback, autonomous catheter insertion in silicone models of major vasculature can be achieved as preliminary step for *in vivo* experiments.

Magnetic tracker probes are 0.8 mm × 8 mm diameter and length, respectively, and they have been used in medical instrumentation tracking projects, as needle guidance [2], bronchoscope navigation [3] and navigation in laparoscopic surgery [4]. Different calibration methods for them using mathematical methods have been compared by Kindratenko [5]. Others using optical trackers for a co-calibration between sensors are proposed by [3, 6, 7]. Here, we navigate catheter in major vasculature models by applying the methodology shown at the beginning of this chapter. For catheter driving devices for telesurgery such as [8–11]; feedback of the catheter motion is provided by encoders or accelerometers placed on the joints of the driving mechanism. For those cases, it is not possible to track accurately the catheter tip position inside a membranous structure, such as blood vessels or silicone models, because catheter is not a rigid body and its shape changes according to the blood vessel wall shape and deformation. In this study an Autonomous Catheter Insertion System (ACIS) was evaluated using micro coils placed on the catheter tip for motion capture and feedback accurately the position and speed to the catheter insertion robot LSM. Then with that information and a sparse LUT as reference, catheter insertion path reconstruction was developed.

Controlling the speed of the catheter tip allows detection of catheter trapping, with that vascular membrane puncture could be avoided. An ACIS should be able to reproduce a predefined insertion path to reach a target and to control catheter speed to avoid puncturing the organ membrane.

7.4.1 Evaluation Field

The ACIS was evaluated for path reconstruction and speed control during a catheter surgery simulation inside silicone models major vasculature and ureter silicone model. Arch of aorta was represented in the region 1 of the model; this region was used for speed control experiments and preliminary experiments with the feedback system. Sub-clavian arteries and a segment of common carotids near aorta arch were represented by region 2 of the model; this region was used for a two-dimensional path reconstruction experiment and the speed control experiment. A ureter silicone membranous model was used for a one-dimensional path reconstruction experiment. A segment of a common carotid was represented by region 3 of the model and was used for path reconstruction experiments. The ACIS was evaluated inside two types of silicone arterial models, block and membranous. Those silicone models were built using CT data of human arteries and ureter, allowing a simulation of endovascular or urological catheter insertion inside silicone models. The shape

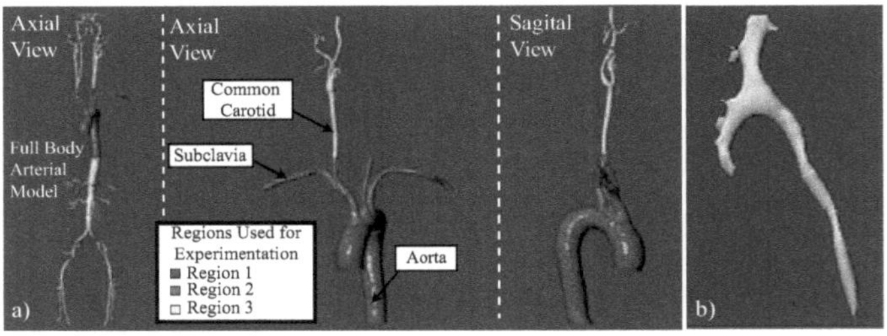

Figure 7.11 (a) CT data of vasculature used for the vasculature models construction, highlighted arteries are the regions where ACIS is evaluated. (b) CT data of ureter used for the silicone model construction.

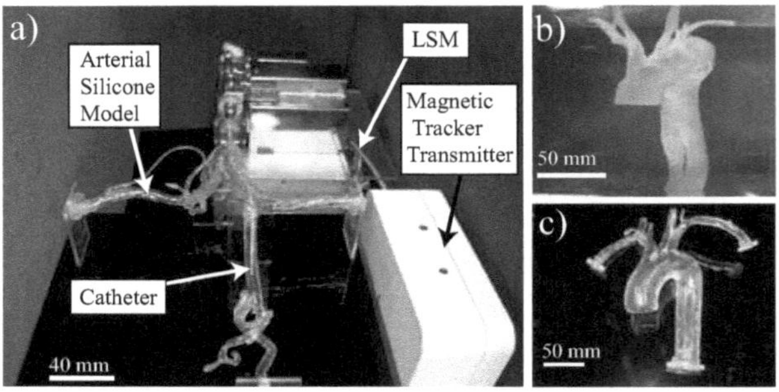

Figure 7.12 (a) Catheter shaped Magnetic Tracker receiver introduced by LSM in silicone membranous model of regions 2 and 3. (b) Silicone block model of regions 1 and 2. (c) Silicone membranous model of regions 1 and 2.

of the silicone block model can be deformed less than 1 mm by the catheter; however, the membranous arterial models admit a more realistic deformation (Figs. 7.11 and 7.12).

7.4.2 Path Planning and Control Software

7.4.2.1 Controller design for ACIS

To apply the studied control method in robot guidance, using the LSM as part of an ACIS, it is needed to design a sparse LUT, M_1, with vectors P'_n.

$$\vec{P}' = (p_{xn}, p_{yn}, p_{zn}, C'_{0n}, C'_{1n}, C'_{2n}, C'_{3n}) \tag{7.7}$$

C'_{0n} is the step length of LSM and refers to the amount of catheter inserted in every cycle by LSM. C'_{1n} indicates the period of the motion of LSM, C'_{2n} indicates to change or not the direction of insertion of the catheter at this point, C'_{3n} indicates the degrees that DC motor will turn to twist the catheter. As spatial points are related to commands to be sent to LSM in M_1, path planning and reconstruction is then possible (Fig. 7.13).

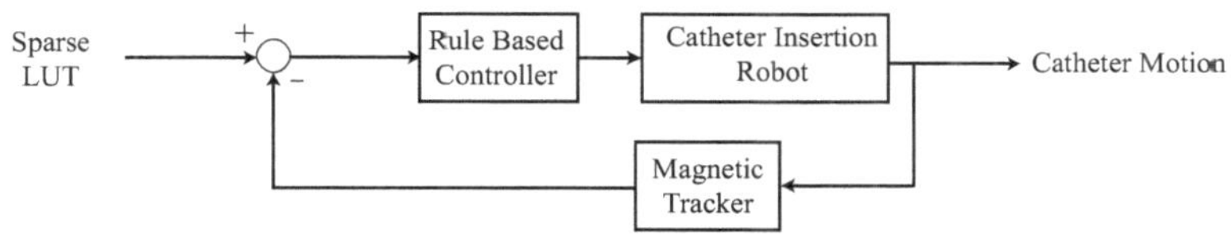

Figure 7.13 Control diagram catheter insertion path reconstruction with ACIS.

For all the experiments with the LSM the detection range varies, as several arterial models are used during experimentation. In those cases, the detection range is set at the arterial cross section diameter to produce a reaction of the LSM as close as possible of the reference points. A software containing the designed rule based controller, interfaces with magnetic tracker and LSM was built. This software allows also collecting points from space with the magnetic tracker, and then to build with them the sparse LUTs to design the catheter insertion paths (Fig. 7.14).

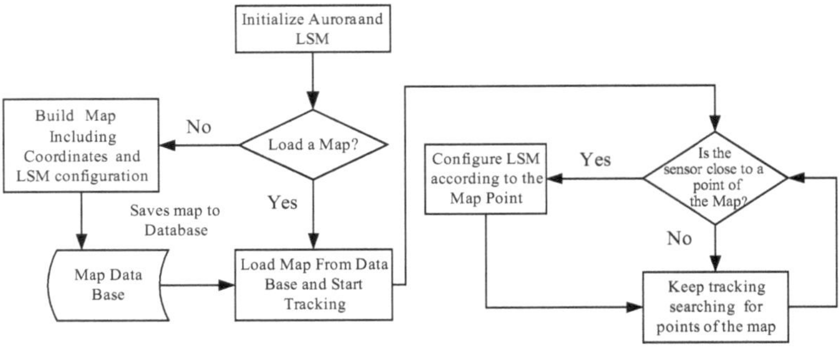

Figure 7.14 Flux diagram of ACIS control software.

7.4.2.2 Magnetic tracker

In some of the experiments, the catheter tip had to been modified in order to obtain better results (Fig. 7.15). For all experiments, a set of reference points was collected using the tracker inside the silicone models. From this set, M_1 was designed containing the planned insertion paths.

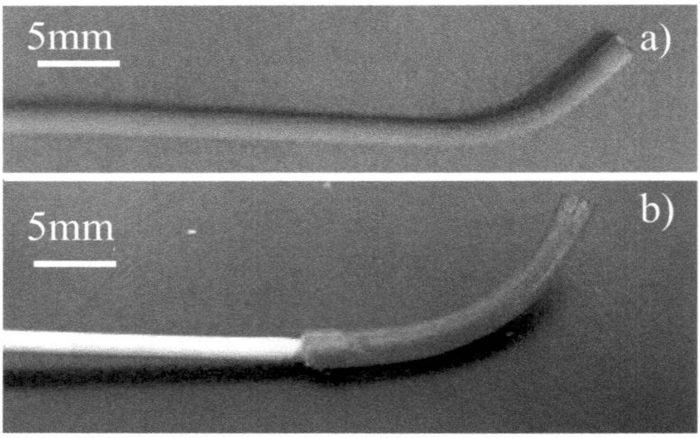

Figure 7.15 (a) Example of a standard 5F catheter. (b) Modified magnetic tracker probe.

7.4.3 Preliminary Experiment for Path Reconstruction

A preliminary experiment was done to evaluate the ACIS reconfiguration accuracy using a sparse LUT built using the model of the artery of regions 1 and 2 as reference. The magnetic tracker was manually inserted along the arterial block model of those regions (Fig. 7.12(b)). Five spatial points were collected during that time. Each collected point was associated with a (C'_{0n}, C'_{1n}, C'_{2n}, C'_{3n}) configuration of the ACIS. The sparse LUT M_1 only modifies the parameter C'_{1n} to change the speed of the catheter insertion. This makes that the catheter trajectory is only modified by the lumen of the silicone model and changes in catheter speed. d_n was set to 20 mm in all reference points, as this is the maximum cross section diameter of the silicone model of vasculature. The magnetic tracker was placed inside the LSM and taken to the entrance of the arterial block model. From the software of the PC the LSM unit and magnetic tracker console were initialized and the feedback system was activated. The system compares then the actual position of catheter with every point of the map using Eq. 7.3. Using this strategy, the ACIS changed LSM's speed along the silicone block model according to M_1.

During the path reconstruction in this configuration, the system changed LSM's catheter insertion speed as specified on M_1 (Fig. 7.16), on a maximum radius of 24.5 mm from each reference point when the specified range was 20 mm. An erratic behavior appeared between P'_0 and P'_1, as their detection range intersected. The system alternated between both defined configurations in this part of the simulation. This phenomena is not desirable as the robot must be guided properly.

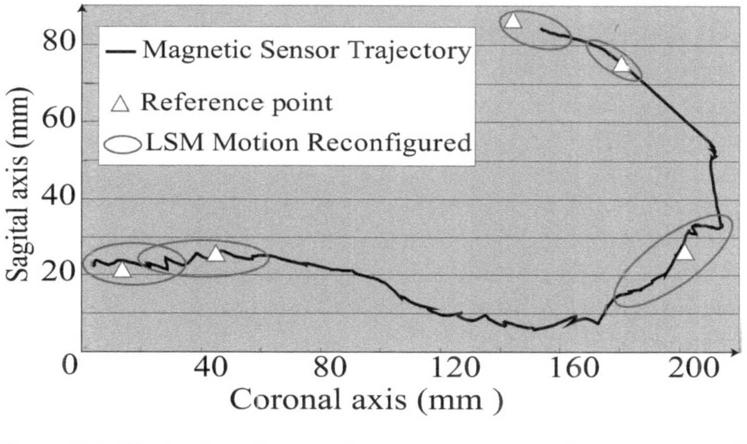

Figure 7.16 Projection of magnetic sensor trajectory during insertion by LSM in arterial block model with reference map feedback.

Then it is deduced that this control could be used to guide a catheter through a predefined path changing the catheter's motion with a sparse LUT. This feedback showed to be useful to reconfigure the system in a neighborhood of a point in space, but its resolution depends on the arterial cross section radius. This can create ambiguities if the points of the map are closer than the cross section diameter of the artery. Therefore a limitation of the system is that the distance between reference points of sparse LUT cannot be less than the cross section diameter of the artery. If two references are closer than a cross section diameter, the system will have more than one configuration in one point of space producing an erratic behavior of the ACIS. In those points, the system may alternate between the defined configurations due to the error introduced by the magnetic tracker.

7.4.4 One-Dimensional Path Reconstruction

A first path reconstruction experiment was done with the purpose to evaluate the effect of the feedback of catheter's tip position on the linear degree of freedom of LSM. Additionally the ACIS's accuracy to change the direction of insertion with respect to a sparse LUT was evaluated. In this case, the sparse LUT M_1, only modifies the parameter C'_{2n} to change the direction of the catheter motion. For that reason, the designed path is said to be one-dimensional. The system was tested inside a silicone model of carotid artery and inside a silicone model of ureter (Fig. 7.17). For the experiment inside the vascular model, four reference points were collected when the magnetic tracker was inserted manually into the straight segment of the silicone model of region 3. From these references a sparse LUT with five vectors was constructed to obtain a trajectory with four directional changes to be reproduced. To reconstruct those trajectories properly, only one reference point is active at the time and the points P'_2 and P'_4 are located in the same coordinates with different configurations.

The detection range d_n was set to 6 mm in all the reference points. The sparse LUT was then loaded on the PC, the magnetic tracker was placed at the start point inside the common carotid model, the magnetic tracker console and LSM unit were initialized by the PC, and the feedback loop was activated. The same methodology was followed for the experiment inside the silicone model of ureter, but the detection range d_n was set to 5 mm in all the reference points. The system compares in both experiments the actual position of catheter with every point of the LUT using Eq. 7.3, the proposed ACIS changes LSM's insertion direction according to the sparse LUT of the arterial model and ureter model.

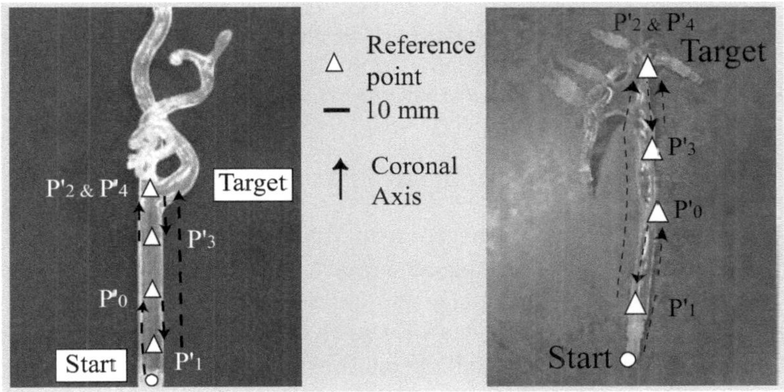

Figure 7.17 Designed one-dimensional trajectory inside silicone membranous model of carotid artery and ureter.

The one-dimensional insertion path was successfully reproduced by the proposed ACIS in both models. The error of the system compared to the first experiment decreased from 24.5 mm to less than 5 mm (Fig. 7.18) showing that the cross section diameter of the model is an important design factor that modifies the error to change system's motion in a specific point of space. In both cases, when the magnetic tracker passes near P'_1 and P'_3 the first time, both points are not active and they did not produced any reaction of ACIS. Also the system reacted in P'_1 and P'_3 when the magnetic tracker passed P'_0 and P'_4, respectively. This kind of configuration could help to detect if the catheter has performed an undesired U turn inside the vascular model, showing that a dynamic sparse LUT that changes the configuration of its components as the trajectory is being reconstructed should guarantee a better path reconstruction and reconstructing more complex paths.

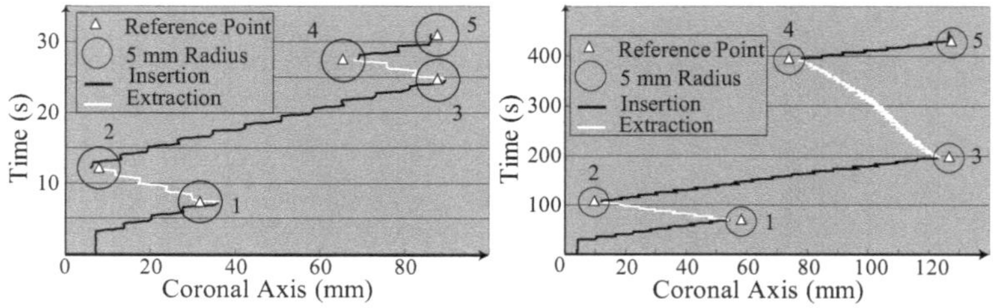

Figure 7.18 Magnetic sensor motion for a one-dimensional path reconstruction with ACIS inside the membranous model of carotid artery (left) and for ureter model, where insertion speed changes are observed at points 3 and 4 (right) [15].

7.4.5 Two-Dimensional Path Reconstruction

It is required that when there is a bifurcation on the path, the system is able to recognize the correct branch, if the insertion path is not being reconstructed properly and go back to the last bifurcation if it identifies that case. To verify those requirements, a trajectory was designed inside an arterial silicone model of regions 2 and 3 (Fig. 7.19). The sparse LUT M_1 only modifies the parameters C'_{2n} to change the direction of the catheter motion and C'_{3n} for twisting the catheter. As only two DOF of the LSM are changed by M_1 to reconstruct the designed path, this path is said to be two-dimensional even if the trajectory is three-dimensional. As all reference points are separated by more than 12 mm, the detection rang d_n was set to 6 mm for both trajectories.

With the model placed near the LSM, the spatial coordinates of each reference point were measured inserting the magnetic tracker into the model. Each set was then associated to configuration C' of the LSM, building that way each vector of M_1. The configuration for each vector was designed with the purpose to control the catheter trajectory inside the silicone model. The computer interfaced the magnetic tracker with the LSM in order to evaluate robot guidance in a closed loop system. In this experiment, LSM moves the magnetic tracker sensor placed on the catheter tip obtaining feedback from it in order to reconstruct automatically a predefined catheter trajectory. The magnetic tracker was placed inside the LSM and taken to the entrance of the arterial model. From the software of the PC, the LSM and magnetic tracker console were initialized. Then, the feedback system was activated. According to the designed trajectory in M_1 the 2 DOF of LSM were automatically modified to reconstruct the trajectories. For this experiment, the trajectory of the catheter was only affected by C'_{2n} and C'_{3n} parameters. The value of the angle of rotation, C'_{3n}, was set manually to 180° and remained fixed during all the experimentation.

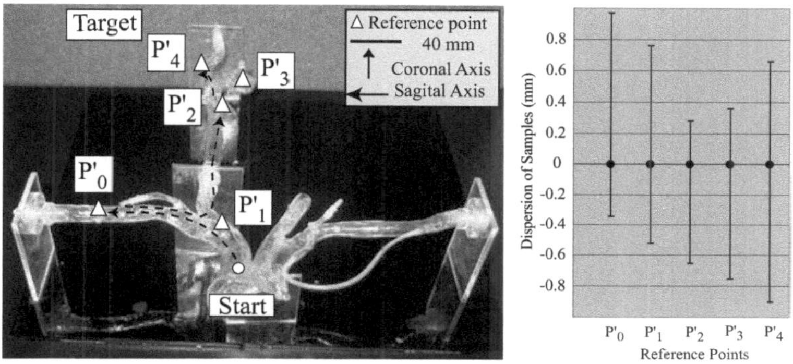

Figure 7.19 Reference points of the planned trajectory (left) with their respective measurement of dispersion (right).

The trajectory is designed to test the branch recognition property of the ACIS, and to show that to ensure a proper path reconstruction, reference points are only needed at least at the target, bifurcations, and marking branches not leading to the target. For this case M_1 includes a target, two bifurcation, and two points marking the branches of vasculature not leading to the target. For this trajectory, the points P'_0 and P'_3 represent a branch not leading to the target. They indicate to the controller to extract the catheter to the last bifurcation. The points P'_1 and P'_2 are the bifurcations of the path. They always indicate a rotation; if the catheter comes from the points P'_0 or P'_3, it indicates a change of direction of insertion too. P'_4 indicates the target of the guidance and stopping the system.

When the catheter shaped sensor was placed in the start point of the trajectory and the feedback loop was activated, the desired path was reconstructed successfully. During the path reconstruction the point P'_3 was not detected, as the rotation was executed properly in P'_2, leading the catheter to the target P'_4. We can observe in Fig. 7.20 that the detection range is set to the maximum for P'_3 and P'_4. Therefore, a specific detection range should be set for each point, in order to set it to the minimum in all cases and avoid erratic behavior of LSM. Detection range is given by the arterial model cross section diameter, because if this limit is not respected there is a chance of missing a reference point during the catheter insertion. Setting the detection range to the minimum for all the reference points, warrantees that reference points placed in different vascular branches will not produce erratic robot behavior.

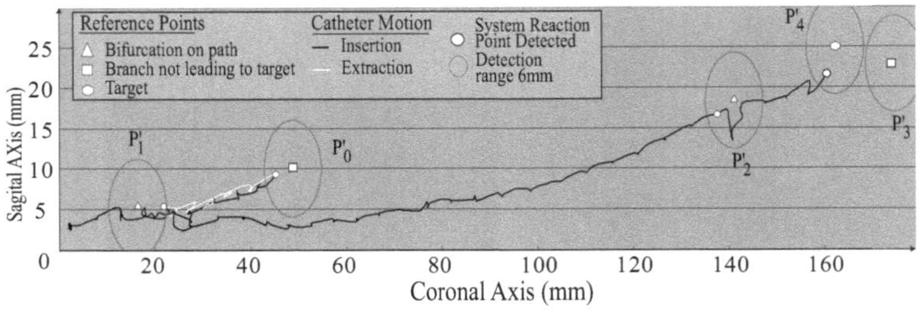

Figure 7.20 Magnetic motion capture sensor trajectory graph for the trajectory reconstruction with ACIS [16].

The detection range d_n is decided when building the sparse LUT, for a probability of 1 to detect a reference point inside a model of cross section diameter d, d_n should calculated from Eq. 7.8:

$$d_n = d - a_c \tag{7.8}$$

where a_c is the catheter cross section radius. As its seen d_n tends to a_c as the cross section diameter of the model is reduced, but this error is the most important one for aorta's model experiment.

LSM's motion of insertion or extraction is not continuous in time. At the one-dimensional path reconstruction experiment, every insertion and extraction cycle is represented on the plot by a step of 6 mm length for all the segments of the path (Fig. 7.18). The length of this step, C_{0n} parameter, affects the accuracy of the system to change the motion when a reference point is detected. Reducing C_{0n} implies a larger time of insertion because less catheter will be inserted per LSM cycle as seen in Fig. 7.18. $C_{0(n-1)}$ is the step length of LSM configured at the point $P_{(n-1)}$. The error introduced by $C_{0(n-1)}$ in P_n is not reduced increasing the sampling frequency of the magnetic tracker because the LSM only allows changing the direction of insertion or twisting the catheter in some specific points of the linear motion cycle without losing synchrony or losing the torsion applied to the catheter, respectively. Therefore, even if several samples are taken during one LSM cycle and the system finds a reference point between them, the LSM will react with a maximum delay of one cycle and a maximum error of $C_{0(n-1)}$ will be introduced to the total error of the ACIS to detect a reference point.

The error introduced by the MMCS, e_m, can be modeled by Eq. 7.9,

$$em = ke_{mt} \tag{7.9}$$

where e_m is the listed error of the MMCS. The parameter k will change according to misalignment of reference frames, distortion of magnetic tracker frame of reference and external interference produced by ferrous metals and oscillating magnetic fields. In this case k will be almost equal to 1 because the control technique proposed at the beginning of the chapter was used to build the controllers. Then the error introduced by the magnetic tracker to the system is minimized to the sensor specifications for position accuracy. As LSM cycle last for one second and changes in its motion are executed once during that cycle, if during that second the Eq. 7.3 is valid then the configuration defined at reference point will be executed at the end of the cycle. Therefore, the system will only be affected inside the detection range, by oscillating magnetic fields with frequency below 0.5 Hz as if the oscillation is faster its magnitude will become zero at least one time during LSM cycle. Therefore, the error of this ACIS at the path reconstruction experiments mainly depends on three factors, the cross section diameter of the model d_n, the step length of LSM $C_{0(n-1)}$, and the error introduced by the sensor e_m. Equation 7.10 is proposed to make a mathematical model of ACIS error to change its configuration in a point P_n of the sparse LUT M,

$$e_A = d_n \pm C_{0(n-1)} \pm ke_m \tag{7.10}$$

where e_A is the total error of this ACIS.

Table 7.4 Detection range and reaction range for the two-dimensional path reconstruction

Reference point	Detection range (mm)	Reaction range (mm)
P'_0	6	4.22
P'_1	6	5.40
P'_2	6	3.96
P'_3	6	—
P'_4	6	2.01

Note: P'_3 is not detected because the catheter insertion trajectory is reproduced properly.

The two-dimensional path reconstruction showed that the proposed ACIS is able to distinguish between two branches in a bifurcation and to pullback the catheter to the last bifurcation to recover the desired trajectory in case it enters to an undesired branch. The parameter C_{3n} define roll rotation of the catheter at the LSM manipulator, the selection of the parameter depends on the shape of the artery and the used catheter tip. This parameter can be set originally based on the motion capture of the catheter inserted by a surgeon inside the silicone arterial model. As observed in Fig. 7.20, during trajectory reconstructions P'_1 had two different configurations. In the first one, if the magnetic tracker detects the reference point no action was taken and when the catheter was coming back from P'_0 it indicates to LSM to twist the catheter and change the sense of insertion of it. Therefore to reconstruct more complex insertion paths dynamic sparse LUTs should be used. When the objective is to correct the distorted frame of reference of the magnetic tracker, the LUT are dense [12]. In this research, a sparse LUT can be used since the micro-coils motion is limited by the walls of the silicone vasculature. In robot manipulation and guidance, the rule-base controller reacted with enough accuracy to control the robotic systems to execute the desired tasks, since the catheter was kept inside the field of view of the camera and catheter insertion trajectory was reconstructed.

The experiments presented in this chapter prove that with an adequate detection range and a sparse LUT that relates the coordinates collected with the magnetic tracker and the commands to be sent to the robot, equations like 7.1–7.5 are enough to achieve control in a restricted space for robot manipulation and guidance. Therefore for reconstructing a catheter insertion path inside a restricted space, such as a silicone vasculature, reference points are needed at least at the following locations: at every bifurcation of the vasculature, at the target to be reached and at all branches not leading to the target. Therefore, this control method may be applied to reconstruct insertion paths inside restricted environments, if bifurcations of the trajectory are not closer than the cross section diameter of the restricted space where the inserted object is traveling. For the system guidance application, controlling the stress produced by the catheter on the vascular model wall is desired to ensure that the catheter stays inside the restricted space. To achieve this, two approaches are possible:

1. To add the feedback of a force sensor on the catheter tip as described [13–14]. The flexibility of the catheter would have to be tested, as well as the effects of this sensor fusion with the magnetic tracker.
2. Another way could be to detect if the catheter is trapped, by controlling the speed of the tip of the catheter, or by using the photo-elastic effect to get feedback of the stress on the vascular wall.

As a first clinical application of the ACIS is to be a medical training tool to store catheter insertion paths done by dexterous surgeons inside silicone models of vasculature. Then the stored knowledge could be reconstructed using the ACIS without the presence of the surgeon, thus providing a new tool for training and knowledge sharing in endovascular surgery. Also such a training system should be used to evaluate the robustness of the system and risks for an *in vivo* autonomous catheter insertion experiment inside ureter or from femoral artery to aorta arch. If sensor fusion is used, the data of the magnetic trackers and intravascular ultrasound probe could be integrated into a system for building the reference trajectory. New techniques to build sparse LUTs should be founded to increase the number of robot control applications of this method. The size of the micro coil of the sensor placed on the catheter tip is a limitation because it cannot be inserted through vascular structures with small cross section diameter and with small curvature radius due to its 8 mm length. If Figs. 7.9 and 7.20 are compared, it can be observed that the LSM moves the catheter in a discrete form compared with the continuous motion when the catheter is inserted by hand. As the insertion path may be reproduced by means of LSM, a similar methodology combined with photo-elastic effect could be used to evaluate the stress produced by different kinds of catheters on the vascular wall. If such experimental setup is constructed, the performance of catheters inside human vascular lumen could be compared using quantitative data.

7.5 Vasculature Imaging Based on Magnetic Trackers and Intravascular Ultrasounds

Real-time three-dimensional vasculature imaging has applications for vascular disease characterization, such as atherosclerotic disease, plaque, and lesion [17], and for providing planning and navigation information for stent graft deployment as well. Conventionally, fluoroscopic images of blood vessels are used for guiding the process of aortic stent graft deployment; for that purpose, a dose 30 mL of contrast media is required along the surgery. However, such technology only provides two-dimensional information of the aortic stent graft position related to the vasculature branches, adding difficulties to the alignment procedure and needs much contrast media for the whole process. Three-dimensional digital subtraction angiography (DSA) technology can provide 3D images of blood vessels; however, it is mainly applied for neurosurgery and requires more than 55 mL contrast media for descending aorta imaging in addition to the 30 mL needed for 2D imaging at the beginning and end of the treatment. Intravascular ultrasounds (IVUS) are used to obtain two-dimensional images of the blood vessel cross sections, providing information of lumen morphology for diagnosis and operation [17–18]. This enables disease diagnosis with less contrast media than 3D DSA, assessment of the interventions, and evaluation of treatments after deploying the stent grafts. During the deployment of stent grafts in descending aorta, the installation of secondary branches of the stent graft consumes a considerable amount of time if the stent graft holes are not aligned properly with the vasculature branches (Fig. 7.21). Such time could be reduced and the effectiveness of the treatment can be increased if the stent graft is navigated to align its holes with the artery branches and active catheters were employed to install the branches.

Sanz-Requena *et al.* made a 3D reconstruction of artery models in [19]. This team captured IVUS images from the video recorded on a VHS tape, and could obtain the distance between cross sections by pulling back IVUS catheter at the speed of 0.5 mm/s and supposed the cross section should be parallel to each other. Therefore, this method can only get one-dimensional depth data of the catheter along blood vessels whose curvatures are relatively small. Other researchers have worked on data fusion of IVUS and angiography [20–22], and they obtained the posture of IVUS tip from angiography images to get the rough localization of IVUS images. Because of the difficulties encountered when trying to track the position of an IVUS transducer using angiography, Godbout et al. proposed a registration approach that relied on a virtual catheter path using a three-dimensional vessel centerline, reconstructed from two calibrated angiograms [22]. However, it is still needed to increase the accuracy for IVUS motion capture and registration.

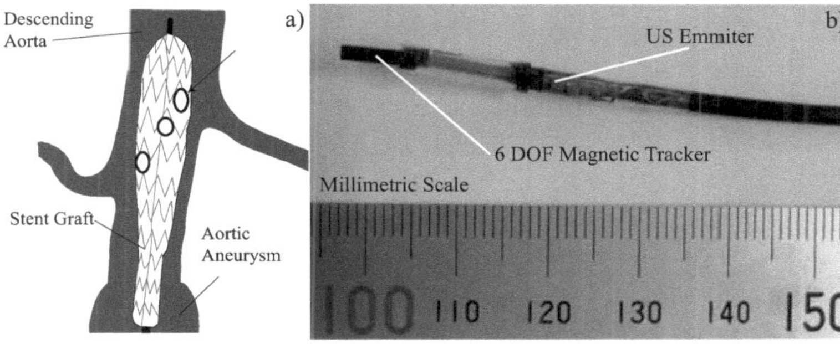

Figure 7.21 (a) Illustration of stent graft in the descending aorta. Vasculature branches and stent graft holes (example pointed with an arrow) must be aligned during the deployment for the graft branches installation. (b) Hybrid probe integrating intravascular ultrasounds with magnetic trackers.

To track the position of catheter tip inside blood vessel models, some researchers installed encoders or accelerometers on the catheter-driving mechanism as source of feedback [24–25]. However, such technology cannot provide precise feedback. Because the catheter is flexible, its shape will be changed easily according to the shape of blood vessel wall and the force exerted on it. For that posture information will be obtained from a magnetic tracker as proposed in [15–16]. In this section, we propose blood vessel 3D imaging relying on the fusion of blood vessel imaging from IVUS and of posture information from magnetic tracker.

7.5.1 Sensor Fusion of IVUS and Magnetic Tracker

An Intravascular Ultrasound Imaging Catheter Visions PV 8.2 (Volcano Corporation, Rancho Cordova, CA, USA) connected to the VH^{TM} IVUS Imaging System were used. The ultrasound frequency is 10 MHz and the detection range is less than 60 mm. NDI Aurora Electromagnetic

Tracking System (NDI Corp., CA), with magnetic sensors with 1.8 mm diameter and 8 mm length, were used for motion capture of the catheter with six degrees of freedom (DOF). The magnetic sensor was installed on the tip of IVUS catheter to obtain its posture. The position precision is 0.30 mm, and that for orientation is 0.20°. After connecting the hybrid probe to the VH™ IVUS Imaging System, a PXC200-AL video capture card was used with sampling rate of 20 frames per second (fps). An additional hybrid probe was built using a 5DOF micro tracker and a Visions PV.018 IVUS (20 MHz) (Fig. 7.21(b)).

7.5.2 Estimation of Disturbance on Magnetic Tracker Measurements

The transducer of the IVUS catheter used in the experiment belongs to the type of multi-element array, which can obtain a 360° cross-sectional image and the ultrasound beam direction and deflection on the display is synchronized [17]. Miniaturized piezoelectric transducers are positioned at the tip to allow small diameter and high-resolution image. Piezoelectric transducers may induce changes on the magnetic field, which may cause dynamic disturbances to the magnetic sensor when close to each other. To quantify the magnitude of this disturbance, measurements with the hybrid probe were done inside a silicone tube filled with glycerin and fixed to a PMMA frame. The disturbance was measured at five locations over a line separated by 20 mm from each other; the measurement was repeated with the silicone tube placed at four different directions in XZ plane. Posture samples from the magnetic tracker were captured during 10 seconds for each location on every direction to measure the dispersion of the samples with IVUS probe enabled and disabled. The disturbance measurements were done with the 5DOF hybrid probe. The dispersion for a single point is shown in Fig. 7.22, the average radius of dispersion r for the four measured directions is shown in Fig. 7.22, r is less than 1.5 mm for all measurements, r increased 1 mm when the IVUS is enabled. This enables imaging of aorta because its diameter is above 20 mm.

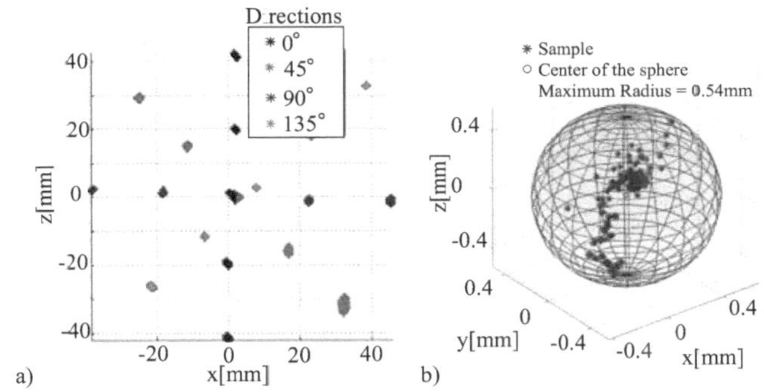

Figure 7.22 (a) Samples dispersion variation along the four studied directions in the XZ plane. (b) Samples dispersion example for one the locations in XZ plane.

7.5.3 Environment for Hybrid Probe Evaluation

To evaluate this imaging method, an *in vitro* environment was constructed by using a silicone membranous model of descending aorta submerged into a water tank. A scan was performed along the descending aorta model at low speed with the 6DOF hybrid probe inserting and extracting the hybrid probe three times. Figure 7.23 shows the membranous silicone model and its original CAD model obtained from the CT images.

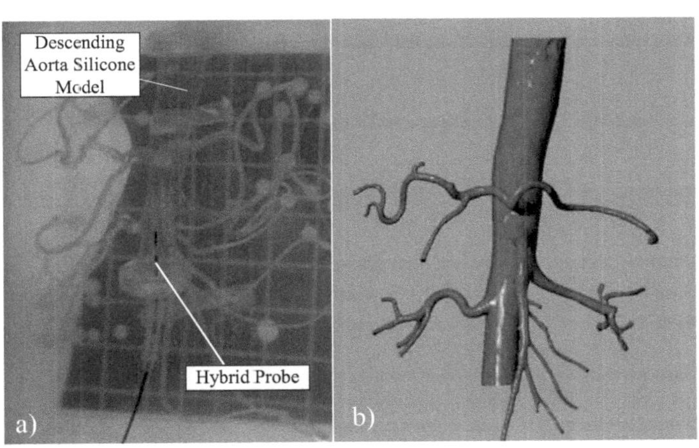

Figure 7.23 (a) Silicone model of descending aorta submerged into the water tank with the hybrid probe inserted. (b) Stereolithographic data used for the silicone model construction.

7.5.4 Image Processing and Kinematics

A sample of a raw IVUS image is shown in Fig. 7.24; the black circular area in the center is the IVUS probe (ultrasounds cannot be detected inside the probe). The image processing was based on Open CV and Visual C++.

The procedure for obtaining the blood vessel contour in 3D is composed of the following steps:

1. Corresponding capture of video frame from IVUS and posture from the magnetic tracker
2. Region of interest (ROI) selection from the raw images
3. Image filtering and conversion to grayscale
4. Catheter center search using Hough circle transformation algorithm
5. Contour detection using Canny algorithm
6. Kinematics for contour posture calculation

Example images for this process and the flux diagram of the image processing algorithm are shown in Figs. 7.24 and 7.25, respectively.

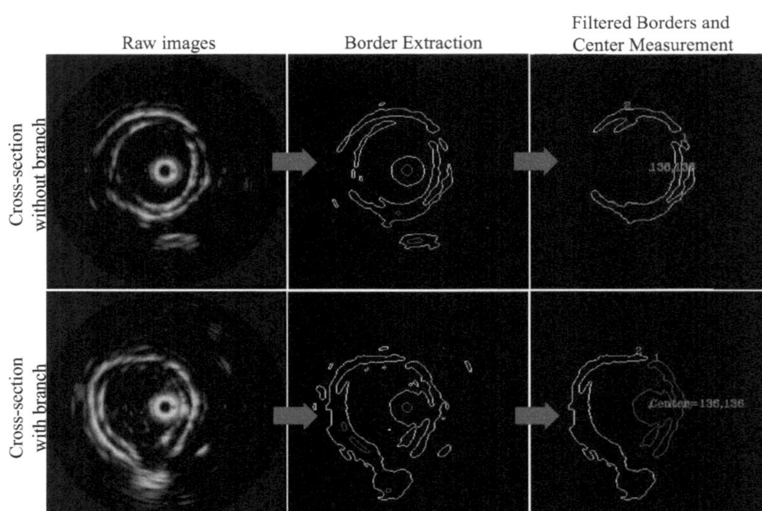

Figure 7.24 Raw image captured from the IVUS and image processing results for: Cross section with and without branch.

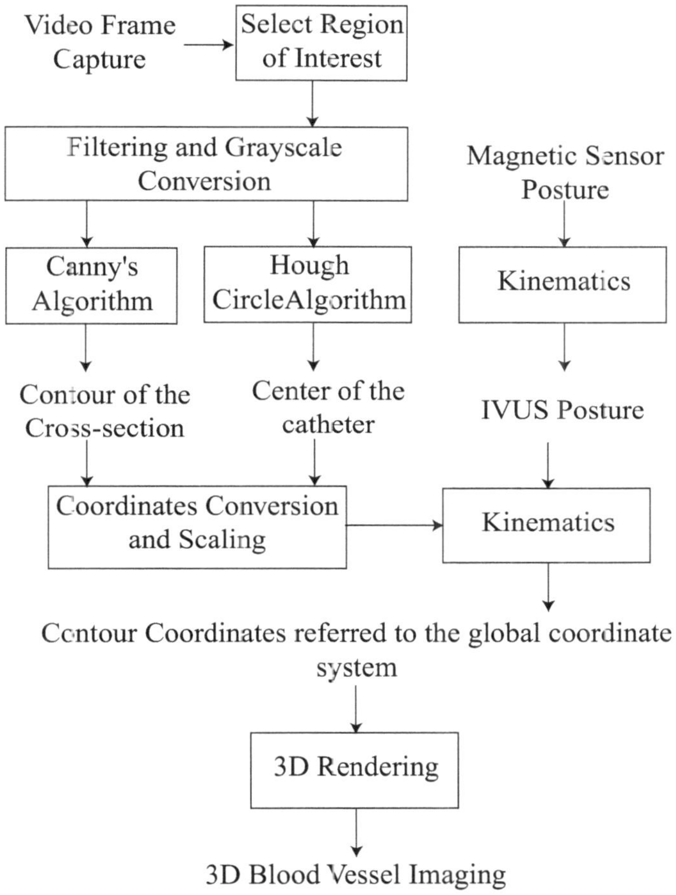

Figure 7.25 Flux diagram for image processing and sensor data integration for vasculature imaging.

As IVUS works in high frequency, therefore a noise reduction filtering should be applied for preserving and enhancing the edges before border detection. For that a median filter with mask size of 5×5 pixels was selected because this filtering is relatively fast. Also a binary threshold is applied to the grayscale images for image segmentation according to the histogram.

After obtaining the contour from Canny Algorithm, the contour coordinates relative the image center can be calculated and expressed with:

$$^{I}P_c(X_P,Y_P,Z_P,1) = {}^{I}P_c(kx,ky,0,1) \tag{7.11}$$

$^{I}P_c$ represents the coordinates of a pixel in millimeters, x and y stand for the coordinates of the contour referred to the image center in pixels. The value of k is 0.217 mm/pixel, which can be obtained from the scale marks on the raw image. This notation enables use these coordinates in the homogeneous coordinate transform for kinematics.

We assume that the two sensors are aligned at all times, sharing the same orientation while moving the integrated sensor slowly inside the silicone model. After obtaining the 6DOF information of the magnetic tracker, we can calculate the posture of IVUS sensor tip, which provides position of the center of catheter that, is known from Hough transform in each image. Figure 7.26 shows the coordinate systems relation among the Aurora emitter, magnetic tracker and IVUS. The Aurora system coordinate can be assumed as the global coordinate. We can calculate the kinematic transformation between Aurora emitter coordinate system and IVUS coordinate system according to:

$$^{O}T_I = {}^{O}T_M \, {}^{M}T_I \tag{7.12}$$

$$^{O}P_I = {}^{O}T_I \, {}^{I}P_C \tag{7.13}$$

$$^{O}P_{MORG} = [T_x, T_y, T_z] \tag{7.15}$$

$$^{M}P_{IORG} = [0,0,-d] \tag{7.16}$$

$^{O}T_M$ stands for the transformation matrix between the magnetic tracker coordinate and the Aurora emitter coordinate and $^{M}T_I$ expresses the transformation matrix between the magnetic tracker coordinate and the IVUS emitter coordinate. $^{O}P_I$ stands for the coordinate of a pixel measured in millimeters in the global coordinate system. $^{O}P_{MORG}$ and $^{M}P_{IORG}$ are the vectors relating the three frames of reference. $^{O}T_M$ is given by

$$^{O}T_M = \begin{bmatrix} 1-2Q_y^2-2Q_z^2 & 2Q_xQ_y-2Q_0Q_z & 2Q_xQ_z+2Q_0Q_y & T_x \\ 2Q_xQ_y+2Q_0Q_z & 1-2Q_x^2-2Q_z^2 & 2Q_yQ_z-2Q_0Q_x & T_y \\ 2Q_xQ_z-2Q_0Q_y & 2Q_yQ_z+2Q_0Q_x & 1-2Q_x^2-2Q_y^2 & T_z \\ 0 & 0 & 0 & 1 \end{bmatrix} \tag{7.14}$$

There are seven parameters T_x, T_y, T_z, Q_0, Q_x, Q_y, Q_z acquired from the posture information of magnetic tracker (T_x, T_y, T_z, are the components of ${}^{O}P_{MORG}$). The first three parameters represent the position, and the others are used to express the orientation by using quaternion's conventions. We need to convert the quaternion information into transformation matrix form [26], and calculate the transformation matrix ${}^{O}T_{M}$. From the previous assumption of sensors alignment, we can define ${}^{M}T_{I}$ with

$$ {}^{M}T_{I} = \begin{bmatrix} 1 & 0 & 0 & 0 \\ 0 & 1 & 0 & 0 \\ 0 & 0 & 1 & -d \\ 0 & 0 & 0 & 1 \end{bmatrix} \tag{7.15} $$

The parameter d stands for the distance between the magnetic tracker and the IVUS emitter (the component in Z_M axis for ${}^{M}P_{IORG}$), which can be measured directly after assembly. After obtaining those parameters, we can obtain the transformation matrix ${}^{O}T_{I}$ and then convert the image coordinate to physical coordinate according to Eq. 7.13.

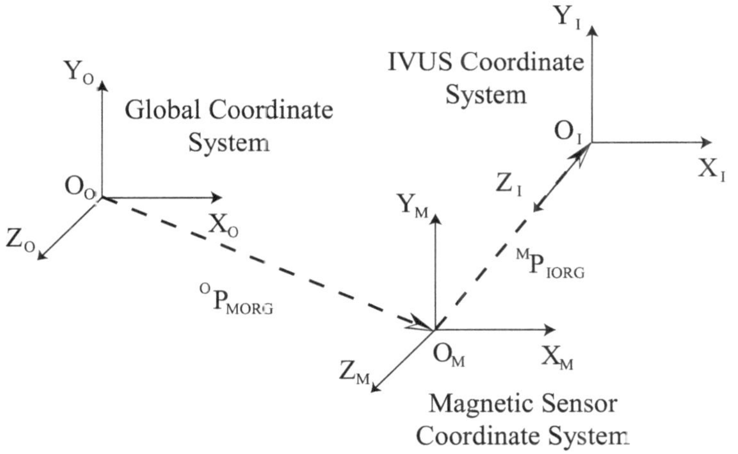

Figure 7.26 Kinematic relation between the frame of reference of Aurora emitter and IVUS probe.

7.5.5 3D Imaging and Rendering

Figure 7.24 shows result images examples for contour extraction process and location of catheter center. Then the contour coordinates are expressed in homogeneous coordinates to be used in the transformation matrix. That way 670 frames obtained from the scan of the blood vessel model were processed. A three-dimensional model was built after assigning 3D coordinates to the pixels identified as part of the blood vessel boundary. Figure 7.27 compares 3D imaging of the blood vessel model constructed using the hybrid probe with the stereo-lithographic data used to build the silicone model.

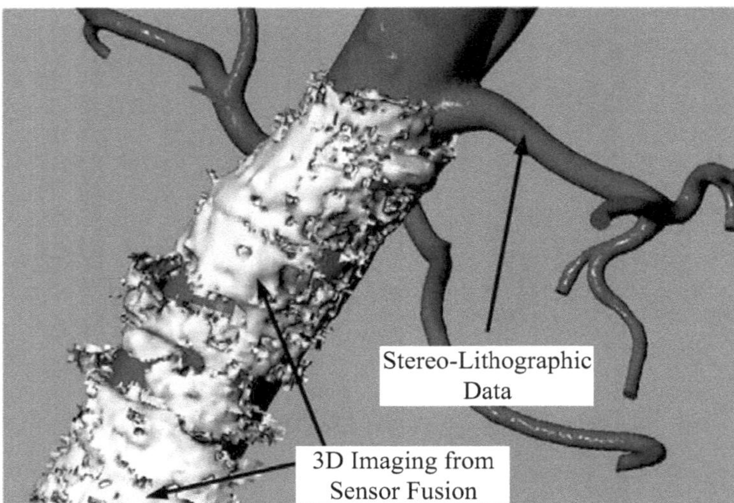

Figure 7.27 3D imaging result for the sensor fusion between magnetic tracker and intravascular ultrasound. 3D rendering for a section of vasculature membrane that corresponds to 670 video frames, is compared with the stereo-lithographic data used for constructing the silicone model.

7.5.6 Error Measurement

For measuring the imaging accuracy, we used a block model of the same vascular structure. The rigid structure of this model enables comparing the 3D imaging of the hybrid probe to the stereo-lithographic (STL) data used for constructing the silicone models. Error was measured by comparing eight points distributed uniformly along respective cross sections of STL data with experimental data. Five corresponding cross sections were selected (Fig. 7.28); the difference between them was averaged at eight locations distributed uniformly around the contour. The maximum error is 1.2 mm and minimum error is 0.735 mm (Fig. 7.29).

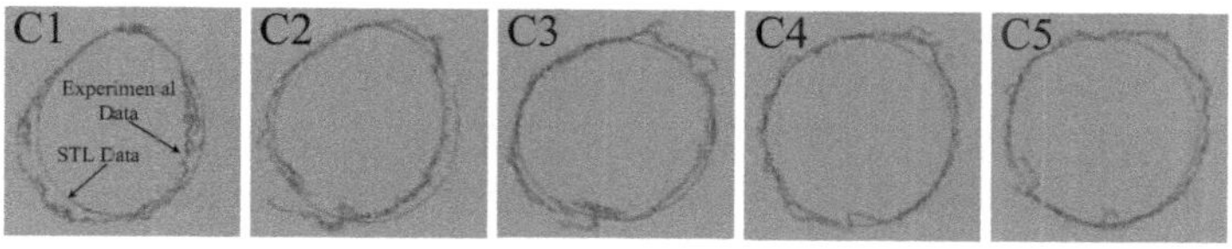

Figure 7.28 Corresponding cross sections of the stereo-lithographic data and 3D imaging data using the hybrid probe, used for error measurement.

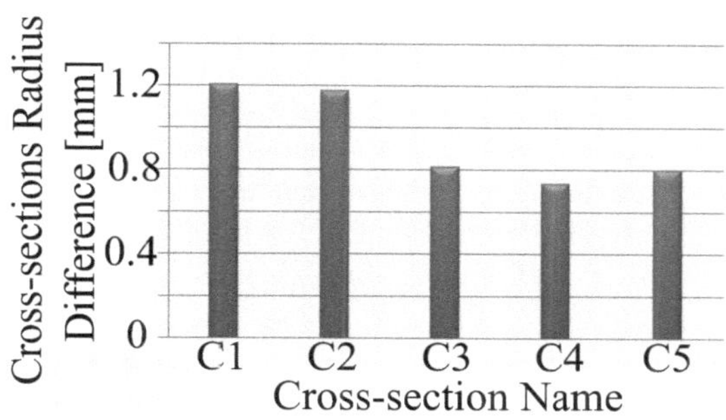

Figure 7.29 Comparison of the average radius difference for the five corresponding cross sections of the stereo-lithographic data and 3D imaging data using the hybrid probe.

References

1. Turski P, Steighorst M, Strother C, Crummy A, Leiberman R, Mistretta C. Digital subtraction angiography road map. *Technical Notes AJR*, 1982, 139: 1233–1234.

2. Krombach G, Mahnken A, Tacke J, *et al.* US-guided nephrostomy with the aid of a magnetic field-based navigation device in the porcine pelvicaliceal system. *J Vasc Intervent Radiol*, 2001, 12: 623–628.

3. Chung A, Edwards P, Deligianni F, Yang G. Freehand co-calibration of optical and electromagnetic trackers for navigated bronchoscopy, in *Proceedings of the Second International Workshop on MIAR*, 2004, 320–328.

4. Ellsmere J, Stoll J, Rattner D, *et al.* A Navigation system for augmenting laparoscopic ultrasound, in *Proceedings of MICCAI 2003, Lecture Notes in Computer Sciences*, 2003, 184–191.

5. Kindratenko V, Bennett A, A survey of electromagnetic position tracker calibration techniques, in *Proceedings of the Eurographics Workshop*, 2000, 13–22.

6. Nakada K, Nakamoto N, Sato Y, *et al.* A rapid method for magnetic tracker calibration using a magneto-optic hybrid tracker, in *Proceedings MICCAI, Lecture Notes in Computer Sciences*, 2003, 285–293.
7. Ikits M, Brederson J, Hansen C, Hollerbach J, An improved calibration framework for electromagnetic tracking devices, in *Proceedings of the IEEE Virtual Reality*, 2000, 63–70.
8. Tanimoto M, Arai F, Fukuda T, *et al.* Telesurgery system for intravascular neurosurgery, in *Proc. MICCAI, Lecture Notes in Computer Sciences*, 2000, 29–39.
9. Arai F, Fujimura R, Fukuda T, Negoro M. New catheter driving method using linear stepping mechanism for intravascular neurosurgery, in *Proceedings of the IEEE-ICRA*, 2002, 2944–2949.
10. Ikeda S, Fukuda T, *et al.* A new remote control device of catheter using linear stepping mechanism for intravascular neurosurgery, in *Proceedings of 20th Annual conference of Robotics Society of Japan*, 2003, 2F17.
11. Jayender J, Patel R, Nikumb S. Robot-assisted catheter insertion using hybrid impedance control, in *Proceedings of the IEEE-ICRA*, 2006, 607–612.
12. Livingstone M, State A. Magnetic tracker for improved augmented reality registration. *PRESENCE: Teleoperators and Virtual Environments*, 1997, 6(5): 14–20.
13. Arai F, Ito M, Kosuga, Fukuda T, *et al.* Concept of intellectual medical treatment support system for active catheter operation. *JSME (Chapter C)*, 1995, 61(587): 372–380 (in Japanese).
14. Tanimoto M, Arai F, Fukuda T, *et al.* Study on micro force sensor for minimum invasive surgery. *Trans Jpn Soc Mech Eng*, 1998, 64(620-C): 150–155.
15. Tercero C, Ikeda S, Uchiyama T, *et al.* Autonomous catheter insertion system using magnetic motion capture sensor for endovascular surgery. *Int J Med Robot Comput Assist Surg*, 2007, 3(1): 52–58.
16. Tercero C, Ikeda S, Fukuda T, *et al.* Robot manipulation and guidance using magnetic motion capture sensor and a rule-based controller. *J Robot Mechatron*, 2008, 20(1): 151–158.
17. Lee JT, White RA. Basics of intravascular ultrasound: An essential tool for the endovascular surgeon. *Semin Vasc Surg*, 2004 (June), 17(2): 110–118.
18. Cardinal M-HR, Meunier J, Soulez G, *et al.* Intravascular ultrasound image segmentation: A three-dimensional fast-marching method

based on gray level distributions. *IEEE Trans Med Imaging*, 2006 (May), 25(5): 590–601.

19. Sanz-Requena R, Moratal D, García-Sánchez DR, *et al.* Automatic segmentation and 3D reconstruction of intravascular ultrasound images for a fast preliminary evaluation of vessel pathologies. *Comput Med Imaging Graph*, 2007 (March), 31(2): 71–80.
20. Wahle A, Prause G, DeJong S, *et al.* Geometrically correct 3D reconstruction of intravascular ultrasound images by fusion with biplane angiography—methods and validation. *IEEE Trans Med Imaging*, 1999 (August), 18(8): 686–699.
21. Slager CJ, Wentzel JJ, Schuurbiers JC. True 3D reconstruction of coronary arteries in patients by fusion of angiography and IVUS (ANGUS) and its quantitative validation. *Circulation*, 2000 (August), 102(5): 511–516.
22. Godbout B, de Guise JA, Soulez G, *et al.* 3D elastic registration of vessel lumen from IVUS data on biplane angiography, in *Proceedings of the 6th International MICCAI Conference*, 2003, 303–310.
23. Godbout B, de Guise JA, Soulez G, *et al.* 3D elastic registration of vessel structures from IVUS data on biplane angiography. *Acad Radiol*, 2005 (January), 12(1): 10–16.
24. Tanimoto M, Arai F, Fukuda T, *et al.* Telesurgery system for intravascular neurosurgery. *Lecture Notes in Computer Sciences*, MICCAI, London, UK, October 2000.
25. Jayender J, Patel R, Nikumb S. Robot-assisted catheter insertion using hybrid impedance control, in *Proceedings of the IEEE ICRA*, 2006, 607–612.
26. Livingstone M, State A. Magnetic tracker for improved augmented reality registration. P*ESCENCE: Teleoperators and Virtual Environments*, 1997, 6(5): 532–546.

Chapter 8

Tailor-Made and Biodegradable Vascular Scaffolds

8.1 Background

Rapid regeneration of injured vascular wall is one of ultimate goals in circulation medicine [1–4]. In most cases, transplant surgery of large or medium-diameter vascular graft has been satisfactory in these days. However, in the case of small diameter artery or veins where the blood flow speed is very low, surgeons still cannot transplant vascular graft with certainty due to low patency rate.

Obstructions of blood vessel at the early stage are due to clot formation, while those at the last stage are due to stenosis caused by the thickening of lumen at inosculation part. In addition, the mismatch of mechanical property between a host vessel and a vascular graft also causes unordinary mechanical stimulation and unordinary tissue growth, and thereby clot or stenosis is formed inside the vascular graft.

When the vascular graft possesses biomimetic structure and function that are very similar to living blood vessel and it can maintain those structure and function, the graft is called to be autonomous. The autonomous vascular graft can coexist with a host living vessel. To create this autonomous graft at the early stage of transplantation, we have to prepare a biomimetic vascular graft *in vitro* and wait for its autonomous regeneration of configuration

Microsurgery: Advances, Simulations, and Applications
Edited by Toshio Fukuda and Carlos Tercero

ISBN 978-981-4364-69-0 (Hardcover), 978-981-4364-70-6 (eBook)
www.panstanford.com

and functions after implantation. To append anti-clot property to a vascular graft, the existence of confluent ECs cultured at stable state along the inner wall is very important. Also, arterial vascular graft must have flexibility and durability to follow the pulsatile motion (70–140 mmHg).

Therefore, the development of vascular graft should contain structural and mechanical design, cell transplantation, tissue regeneration based on cellular and tissue biomechanics (Fig. 8.1).

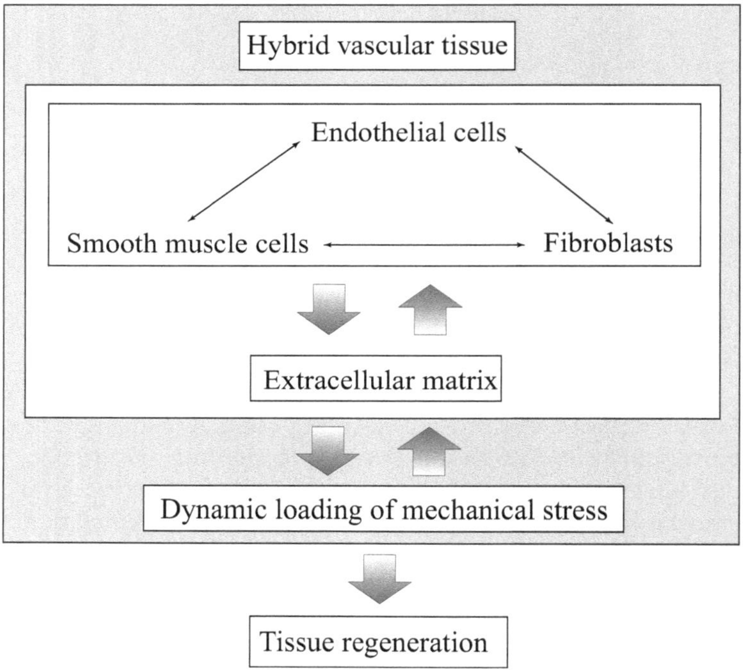

Figure 8.1 Strategy of vascular wall regeneration.

So far, various types of vascular scaffold have been developed. In this section, we introduce several representative vascular grafts. *TERUMO* developed vascular grafts made of polyester fiber or ePTFE. They fabricated grafts with large curvature or bifurcations using biocompatible plastics. However, the vascular wall does not seem to be porous in this image and biodegradable materials are preferred because they disappear as time goes by after transplantation. To fabricate biodegradable and porous vascular grafts, Matsuda *et al.* has been continuing the development of mechanoactive scaffold by electrospinning [5–10]. Electrospinning is suitable for the short-time fabrication of sheet-like or straight cylindrical scaffolds. It is not suitable for scaffolds with complex configurations such as curvature or bifurcations.

Matsuda *et al.* also developed a vascular graft with biomimetic elasticity especially in its pressure–diameter relationship. The main technological feature of this mechanoactive graft is the use of a durable synthetic elastomer (SPU) coupled with laser-ablated microfabrication with CAD and CAM [11, 12]. Circular pores were generated on the surface of SPU tubes by excimer laser (KrF: 248 nm) with various pore densities (wall thickness: 100 μm, length 2 cm, inner diameter: 1.5 mm, pore diameter: 100 μm). The high pore-density tube produced compliance similar to that of human coronaries within the physiological pressure range [13]. They extended this research to the imitation of pressure–diameter property of native artery. Artery indicates unique biomechanical properties: large inflation in the low-pressure regions, gradually reduced inflation in the physiological pressure regions, and little inflation in the high-pressure regions, which is termed the "J curve" in the pressure–diameter plot.

They achieved this property *in vitro* by developing a coaxial double-tubular compliant graft. The coaxial compliant graft was assembled by inserting the high-compliance inner tube into the low-compliance outer tube. By increasing the intraluminal hydrodynamic pressure, the inner tube inflates in the low-pressure regions. After the inner tube made contact with the outer tube, both tubes inflated together gradually in the high-pressure regions. This fabricated coaxial double-tubular graft exhibited the J-curve mimicking that of canine carotid arteries.

Another major way to fabricate biodegradable and porous scaffolds is porogen leaching method using a casting mold. Recently, Gao *et al.* developed a way to fabricate small-diameter vascular graft using PGS [14]. At this moment, we presume that this method has the highest potential to meet all the criteria of (i) biodegradability, (ii) porosity, (iii) complex configuration such as bifurcations, owing to the use of the casting mold.

However, none of previous technologies focused on the precise reconstruction of vascular configuration. As a requirement for scaffolds, it is very important to recreate the configuration of the patient's vascular structure. Vascular scaffolds with biomimetic configuration will fit surrounding vascular tissues at early stage. Recently, our group established a unique method for fabricating 3D elastomeric models of patient-specific arteries to simulate intravascular surgery [15] and to evaluate the stress condition inside a blood vessel [16]. The introduction of these reconstructive techniques into tissue engineering enables the development of tailored biodegradable scaffolds based on CT data of patients. Based on this motivation, in this research, we aim to fabricate biodegradable and porous scaffolds with patient-specific arterial configuration.

8.2 Preparation of Polymer Solution Including Salt Microparticles

High-molecular-weight equimolar PLCL possesses elasticity [7], mechanoactivity [9], and bioactivity [10] and is an appropriate biodegradable material for artificial vascular constructs. Based on these studies, a 5 w% solution of PLCL (molar ratio: 50:50, molecular weight: 4.05×10^5, BMG Inc., Japan) in chloroform (Wako Pure Chemical Industries, Ltd., Japan) was prepared. Then NaCl microparticles (diameter: 90–106 μm) ground with a mortar were added to the polymer solution. The diameter of the NaCl microparticles was maintained between 90 and 106 μm by passing them through two sieves. A 3.3 w% solution of NaCl microparticles in the polymer solution (in chloroform) was used for the fabrication of carotid artery scaffolds, while a 0–8 w% solution in chloroform was used for the tensile tests (Table 8.1).

Table 8.1 Average Young's modulus of scaffolds at different polymer compositions

Porosity	PLCL 3% in $CHCl_3$	PLCL 5% in $CHCl_3$	PLCL 10% in $CHCl_3$
NaCl : PLCL = 0 : 10	2.4 MPa	4.0 MPa	5.2 MPa
NaCl : PLCL = 2 : 8	2.2 MPa	2.6 MPa	3.3 MPa
NaCl : PLCL = 4 : 6	1.6 MPa	1.8 MPa	2.0 MPa
NaCl : PLCL = 6 : 4	1.0 MPa	1.3 MPa	1.8 MPa
NaCl : PLCL = 8 : 2	0.6 MPa	0.8 MPa	MPa

8.3 Fabrication of Carotid Artery Scaffold

Figures 8.2(a–c) show the fabrication process of patient-specific scaffolds. The 3D configuration of the carotid artery (Fig. 8.2(a)) was developed using a method described previously [17]. First, the cervical portion was scanned by radiographic X-ray equipment (64-row CT scanner, helical scan) and the image slices obtained were saved in a personal computer (PC). Each slice image was laminated to create a 3D CAD image of the carotid artery. This 3D image was materialized by rapid prototyping at 13 μm of laminating pitch [15]. To coat the PLCL membrane as uniformly as possible, dip coating of polymer solution onto the 3D model of the blood vessel was repeated three times in one direction and three times in the opposite direction by setting the model upside down (Fig. 8.2(b)). Each coating was followed by drying at room temperature for 2 minutes. The PLCL-coated models then were soaked in deionized (DI) water followed by dissolution of the NaCl microparticles. After the NaCl microparticles were completely removed from the PLCL, the salt-leached PLCL scaffold was dried for 24 hours at room temperature (Fig. 8.2(c)). A digital salinometer (Sekisui Chemical, Ltd., Japan) was used to measure the salt concentration.

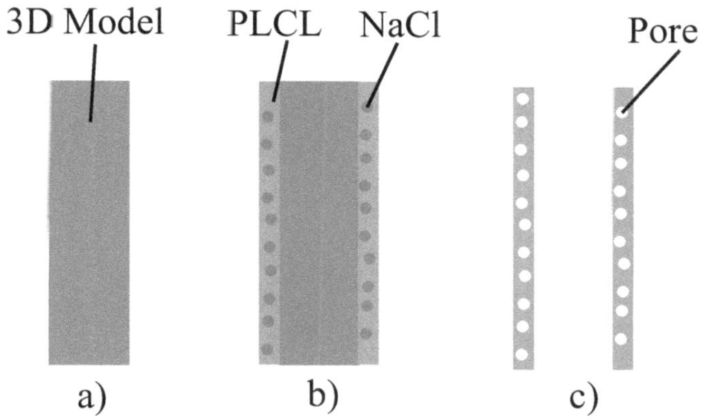

Figure 8.2 General fabrication process of patient-specific, porous and biodegradable scaffold.

Figure 8.3(a) shows the 3D CAD image of a reconstructed carotid artery. From these data, a porous biodegradable scaffold made of PLCL was successfully developed based on the configuration of a patient-specific carotid artery (Figs. 8.3(b) and 8.4). Selective dissolution of the 3D model was successful (Fig. 8.2), in fact, DI water dissolved NaCl microparticles. Figure 8.3(c) shows an SEM image of a porous structure on the surface of a scaffold. Due to the removed salt, the surface of the PLCL scaffold possesses a porous structure. Figure 8.3(d) shows a close-up view of the porous structure. A pore diameter of 100 μm was created due to the NaCl microparticles; interconnected smaller pores also were created. Pore size was less than 100 μm across the scaffold.

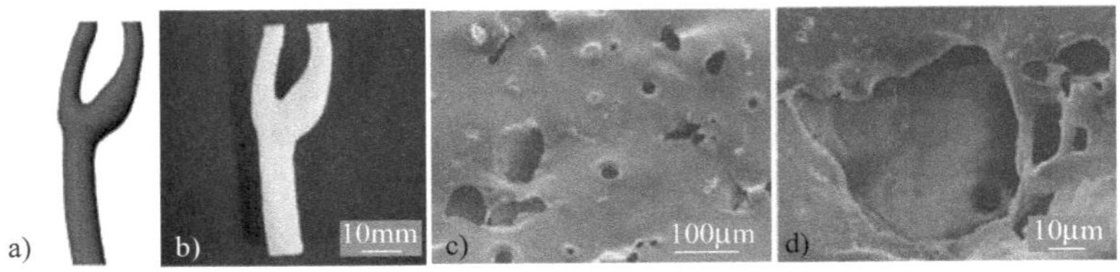

Figure 8.3 (a) CAD image of the blood vessel model. (b) biodegradable PLCL scaffold. (c–d) SEM micrographs of porous structure of the PLCL scaffold.

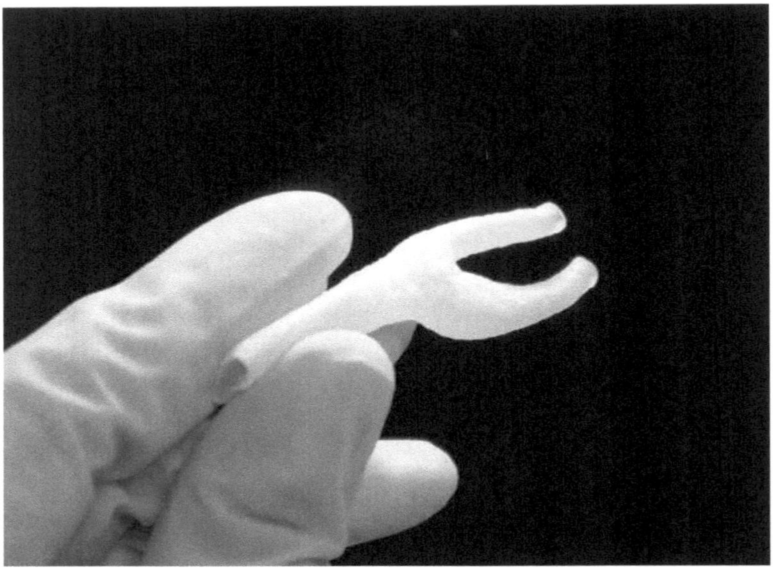

Figure 8.4 Patient-specific, porous and biodegradable vascular scaffold of carotid artery.

8.4 Evaluation of Wall Thickness and Young's Modulus

8.4.1 Measurement of the Thickness of PLCL Membranes after Dip Coating

For simplicity, the models tested were straight cylinders (inner diameter: 5 mm, outer diameter: 5.2 mm, length: 40 mm). During the fabrication of PLCL scaffolds, dip coating of PLCL solution onto the cylinder model was repeated three times in one direction and three times in the opposite direction by setting the model upside down. Besides the fabrication of carotid artery scaffolds, the thickness of PLCL membrane was measured at dip coating by changing the pull-up velocity of cylinder models (1 mm/s, 2 mm/s, 3 mm/s, 4 mm/s) and the PLCL concentration in chloroform (3 w%, 5 w%, 10 w%). The weight ratio of NaCl microparticles to PLCL was fixed for this experiment (NaCl : PLCL = 4 : 6 at w%). After dip coating six times, the thickness of the coated PLCL membrane at the middle of the cylinder models was measured. Models were sliced into rings and the thickness of PLCL membranes was observed using an inverted optical microscope (IX71, Olympus Co. Ltd., Japan).

At each experimental condition, five samples were measured to calculate standard deviation (SD). Overall, 60 ($4 \times 3 \times 5$) samples were tested. Figure 8.5 shows the relation between the thickness of PLCL at dip coating and pull-up velocity and the concentration of PLCL in chloroform ($n = 5$ for each bar, vertical axis: logarithmic scale). This graph indicates the following results: (i) PLCL thickness was controllable from 40 μm to 1200 μm; (ii) a higher PLCL concentration in chloroform yielded a thicker membrane; and (iii) pull-up velocity during dip coating had little effect on the thickness, but higher velocity yielded a thicker membrane.

8.4.2 Tensile Test

Sheet-like rectangular PLCL scaffolds at compositions listed in Table 8.1 were used as samples for the tensile test. The 15 types of polymer solutions were dried on a glass dish followed by leaching of the NaCl microparticles by DI water. The DI water was changed every 4 hours until no NaCl microparticles could be detected. The sheet-like PLCL scaffolds obtained were stretched by a tensile testing machine (stretching velocity: 0.3 mm/s, shape of specimen: JIS Z2201). After measuring stresses and strains of 15 scaffolds, Young's modulus of each scaffold was calculated using

$$E = \frac{P \times l}{\delta l \times t \times d} \tag{8.1}$$

where E, P, l, δl, t, and d are the Young's modulus, stress, length, variation of length, thickness, and width of the sheet, respectively. At each experimental condition, six samples were measured to obtain standard deviation (SD). Therefore, 90 (5 × 3 × 6) samples were tested.

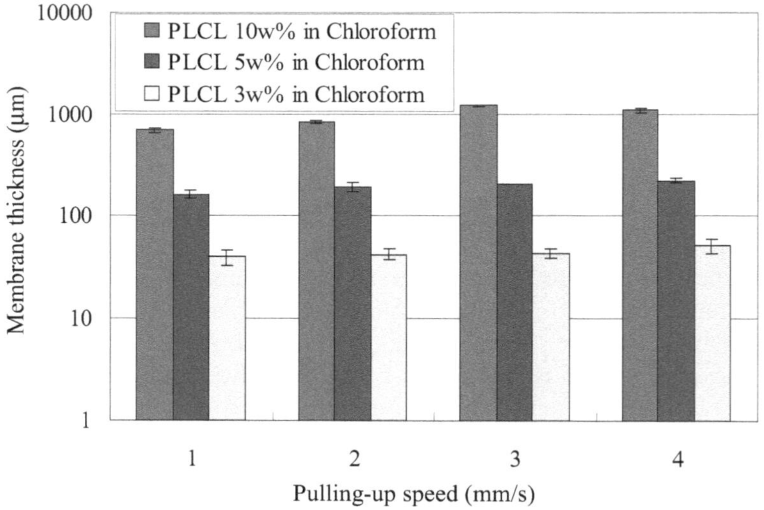

Figure 8.5 Relation between pull-up velocity and thickness of dip-coated PLCL membrane (n = 5 for each bar, vertical axis : logarithmic scale, PLCL : NaCl = 6 : 4 wt%).

Figure 8.6 shows an experimental setup used to measure Young's modulus. A commercial dip coater was modified to be a tensile tester by attaching a load cell. The load cell was connected to a force gage, which was also connected to a laptop. The pull-up motion of a dip coater was used to stretch specimens of PLCL sheet-like scaffolds (pull-up speed: 0.1 mm/s). Time series variation of tensile force was recorded in PC and was subsequently converted into a stress–strain curve. Specimens were stretched until the tensile force reached 5 N.

For porous PLCL scaffolds, we confirmed that Young's modulus depends on the porosity and concentration of PLCL in chloroform. Table 8.1 and Fig. 8.6 show the relationship between Young's modulus and porosity (i.e., the weight ratio between NaCl and PLCL) and the concentration of PLCL in chloroform. These data indicate that Young's modulus for PLCL scaffolds was larger in scaffolds with lower porosity and higher PLCL concentration, and was controllable from 0.6 to 5.2 MPa.

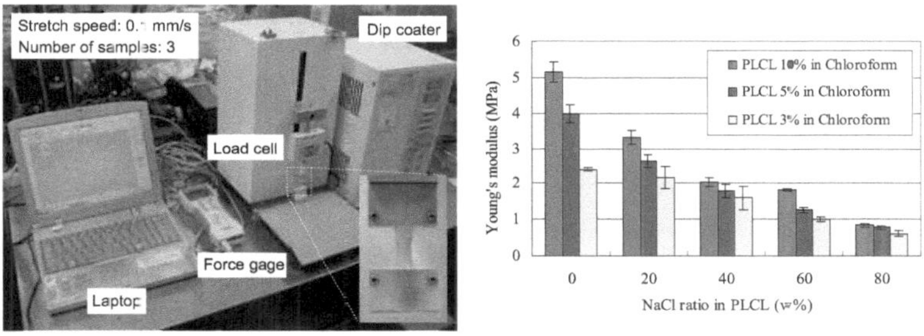

Figure 8.6 Environment of tensile tests by using a dip-coater as a stretching apparatus (Left). Relationship between polymer composition, porosity and Young's moduli of sheet-like scaffolds (Right).

8.5 Spatial Distribution of Pores Inside Scaffolds

We also evaluated the spatial distribution of pores inside four kinds of scaffolds shown in Fig. 8.7. Scaffolds were fabricated from polymer solution (PLCL 5% in chloroform) with four different composition of sodium chloride (NaCl : PLCL = 2 : 8, 4 : 6, 6 : 4, 8 : 2 in weight %). Dip coating of polymer solution on the tube models was operated 12 times (6 times to one direction and another 6 times to inverse direction) at the pull-up speed of 1 mm/s. Based on the schematic shown in Fig. 8.8, the fabricated scaffolds were cut and divided into three parts to evaluate spatial distribution of porosity. Figure 8.9 shows optical images of porous configuration at four kinds of scaffolds. These images show cross-sectional and surface configuration at the middle of scaffold (Part B in Fig. 8.9). Porous structure was confirmed all over the scaffold at every kind of scaffolds.

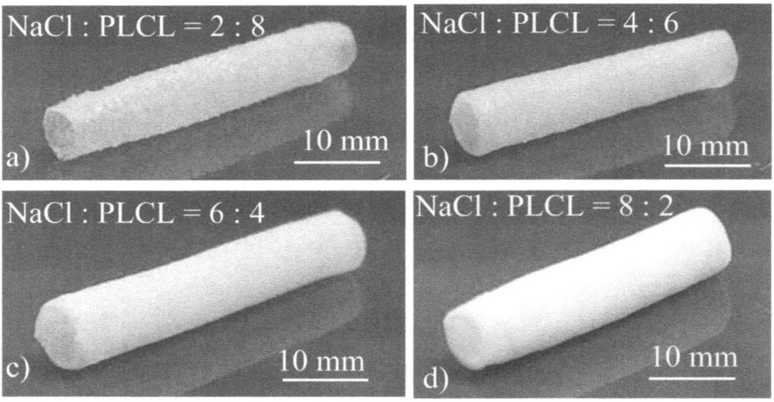

Figure 8.7 PLCL tubular scaffolds with four different porosities.

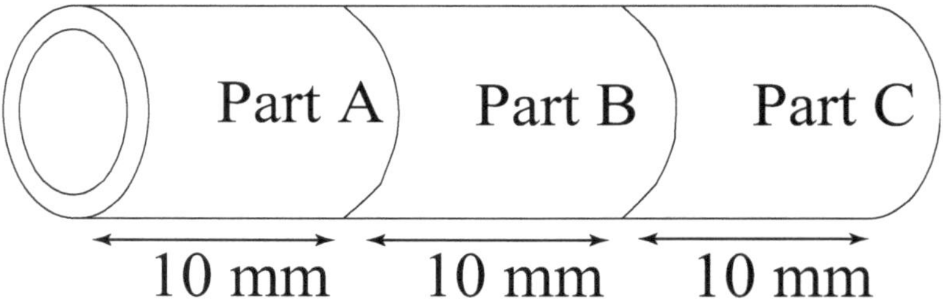

Figure 8.8 Evaluation of porosity at three parts of each scaffold.

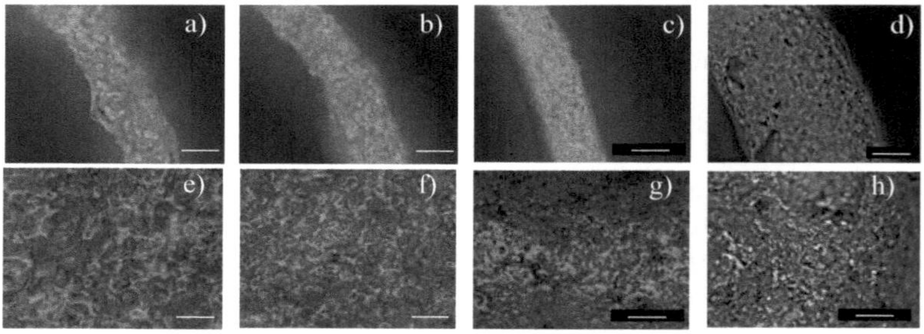

Figure 8.9 (a–d) Cross section images of four scaffolds at part B (NaCl : PLCL = 2 : 8, 4 : 6, 6 : 4, 8 : 2), (e–h) inner wall images of four scaffolds at part B (NaCl : PLCL = 2 : 8, 4 : 6, 6 : 4, 8 : 2). Scale: green bar = 250 μm.

For quantitative measurement of porosity, we determined to use following equation:

$$\text{Porosity}(\%) = (\rho_{\text{PLCL}} - \rho_{\text{d}})/\rho_{\text{PLCL}} \times 100 \quad (8.2)$$

where ρ_d is the density of PLCL scaffolds and ρ_{PLCL} is the density of nonporous PLCL polymer, which was 1.43 g/cm^3 at 26°C. This evaluation method of porosity is quite simple but very reliable in terms of using three-dimensional information of scaffolds (volume and weight).

To calculate the density, it is necessary to calculate the volume of scaffold. First, we measured the wall thickness of each scaffold by an optical microscope. Figure 8.10 shows the distribution of wall thickness at three parts of scaffold. The thickness was measured at four different points and averaged (n = 4). The tendency of thicker wall at higher concentration of sodium chloride was confirmed. The wall thickness changed from 300 μm to 800 μm in these conditions. The average wall thickness was used to calculate the volume of 12 samples (3 parts × 4 compositions).

After the measurement of scaffold volume and weight, we calculated the density of each scaffold at three different parts (parts A, B, and C in Fig. 8.8). Figure 8.10 shows the distribution of porosity at four different polymer composition and three different parts. Good correlation of higher porosity at higher concentration of sodium chloride was confirmed. Porosity changed from 5% to 40% as the NaCl ratio increased. The deviation of porosity depending on parts was less than 15% (NaCl : PLCL = 6 : 4).

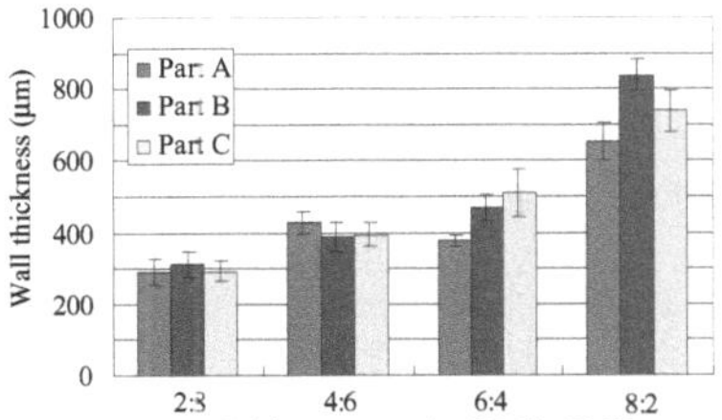

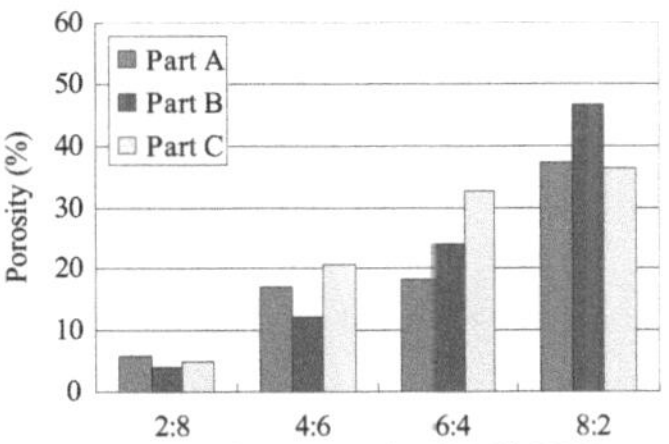

Figure 8.10 (Left) Relation between porosity and wall thickness at three parts. (Right) Variation of porosity according to the increase of NaCl ratio to polymer.

8.6 Cell Culture for Confirmation of Biocompatibility and Safeness of Fabrication Methods

8.6.1 General Cell Culture

HUVECs, growth medium (EGM-2), and subculture kits were obtained from Cambrex Bio Science Walkersville, Inc. (Maryland, USA). The subculture kits are composed of HEPES buffered solution (HEPES-BSS), trypsin neutralizing solution, and trypsin/EDTA. A humidified incubator at 37°C and 5% CO_2 was used for cell culture. HUVECs were cultured on plastic culture dishes in EGM2 containing 2% FBS and other growth factors. Figure 8.11 shows fluorescent image of HUVECs on a tissue culture dish. Nuclei and actin filaments of HUVECs at the same position were observed. Confluent cells were washed in HEPES-BSS and detached from dishes with trypsin/EDTA. These cells were seeded on scaffolds and maintained with the same media.

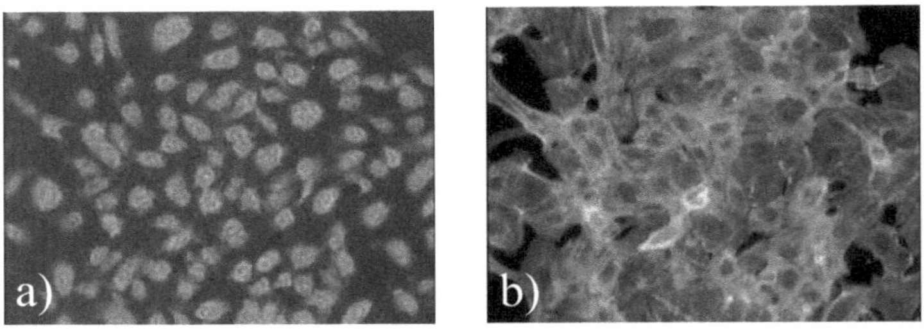

Figure 8.11 Fluorescent micrographs of (a) nuclei, and (b) actin filaments of HUVECs.

8.6.2 Cell Seeding and Culture on Scaffolds

Round sheet-like PLCL scaffolds (diameter = 15 mm) were developed from a polymeric solution (PLCL 5 w% and NaCl 3.3 w% in chloroform, PLCL : NaCl = 6 : 4 w%) in the same manner as previously presented. After ultrasonic washing and treatment with 70% ethanol, the scaffolds were dried. During drying, the growth medium, carbonic acid buffer and HEPES, were added to neutralize the 0.5% type I collagen solution (pH 3, Koken Co. Ltd., Japan). The dried scaffolds were coated with neutral collagen under decompression for 5 minutes. The collagen-coated scaffolds were kept at 37°C for 1 hour to initiate the formation of collagen gel, after which scaffolds were fixed onto the wells of a 24-well plate. Over 4 days, HUVECs were seeded in wells with and without scaffold (on the first and third days, number of cells: 0.5×10^4/well at each seeding, 5% CO_2).

8.6.3 Observation of HUVECs on Scaffolds

An optical microscope (BX62, Olympus Co. Ltd., Japan) was used to observe HUVECs on scaffolds with a fluorescent system including two mirror units (blue and green excitation, Olympus), a conventional broadband illumination source, and an electron multiplying CCD (Luca^{EM}, Andor Technology Plc., USA). The scaffolds with seeded HUVECs were stained after being cultured for 4 days. After cell staining, the scaffolds were exposed to excitation light under the fluorescent microscopic system (objective lens: 40×). SYTO 13 Green-fluorescent nucleic acid stains (Molecular Probes, Inc., USA) and phalloidin–rohdamine X (Wako Pure Chemical Industries, Ltd., Japan) conjugate were used to observe cellular status on the scaffolds. In general, SYTO dyes stain nucleic acids while phalloidin conjugates stain structures with F-actin behavior such as actin fibers. The excitation and emission wavelengths of SYTO 13 dye in the presence of DNA were 488 and 509 nm, respectively. The excitation and emission wavelengths of the phalloidin–rohdamine X conjugate in the presence of F-actin were 556 and 574 nm, respectively (Fig. 8.12).

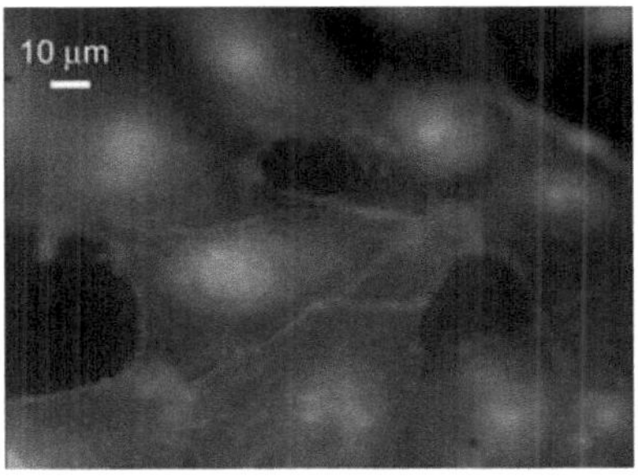

Figure 8.12 Fluorescent micrograph of HUVECs on a scaffold. The nucleic acids and F-actin were stained by SYTO 13 and phalloidin–rohdamine X conjugate, respectively, after cell culture (4 days). Green sites indicate nucleic acids of HUVECs and red sites indicate F-actin.

Results indicate that patient-specific biodegradable scaffolds could be successfully fabricated by the proposed method. Inside the PLCL scaffolds, pores of less than 100 μm in diameter were created as shown in Figs. 8.3(c–d), indicating that NaCl microparticles were successfully leached out.

As shown in Fig. 8.5, the thickness of the PLCL membrane was mainly affected by the concentration of PLCL in chloroform. Because polymer solutions with higher PLCL concentrations have higher viscosity, they are less affected by gravity during pull-up. The thickness of a human artery is on the order of hundreds of microns. Figure 8.5 shows that the thickness of PLCL could be controlled from tens of micron to a millimeter.

Young's modulus for human blood vessel is between 1 and 3 MPa. As shown in Fig. 8.6, the results of tensile tests indicated that Young's modulus for these PLCL scaffolds covers this range. For the fabrication of artificial carotid arteries, a composition of 5 w% PLCL and 3.3 w% NaCl in chloroform (PLCL : NaCl = 60 : 40 w%, Young's modulus = 1.8 MPa) was selected. This Young's modulus is close to that of the human artery. In addition, the trend for a higher Young's modulus at lower porosity and higher PLCL concentration is reasonable. When implanting artificial blood vessels, it is desirable that the artificial blood vessel matches that of the internal blood vessel. Young's modulus of blood vessel walls varies among individuals due to age or artery disease. Thus, the ability to control the configuration and Young's modulus of scaffolds is extremely useful for patient-specific surgery.

The size limitation of this method is the minimum reproduction of 1 mm of inner diameter. However, this coverage is adequate for vascular treatment because most vascular regions treated during surgery have diameters greater than 1 mm. The laminating pitch of rapid prototyping [18] results in rough resolution relative to soft-lithography [19]. However, in scaffold fabrication, a laminating pitch of 13 μm is small enough compared to the diameter of the NaCl microparticles (100 μm). Therefore, rapid prototyping is appropriate for the development of porous scaffolds.

Here, we presented a novel method for scaffold fabrication using rapid prototyping, lost wax, dip coating, selective dissolution, and salt leaching processes. This allowed the successful development of a patient-specific biodegradable scaffold based on CAD data of a carotid artery obtained from CT images. Young's modulus of the sheet-like scaffolds and membrane thickness of the cylindrical scaffolds were measured by adjusting parameters of porosity, polymer concentration, and pull-up velocity. The biocompatibility of scaffolds was confirmed by the growth of HUVECs on the scaffolds.

The proposed fabrication technique has three features that make it attractive:

(i) It can create patient-specific arterial scaffolds with >1 mm inner diameter.
(ii) It can reproduce complex configurations other than simple cylindrical structures, such as bifurcations.
(iii) It can reproduce a precritical original vascular structure from a patient's damaged vascular structure (e.g., we can remove aneurysms or widen an area of stenosis by computer). Although maintaining a uniform thickness of the PLCL membrane is difficult due to the limitations of dip coating, this fabrication method has great potential to reconstruct tailor-made biodegradable scaffolds based on CT images of vascular structure. The relation between pore size and growth of HUVECs (size: 10–60 μm) requires additional investigation.

8.7 Development of Biodegradable Scaffolds by Casting from Magnetically Assembled Sugar Particles

Recent progress in regenerative medicine is receiving attention as an alternative treatment for organ transplantation. One main approach to regenerating tissues is to reproduce an artificial organ *in vitro* first and then transfer it into a patient's body [20, 21]. For *in vitro* preparation of tissues, tissue engineering is very important in terms of developing scaffolds that restore, maintain, and improve functions of damaged tissue [22]. The development of appropriate 3D scaffolds for the maintenance of cellular viability is important and has been a challenge in the field of tissue engineering. While scaffold materials must meet certain biological criteria for implantation, an ideal scaffold should have 3D well-connected macro and micro porous structure to accommodate cells and to facilitate the transportation of nutrient. In general, scaffolds with high porosity and interconnected pores foster tissue regeneration as long as they have appropriate strength.

In previous studies, microscopic and macroscopic 3D porous structures have been fabricated by various methods such as particulate leaching [23, 24], electrospinning [25, 26], multilayer lamination [27, 28], molding [29, 30], 3D printing [31, 32], and phase separation [33]. Above all methods, particulate leaching has the following absolute advantages compared with other techniques:

(i) The size of particulates can be controlled easily by using sieves, thereby ensuring pores with controlled size.
(ii) Particulates can be packed in complex 3D structure, thereby ensuring 3D porous scaffolds with arbitrary shape [34, 35].
(iii) Particulates such as salt or sugar do not solve in general organic solvents and can be removed easily by water.

Two main techniques have been used to achieve well-connected pores and high porosity inside the scaffolds by particulate leaching. Murphy *et al.* proposed a way to improve pore connectivity by salt fusion under moderate humidity and temperature [36]. On the other hand, Wei *et al.* proposed another unique way to enhance the pore connectivity by emulsified sugar spheres [37]. Well-connected pores with uniform diameter should be very useful for effective and fast regeneration of tissue.

However, in addition to all of these previous efforts, self-assembly of micro-scale particulates is also interesting and has great potential to achieve high porosity and pore-connectivity in a scaffold. While self-assembly of spherical particles is very attractive to generate ordered array of pores inside a scaffold [38, 39], magnetic self-assembly is very simple and efficient way to achieve ordered 3D constructs [40, 41]. Here, we propose a way to develop self-assembled templates for particulate leaching by encapsulating ferrite microparticles inside sugar microspheres. The fabricated and self-assembled magnetic sugar particles (MSPs) were utilized as a template for magnetic sugar leaching (MSL). The developed scaffolds after MSL had ordered, well-connected micropores. The mechanical strength of a sheet-like scaffold was evaluated by tensile test and its biocompatibility was confirmed by the viability of human umbilical vein endothelial cells (HUVECs).

8.7.1 Preparation of Magnetic Sugar Particles

Poly(L-lactide-co-ε-caprolactone, PLCL) (C_{LA} : C_{CL} = 52:48, Mw = 540,000) was purchased from BMG (Japan). D-fructose (m.p. 102–

104°C), sorbitanmonooleate (Span 80, surfactant) were purchased from Sigma (St. Louis, MO). Mineral oil and hexane were from Wako Pure Chemical (Japan). Ferrite particles (average size: 5 μm) were kindly donated from Toda Kogyo (Japan). Neodymium magnets were purchased from NeoMag (Japan).

HUVECs and HuMedia EG2 were purchased from Kurabo (Japan). D-PBS, carbonic acid buffer, HEPES were purchased from Wako Pure Chemical. 0.5% type I collagen solution and SYTO 13 were purchased from Koken (Japan) and Invitrogen (USA), respectively.

Magnetic sugar particles (MSPs) as particulates for leaching were prepared by an emulsion technique. Figure 8.13 shows the fabrication process of MSPs. Typically, 60 g of D-fructose were melted at 120°C until clear yellow liquid was obtained. Then 3 g of ferrite particles was added in the solution to append magnetism. The brown-colored liquefied fructose with ferrite particles was emulsified into 50 mL mineral oil with 1.3 mL Span 80 at 120°C under stirring. The suspension was cooled down using an ice-bath with effective stirring to solidify MSPs. After discarding the mineral oil, MSPs were washed in hexane to remove mineral oil and surfactant completely. The diameter of MSPs was controlled to desired range by using sieves in hexane. MSPs were stored in hexane at low temperature (<10°C) to prevent humidity and deformation. In this report, we controlled the size in the range of 53–106 μm and 250–500 μm to demonstrate the controllability of diameter. In addition, we controlled the size in the range of 150–500 μm (MSPs 1), 300–500 μm (MSPs 2) and 425–500 μm (MSPs 3) to demonstrate the controllability of standard deviation in diameter.

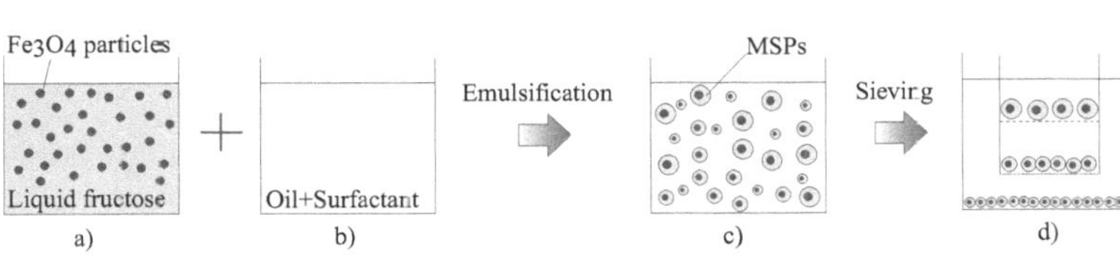

Figure 8.13 Fabrication process of magnetic sugar particles (MSPs) with controlled diameter.

Owing to the activity of surfactant, liquefied D-fructose containing ferrite particles formed emulsion with mineral oil at 120°C. After the emulsification, MSPs were successfully solidified with little damage on their spherical shape. Figure 8.14 shows an optical microscope image of MSPs with controlled diameter suspended in hexane. Ferrite particles were successfully encapsulated in sugar microspheres. Even without external magnetic force, MSPs formed self-assembled structure because each ferrite particle itself has slight magnetic force. The average diameter and standard deviation (SD) of large and small MSPs were 294 ± 41 mm and 69 ± 18 mm, respectively. In addition to the size, SD in diameter of MSPs was also tunable. Figure 8.15 shows optical microscope images of three kinds of MSPs. As shown here, SD in diameter is large in Fig. 8.15(a) while it is small in Fig. 8.15(c). By changing the width of sieving range, degree of variance was controlled successfully.

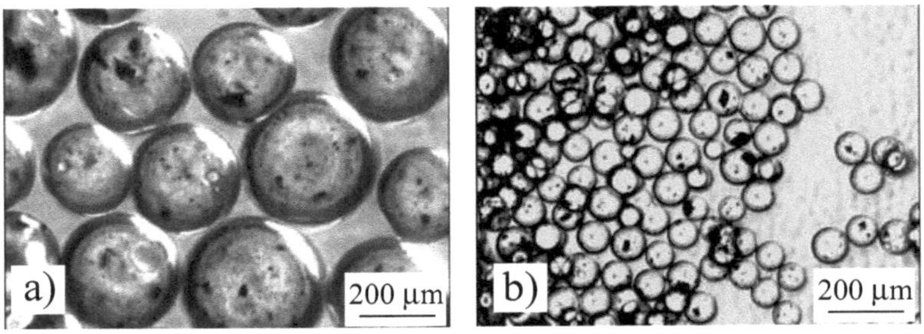

Figure 8.14 Emulsified, solidified and sieved MSPs in hexane (a) large MSPs sieved between the range of 250–500 μm (b) small MSPs sieved between the range of 53–106 μm.

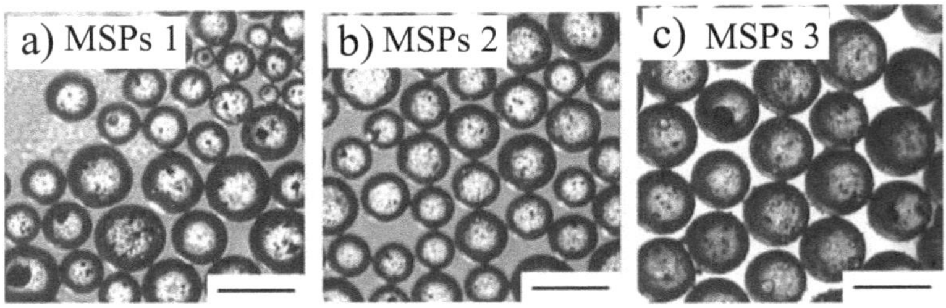

Figure 8.15 Magnetic sugar particles with tunable standard deviation of diameter. Standard deviation is large in (a) while the size is almost uniform in (c). Numerical information is shown in Fig. 8.16.

Figure 8.16 shows numerical information of Fig. 8.15. The average diameter and SD of MSPs 1, MSPs 2, MSPs 3 were 295 ± 105 μm, 369 ± 59 μm, 456 ± 24 μm, respectively. This plot indicates the following results:

(i) Larger sieving range yielded MSPs with larger SD of diameter while smaller sieving range yielded almost uniform size of MSPs.
(ii) The diameter of all the MSPs were truly in the sieving range.

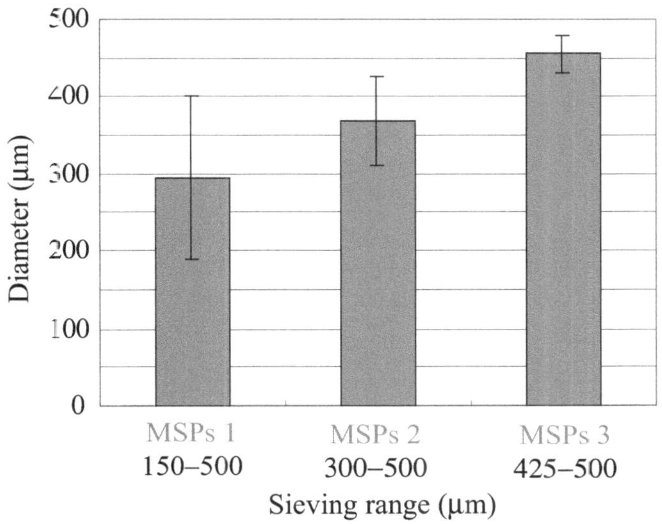

Figure 8.16 Relation between sieving range and diameter of MSPs. The diameter of all the MSPs was truly in the sieving range.

8.7.2 Fabrication of Porous PLCL Sheet-Like Scaffolds by Magnetic Sugar Leaching

Figure 8.17 shows the fabrication process of PLCL sheet-like scaffolds. First, a specially coated silicon wafer was dried in a hot oven (Fig. 8.17(a)). MSPs fabricated in the previous section were then attracted to form an assembly by a neodymium magnet (magnetic flux density: 0.122 T) located under the wafer (Fig. 8.17(b)). Then polymer solution (PLCL 10 w% in chloroform) was dropped on MSPs and vacuum was applied to fill the interspace between each particle (Fig. 8.17(c)). After the complete evaporation of chloroform and removal of the magnet, the composite with silicon wafer was soaked in water to dissolve sugar. Finally, we obtained porous and biodegradable scaffolds with well-connected pores (Fig. 8.17(d)) [42].

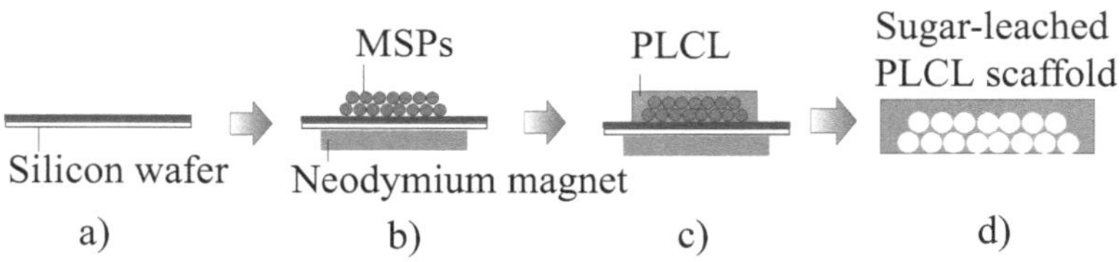

Figure 8.17 Fabrication process of a sheet-like PLCL scaffold.

MSPs were used to fabricate a sheet-like scaffold. Figure 8.18 shows an overview and surface configurations of fabricated scaffold. The self-assembly of MSPs with controlled diameter contributed to generating effective interspaces where polymer solution infiltrated. The vacuum-assisted infiltration of PLCL solution was successful and polymer structure was maintained after whole fabrication process. The porous configuration was visible to naked eyes owing to large pore size. Circular holes observed in Figs. 8.18(b–c) were due to the contact between MSPs and silicon substrate at the stage of magnetic attraction. Pores of less than 50 μm were also scattered sparsely on the surface (Fig. 8.18(c)). These small pores are probably due to the reaction of phase separation among PLCL, chloroform and some residual hexane.

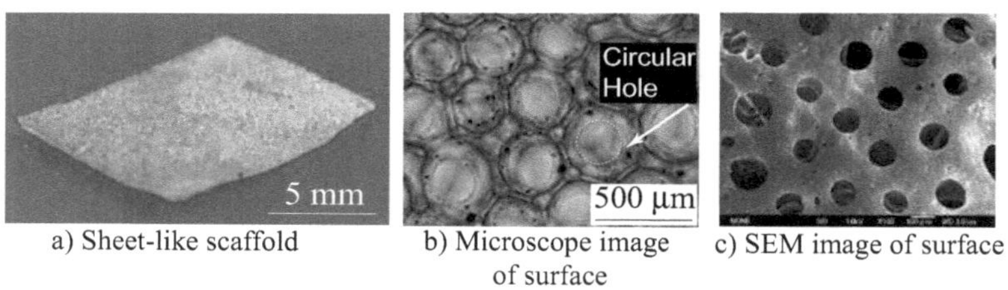

Figure 8.18 Surface configuration of sheet-like PLCL scaffolds captured by optical and electronic microscope.

MSPs 1, MSPs 2 and MSPs 3 shown in Fig. 8.15 were also used to fabricate PLCL scaffolds. In the same way on previous method, those sugar particles were cast to PLCL polymer. Figure 8.19 shows the surface configurations of PLCL scaffolds cast from MSPs 1, MSPs 2 and MSPs 3. The average diameter and SD of pore size is qualitatively very close to those of MSPs.

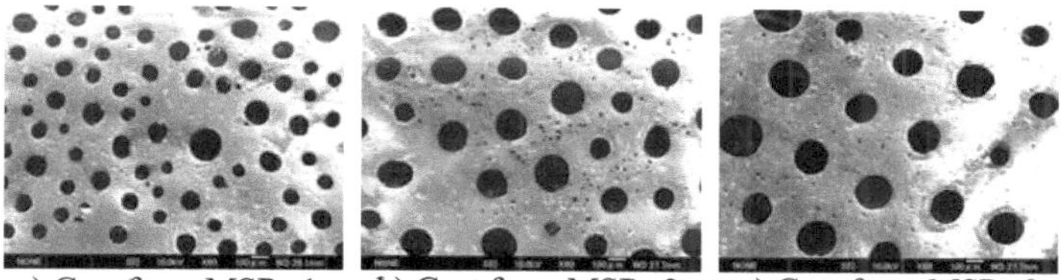

a) Cast from MSPs 1 b) Cast from MSPs 2 c) Cast from MSPs 3

Figure 8.19 Surface configuration of PLCL scaffolds cast from MSPs 1 (295 ± 105 mm), MSPs 2 (369 ± 59 μm) and MSPs 3 (456 ± 24 μm). Small pores less than 20 μm were generated from co-existing phase of PLCL, chloroform, and hexane.

Interior configuration of scaffolds was observed to confirm 3D porous structure by a SEM and a stereomicroscope (Fig. 8.20). Figure 20(b) synthesized from 2D images captured at seven different focal depths using VHX-900. Therefore, the configuration of the scaffold to the direction of depth can be observed very well. Well-connected pores owing to magnet-assisted assembly of MSPs and successful infiltration of polymer can be confirmed from these images.

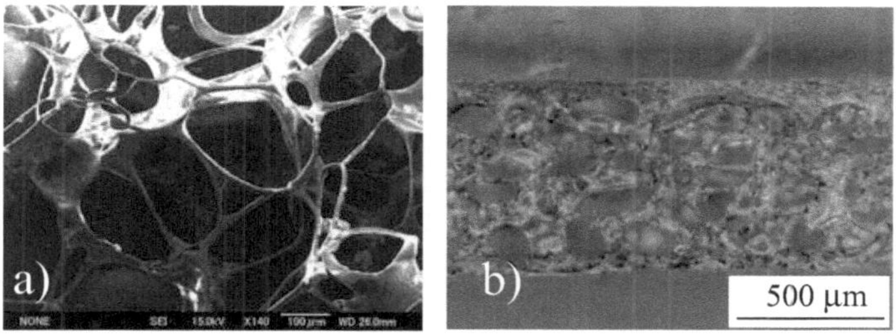

Figure 8.20 Interior configuration of sheet-like scaffolds captured by (a) SEM, and (b) stereomicroscope.

8.7.3 Evaluation of Young's Modulus and Porosity

Strength and compliance of sheet-like scaffolds were evaluated in terms of Young's modulus and porosity. Those parameters are very important for scaffolds to maintain their shape and fit to surrounding tissue.

8.7.3.1 Evaluation of Young's modulus

Tensile tests were performed to measure Young's modulus of sheet-like scaffolds fabricated in the previous section and nonporous PLCL.

The experimental setup described in Fig. 8.6 was used to measure Young's modulus. A commercial dip coater was modified to be a tensile tester by attaching a load cell. The load cell was connected to a force gage, which was also connected to a laptop. The pull-up motion of a dip coater was used to stretch specimens of PLCL sheet-like scaffolds (pull-up speed: 0.1 mm/s). Time series variation of tensile force was recorded in PC and was subsequently converted into a stress–strain curve. Specimens were stretched until the tensile force reached 5 N.

Evaluation of porosity

Porosity of the scaffold was also evaluated in a following way. The apparent density (ρ_d) of the scaffold was calculated from its dry weight and volume. Here, ρ_d was calculated as 0.34 g/cm^3. The porosity of the scaffold was determined according to the following equation [4]:

$$\text{Porosity}(\%) = (\rho_{\text{PLCL}} - \rho_d)/\rho_{\text{PLCL}} \times 100, \tag{8.3}$$

where ρ_{PLCL} was the density of nonporous PLCL polymer, which was 1.43 g/cm^3 at 26°C.

Four types of PLCL sheets were evaluated (PLCL nonporous sheet, PLCL porous sheets cast from MSPs 1, MSPs 2, and MSPs 3). Three samples were tested for each condition to consider SD. Figure 8.21 shows stress–strain curves of PLCL sheets obtained by tensile test. All the curves had inverse J shape that is peculiar to typical elastomers. Elastic (Young's) modulus of biodegradable elastomer is generally estimated by the gradient of initial linear part. We calculated Young's modulus from the strain range of $0 < \varepsilon < 0.1$. Figure 8.22 shows a quantitative data of Young's modulus and porosity for nonporous PLCL sheet and three porous scaffolds cast from MSPs. The Young's modulus of nonporous PLCL sheet was 5.2 MPa. Regarding porous PLCL scaffolds, those cast from MSPs with larger SD of diameter yielded higher Young's modulus and lower porosity. Good correlation of higher porosity and lower Young's modulus supports the credibility of this data. Table 8.2 shows precise values of Young's modulus and porosity shown in Fig. 8.22. The elasticity of the entire porous scaffold was quite close to the elastic modulus of human artery (1–3 MPa).

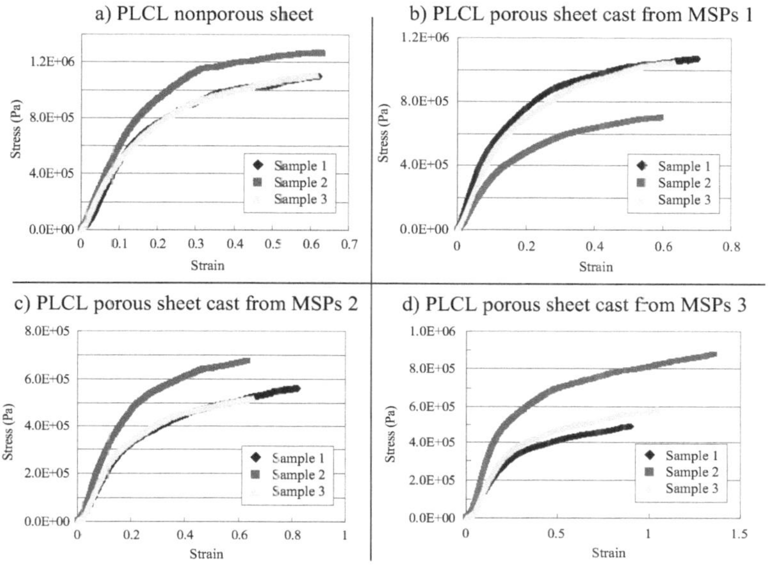

Figure 8.21 Stress-strain curves of four types of PLCL sheet: (a) nonporous, (b–d) porous.

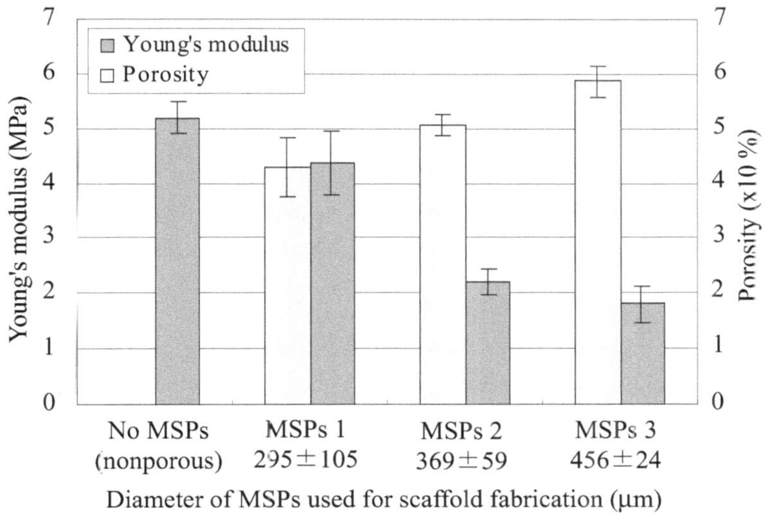

Figure 8.22 Young's modulus and porosity of four types of PLCL sheets. Number of samples: 3.

Table 8.2 Porosity and Young's modulus of each PLCL scaffold

Type of scaffold	Porosity (%)	Young's modulus (MPa)
PLCL nonporous sheet	0	5.2 ± 0.6
PLCL porous sheet (cast from MSPs 1)	42.94 ± 11.08	4.37 ± 1.13
PLCL porous sheet (cast from MSPs 2)	50.75 ± 3.73	2.22 ± 0.47
PLCL porous sheet (cast from MSPs 3)	58.62 ± 6.13	± 0.65

8.7.4 Magnetic Manipulation of Particles

Here, we demonstrate the magnetic manipulation of MSPs using an external magnetic force and a conventional patterning device. Figure 8.23(a) shows a schematic of patterning experiment. A patterning device comprising acryl and steel sheets/wires (Figs. 8.23(b, c)) was placed on a neodymium magnet (magnetic flux density: 0.38 T). Coated silicon wafer as a substrate was placed on the patterning device. MSPs suspended in hexane formed an assembled structure owing to the patterned magnetic field. Figure 8.23(d) shows the analytical result of three-dimensional distribution of magnetic flux density on the patterning device shown in Fig. 8.23(b). A free software "Super Moment" was utilized in the magnetic moment analysis. Figures 8.23(e, f) show assembled MSPs on the steel sheet. In Fig. 8.23(f), three-dimensional assembly of MSPs can be confirmed.

In addition to linear pattern, we also tried to assemble MSPs on a polka-dot pattern. Figure 8.23(c) is a device comprising acryl and steel wires with four different diameters (0.9 mm, 0.7 mm, 0.45 mm, 0.28 mm). The magnetic flux density on this device was calculated as shown in Fig. 8.23(g). The strength of magnetic field was dependent on the diameter of steel wires. Figure 8.23(h) shows assembled MSPs on the surface of patterning device. Polka-dot pattern of MSPs with gradient degree of assembly was accomplished. From these experiments, we confirmed the good correlation between magnetostatic analysis and assembled MSPs. We also confirmed that MSPs can be controlled by external magnetic force.

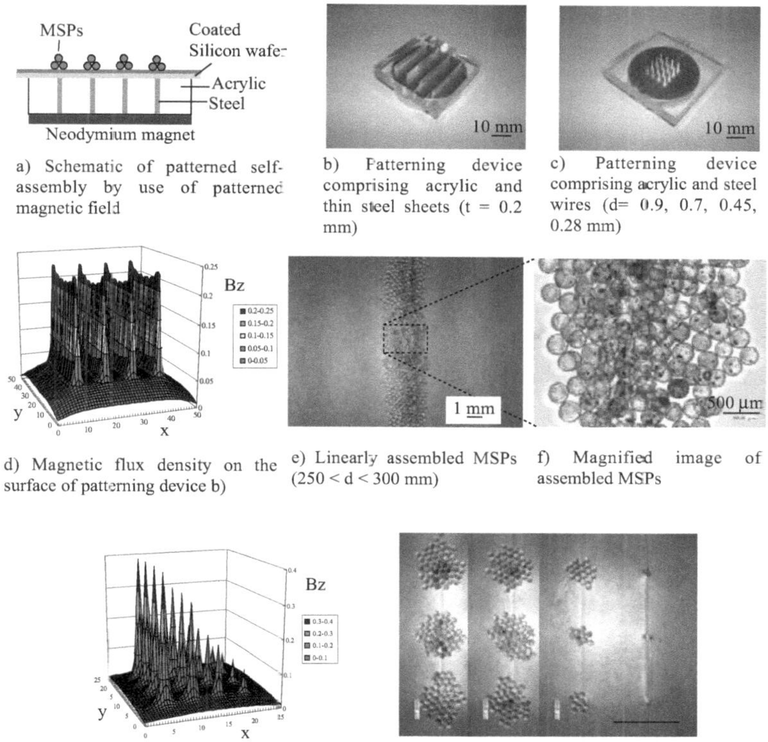

a) Schematic of patterned self-assembly by use of patterned magnetic field

b) Patterning device comprising acrylic and thin steel sheets (t = 0.2 mm)

c) Patterning device comprising acrylic and steel wires (d= 0.9, 0.7, 0.45, 0.28 mm)

d) Magnetic flux density on the surface of patterning device b)

e) Linearly assembled MSPs (250 < d < 300 mm)

f) Magnified image of assembled MSPs

g) Magnetic flux density on the surface of patterning device c)

h) Assembly of MSPs on a polka-dot pattern

Figure 8.23 Magnetic manipulation and patterning of MSPs by use of external magnetic force and patterning device.

By using the linear assembly of MSPs 2 based on the schematic of Fig. 8.23, polymer solution of PLCL was cast on those patterned particles similarly described in Fig. 8.17. We call this combination method of magnetic patterning and sugar leaching as "patterned sugar leaching." Figures 8.24(a, b) show a pore network running to vertical direction from a macroscopic viewpoint. Although there was variance of line width depending on sites, usage of the magnetic patterning device could produce a patterned pore network with the line width of several millimeters. By the combination of magnetic porogens and a 2D patterned magnetic field, the positioning of pores (ϕ > 50 μm) could be controlled two-dimensionally in the order of millimeters.

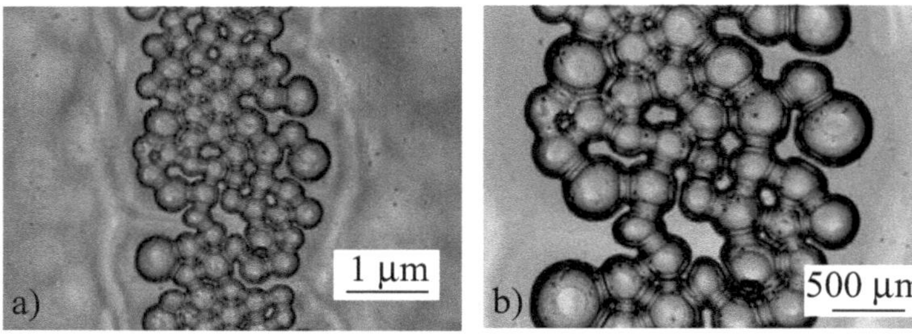

Figure 8.24 Spherical pore network inside a PLCL 2D substrate.

8.7.5 Fabrication of a Tubular Scaffold

Figure 8.25 shows a fabrication process of a PLCL tubular scaffold. A tube with arbitrary wall thickness (inner diameter >1 mm) as a sacrificial model can be prepared as described previously (Fig. 8.25(a)), [42]. Then a PLCL sheet-like scaffold was rolled around the tube with several millimeters' overlap (Figs. 8.25(b–c)). After winding, the boundary of inner and outer sheet was welded by original polymer solution (PLCL 10 w% in chloroform [Figure 8.25(d)]). After the dissolution of sacrificial tube model, we obtained a PLCL cylindrical scaffold with controlled pore layouts (Fig. 8.25(e)).

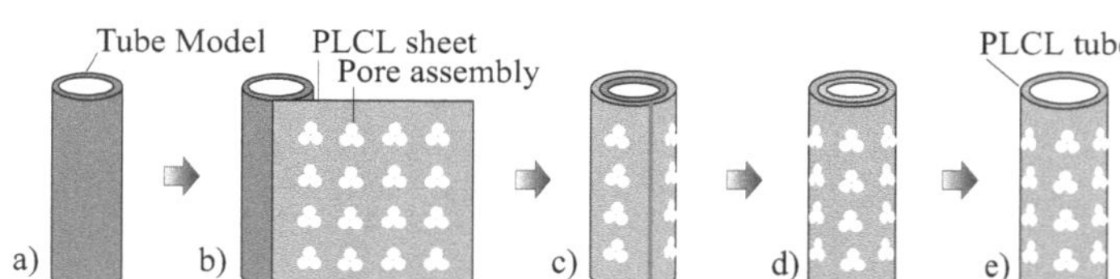

Figure 8.25 Fabrication process of a tubular scaffold (a) Tube model was obtained by lost-wax method (b–c) PLCL sheet with pores was rolled around the tube model. (d) The edge line of PLCL sheet was welded by original PLCL solution. (e) After the dissolution of the tube model, PLCL tubular construct with porous configuration was obtained.

Sheet-like scaffolds of the previous section were used to fabricate tubular scaffolds. Scaffolds were rolled around the tube models and the edge line between inner and outer layer was welded by original PLCL solution. After drying and dissolution of the tube model, tubular scaffolds were successfully obtained as shown in Fig. 8.26.

We chose PLCL in terms of bioactivity [43] and mechanoactivity [44] for future application in vascular tissue engineering. Initially, we tried to assemble MSPs around the cylindrical surface by using a cylindrical bar magnet. However, theoretically it is almost impossible to generate a uniform magnetic field all over the surface, especially if its axial length becomes longer than 1 cm. When we consider the application in vascular tissue engineering, this method of winding a sheet with use of MSPs will be the better way to control the pore layout and porosity inside a small-diameter vascular graft.

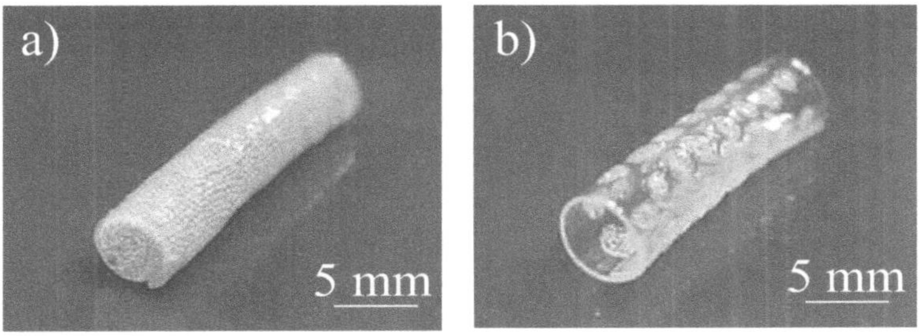

Figure 8.26 (a) PLCL tubular scaffold with porous structure all over the inner surface (b) PLCL tubular scaffold with polka-dot pattern of pores.

So far, straight cylindrical vascular grafts have been fabricated by electrospinning [25, 26, 43, 44] or excimer laser ablation. In the case of electrospinning, it is very difficult to control pore layout to the axial direction. In the case of excimer laser ablation, they spent 8 hours to fabricate a 2 cm-long vascular graft. Compared with these methods, our method can create patterned porous configuration in 2D plane by use of patterned magnetic field and 2D sheets can be easily rolled to tubes in a short time regardless of tube size (total fabrication time <4 hours). The maximum axial length and minimum inner diameter available by this method are 4 cm and 1 mm, respectively.

8.7.6 Cell Culture for Confirmation of Biocompatibility and Safeness of Fabrication Methods

8.7.6.1 General cell culture

A humidified incubator at 37°C and 5% CO_2 was used for cell culture. HUVECs were cultured on plastic culture dishes in Humedia-EG2. Sub-confluent cells were washed in PBS and detached from dishes with trypsin/EDTA. These cells were seeded on scaffolds and maintained with the same media.

8.7.6.2 Cell seeding and culture on scaffolds

Sheet-like PLCL scaffolds were developed in the same manner as described using the MSP. After oxidization and washing with 70% ethanol, the scaffolds were dried. During drying, HuMedia EG2, carbonic acid buffer and HEPES, were added to neutralize the 0.5% type I collagen solution. The dried scaffolds were coated with neutral collagen under decompression for 5 minutes. The collagen-coated scaffolds were kept at 37°C for 1 hour to initiate the formation of collagen gel, after which scaffolds were fixed onto the wells of a six-well plate. Over 4 days, HUVECs were cultured in wells with scaffolds. (Cells were seeded on the first and third day. Number of cells: 2.67×10^5 /well, 5% CO_2)

8.7.6.3 Observation of viable cells on scaffolds

An inverted optical microscope (IX 71, Olympus Co. Ltd., Japan) was used to observe HUVECs on scaffolds with a conventional fluorescent system. The scaffolds with seeded HUVECs were stained after being cultured for 4 days. After cell staining, the scaffolds were exposed to excitation light under the fluorescent microscopic system (objective lens: 10×). SYTO 13 Green-fluorescent nucleic acid stains was used to observe cellular status on the scaffolds. The excitation and emission wavelengths of SYTO 13 dye in the presence of DNA were 488 nm and 509 nm, respectively.

Figure 8.27 shows a fluorescent micrograph of nucleic acids of HUVECs on a scaffold. White sites indicate nucleic acids of HUVECs. Because the sheet-like scaffold has 3D porous structure, some cells were observed out of the focal plane. Although not as clear as SEM images, contours of pores can also be observed from this image. Black sites around the center of the figure indicate residual ferrite particles. We believe these ferrite particles have little damage on cell growth and morphology because some groups have been encapsulating ferrite particles inside living cells [40, 45]. From this image showing coherent and viable HUVECs all over the scaffold, at least we can say that that scaffolds developed by our fabrication method is safe for cell culture.

The functionalization of porogen will be very useful to expand and break through the fabrication range of biodegradable scaffolds and biological substrates. In this chapter, we proposed a novel method to develop porous biodegradable scaffolds by use of magnetic sugar particles (MSPs) as novel type of porogens.

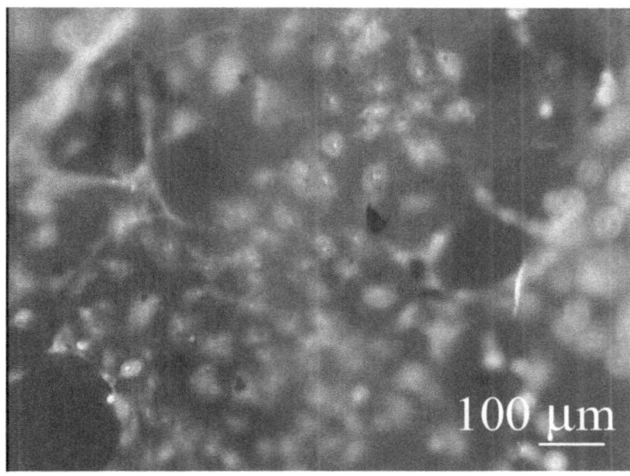

Figure 8.27 A fluorescent micrograph of HUVECs attached on a sheet-like scaffold. White sites indicate the location of viable HUVECs. The diameter of MSPs used for the scaffold varies from 300–500 μm (MSPs 2).

MSPs were successfully manipulated and assembled to form a casting template by external magnetic force. The diameter and its standard deviation (SD) of MSPs were highly tunable by using sieves in hexane. Furthermore, the diameter was truly in the range of mesh sizes of used sieves. Those diameter-controlled particles were cast in PLCL and pores with controlled size were successfully connected and generated inside scaffolds. Young's modulus and porosity as important parameters for scaffolds were also tunable and good correlation between those parameters were confirmed. We believe that MSPs will be important tools for the fabrication of biodegradable scaffolds with controlled pore size, pore layout and total scaffold porosity. Fabrication of patterned spherical pores on a substrate is very difficult to achieve by previous lithographic techniques even if the substrate is 2D. The advantages of using magnetic porogens are that (i) porosity and pore layouts in scaffolds can be controlled by patterning the magnetic field and (ii) magnetic assembly ensures well-connected pores inside scaffolds.

8.7.7 MSP Steering Principle, Modeling and Evaluation

8.7.7.1 MSP magnetically steering principle

Three-dimensional, *in vivo* remote magnetic manipulation of small object have been demonstrated in cardiac catheterization, active exploration of human gastrointestinal tract and for delivering drugs, hyperthermia and other therapies within biological systems [46–49]. Here, a combined structure of Maxwell coils and Helmholtz coils is used to generate gradient magnetic field at arbitrary direction in Y-Z plane of working area. As the central area of Maxwell coils is a low field area, where magnetic torque on MSP can be neglected but the intensity of magnetic field increases and the direction changes when the MSP is away from the central area, this may result in unnecessary rotation on MSP and malfunction of actuating MSP moving to the opposite direction. Thus Helmholtz coils is introduced to add a uniform magnetic field to augment the magnetic field intensity in target area. For achieving a robust system to realize effective controllability of MSP in hexane, a magnetic field generator structure combined Maxwell coils and Helmholtz coils is proposed as shown in Fig. 8.28. Coils with the radius of R1 and R2 are Maxwell coils and Helmholtz coils, respectively, with the axial direction aligned with Y axis. Similarly coils with the radius of R3 and R4 are Maxwell coils and Helmholtz coils, respectively, with the axial direction aligned with *Z* axis. Arbitrary direction gradient magnetic field can be achieved at Y-Z plane by properly configuring the input current for each coil.

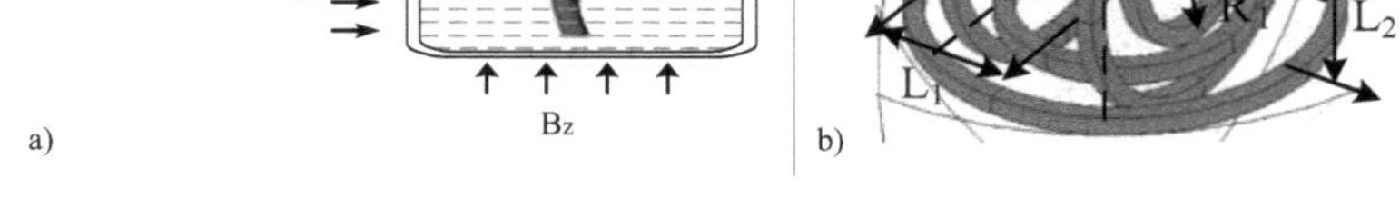

Figure 8.28 (a) Schematic of magnetic steering system for MSP. (b) Combined structure of Maxwell coils and Helmholtz coils for 3D manipulation of MSP.

A magnetic force will be obtained when magnetized MSP exposed to the magnetic field. Since MSP is generally stable in hexane, the whole motion control process is conducted in hexane solution. MSP's gravity force, viscous resistance and dynamic pressure force between the front part and rear part of the MSP, buoyant force are the main contributors influencing the motion of MSP in hexane. Magnetic force is introduced by the combined coils to balance the gravity force, overcome the drag force from the hexane and drive the MSP moving into the desired position toward the scaffold. Figure 8.29 shows the schematic of MSP moving in hexane.

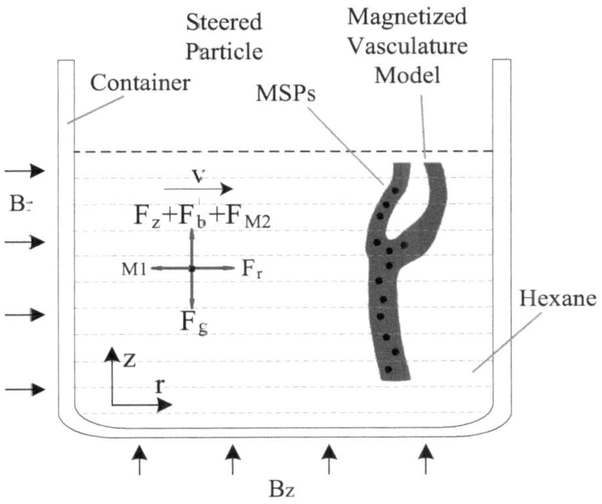

Figure 8.29 Dynamic mechanic model for MSP moving in hexane.

8.7.7.2 Fluid dynamics of MSP in hexane

For better control the movement of MSP in hexane, the dynamics model is established first.

The movement of MSP in hexane is governed by the Navier–Stokes equations:

$$\rho\frac{\partial u}{\partial t}-\Delta\eta(\Delta u+(\Delta u)^{\mathrm{T}})+\rho u\cdot\Delta u+\Delta p=F \tag{8.5}$$

where ρ denotes the fluid density (kg/m^3), u the velocity vector (m/s), η the dynamic viscosity (Ns/m^2), and $\boldsymbol{p}$ the pressure (Pa). The fluid is hexane with a dynamic viscosity of 0.294×10^{-3} Ns/m^2 and a density of 654.8 kg/m^3. F is the volume force of MSP.

At the direction of gravity force, when no external force is applied, the ordinary differential equation that describes the force balance is

$$m\ddot{z} = F_g + F_z + F_b \tag{8.6}$$

where m denotes the mass of the particle, $z(m)$ the position of the particle, F_g the gravitational force, F_b the buoyancy force and F_z the z-component of the force that hexane exerts on the MSP. The difference of gravitational force and buoyancy force is given by

$$F_g - F_b = (\rho_M - \rho)V_m g \tag{8.7}$$

where V_M is the volume of the MSP and ρ_M is the density. is $g = 9.81$ m/s^2 the acceleration due to gravity. The force that hexane exerts on the MSP is calculated by integrating the normal component of the stress tensor over the surface of the particle. The force's z-component is given by

$$F_z = 2\pi \int_S rn(-pI + \eta(\Delta u + (\Delta u)^T))ds \tag{8.8}$$

where r is the radial coordinate and n is the normal vector on the surface of the MSP, and I is the identity 3×3 tensor (or matrix). The initial values for the position and velocities are zero. Similar result of r-component F_r also can be obtained by Eq. 8.8.

8.7.7.3 Electromagnetic analysis for magnetic steering

When placing a magnetized particle in magnetic field, surface magnetization current and volume magnetization current inside the ferromagnetic material will be generated after its magnetization in the magnetic field, the magnetic force is interpreted as the magnetic force on molecular current. For isotropic medium it can be calculated as:

$$F_M = \iiint_V (\nabla \times M) \times B dV \tag{8.9}$$

And for the transient analysis of gradient magnetic field and uniform magnetic field, electromagnetic conversion problem is governed by Biot–Savart law:

$$dB = \frac{\mu_0}{4\pi} \times \frac{Idl \times r}{|r|^3} \tag{8.10}$$

where μ_0 is the permeability of the vacuum, magnetic constant, r is the displacement unit vector in the direction pointing from the wire element towards the point at which the field is being computed, $d\vec{l}$ is a vector, whose magnitude is the length of the differential element of the wire, and whose direction is the direction of conventional current, and B is the magnetic flux density.

Finite-element methods are used to simulate the motion of MSP moving in hexane, and magnetic field distributions of the proposed magnetic steering system. Gauss meter (MG-3002, Sato Inc. Japan) is used to measure the actual values of magnetic field.

8.7.7.4 Motion simulation of MSP in hexane

Simulation results supported the feasibility of proposed method for propelling MSP/MSP cluster moving at desired trajectory to form specific shape of MSP model. And the model with MSPs attached on this surface will be used to fabricate vascular scaffold with porogen leaching method. Figure 8.30 is a visualization of the solved velocity distribution of MSP when passing through hexane. The arrows show the direction of velocity and the background color shows the distribution of the velocity field. When MSP moves in hexane, the viscous resistance force and dynamic pressure force between the front part and rear part of the MSP can be calculated by integrating stress tensor on MSP's boundary. Figure 8.31 shows the relation between the velocity of MSP and drag force from Hexane. In this figure, the solid line plots the numerical calculation result. We use MSP with the diameter of 200 μm in this calculation. The viscous force and pressure difference drag force at MSP's movement direction increase in proportion to the velocity of MSP and are almost with equal magnitude at same velocity. The total drag force is about 0.001 μN when MSP with a velocity of 2 mm/s.

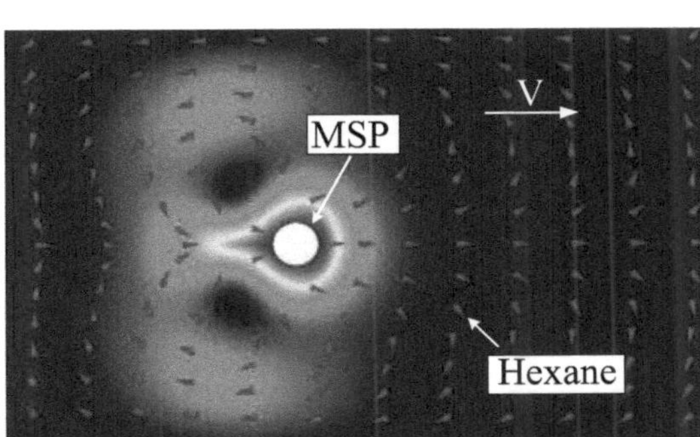

Figure 8.30 Solved velocity distribution of MSP passing through hexane.

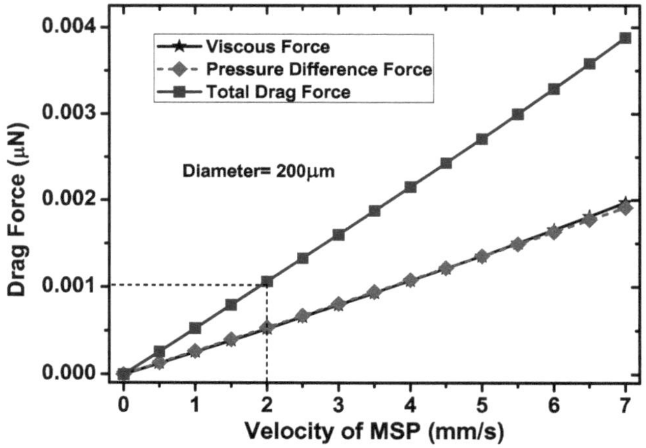

Figure 8.31 Relation between velocity of MSP and drag force.

8.7.7.5 Calculation of magnetic field

The distribution of magnetic flux density of Maxwell coils along the axial direction is illustrated in Fig. 8.32. The magnitude of the magnetic flux density varies linearly inside the two coils. A group of Maxwell coils is used to generate the required gradient magnetic field; the parameters are 55 mm in diameter and 80 mm^2 in the area of cross section. The calculation results shown in Fig. 8.32 show the magnetic flux density increase in proportion to the input current density. Different magnetic force can be obtained by adjusting the input current of Maxwell coils to generate different gradient magnetic field. The value of J_0 is 4.0×10^6 A/m^2, which is the maximum input current density for general copper wire. When the value is set to J_0, the gradient of magnetic flux intensity in working area is approximately equal to 330 mT/m, which is sufficient to drive the MSP moving in the hexane at a desired trajectory.

A group of Helmholtz coils is used to generate required uniform magnetic field. Figure 8.33 illustrates the magnetic flux density of Helmholtz coils in axial directions. The magnitude of magnetic flux density varies with the differences of coordinate in axial direction, and a high degree of uniformity is achieved in the central region of −0.02 to 0.02 m. The parameters are 95 mm in diameter and 80 mm^2 in the area of cross section. Figure 8.33 shows the magnetic flux density increase in proportion to the input current density. When the value is set to 5×10^5 A/m^2, the magnitude of magnetic intensity in uniform area is approximately equal to 0.8 Gs, which is sufficient to increase the magnetic field generated by Maxwell coils to be in one direction.

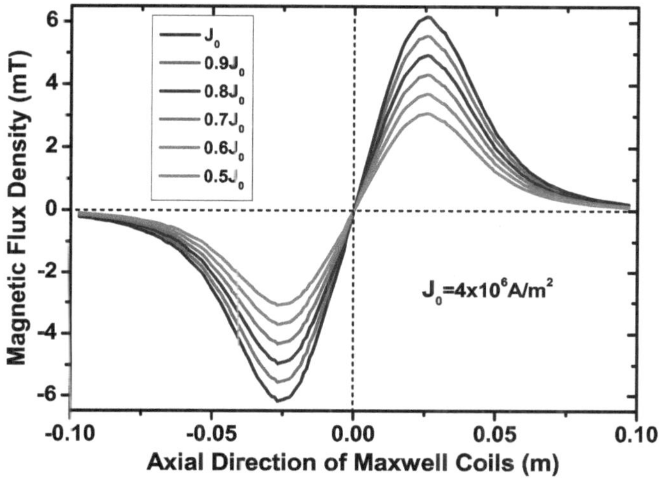

Figure 8.32 Magnetic field distribution inside Maxwell coils at six different input currents density of coils.

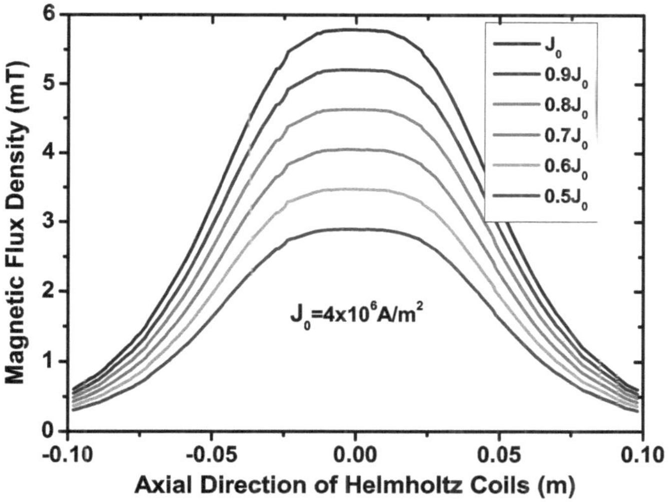

Figure 8.33 Magnetic field distribution inside Helmholtz coils at six different input currents density of coils.

Figure 8.34 shows the relation between input current density and magnetic field gradient of Maxwell coils and the magnetic force as a function of input current density. Different MSPs with magnetization of 2.5×10^4 A/m and 5.0×10^4 A/m, respectively, are deployed to calculate magnetic force at different input current density. The magnitude of magnetic flux density for Helmholtz coils must correspond well with the gradient magnetic field. As the figure above illustrates, the magnetic force is proportional to input current density. Compared with Fig. 8.31, it is possible to determine motion control parameters of MSP. Once the required velocity of MSP is determined according to the requirement of different motion occasions, the drag force from hexane can be obtained according to the blue square line in Fig. 8.31. In order to overcome the drag force, the magnitude of gradient magnetic field must be adjusted to an appropriate value for generating enough magnetic force F_M, which can be obtained in terms of triangle plot line.

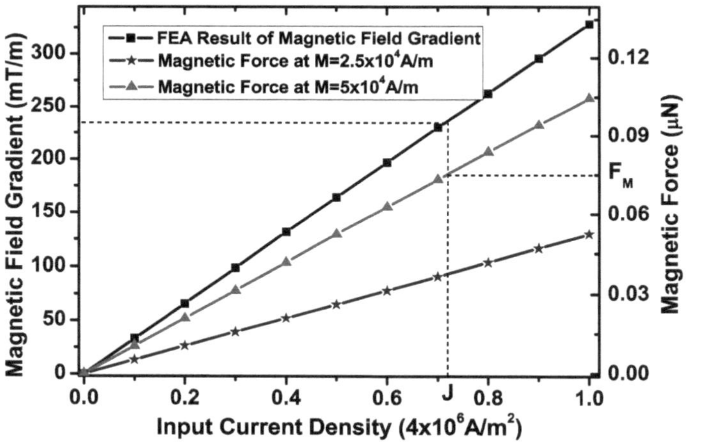

Figure 8.34 Relation between input current density and magnetic flux density of Maxwell coils.

8.7.7.6 MSP steering evaluation

Magnetic sugar particles as controllable units are prepared by an emulsion technique. First, 60 g of D-fructose are melted at 120°C until a clear yellow liquid is obtained. Then 6 g of magnetic powder (MQFP-B-20076, Magnequench Inc., Singapore) are added in the solution to append magnetism. The brown-colored liquefied fructose with magnetic powders is emulsified into 50 mL mineral oil at 120°C under stirring. The suspension is cooled down rapidly using an ice-bath with effective stirring to solidify MSPs separately. Solidified MSPs are washed in hexane to remove mineral oil and surfactant completely. The diameter of MSPs is controlled between the different mesh sizes of two sieves in hexane.

In order to prove the feasibility of proposed method, a magnetic motion control system for MSP is developed as shown in Fig. 8.35(a). Experiments of motion control are conducted in a well chamber (Figure 8.35(a)) CCD camera (PRO5, RF System Lab., Japan) and stereoscope (L-50, Hozan Co., Japan) collaborate to record the whole motion process, and specific software is developed for input current control and image analysis [50]. The system can generate controllable gradient magnetic field at the axial direction of well chamber with the work area of 40 mm × 35 mm, and realize a resolution of 1 μm in position capture, and with a vision scope of 20 mm in diameter.

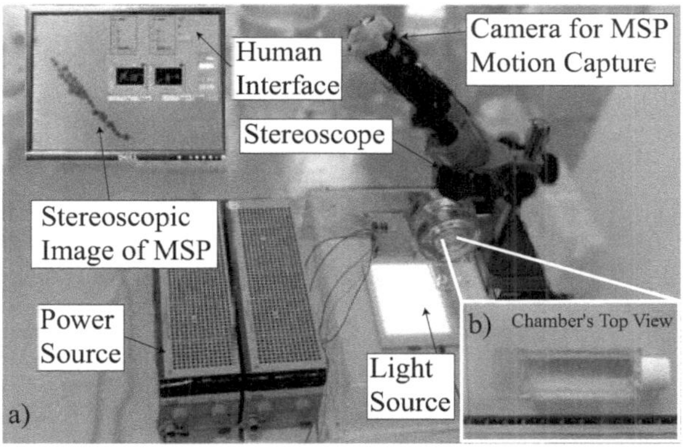

Figure 8.35 (a) Magnetic motion control system for MSP, working at room temperature; (b) Top view of well chamber with the length, width and height of 50 mm, 20 mm, and 20 mm, respectively, the chamber is filled with hexane and MSPs are injected for motion evaluation experiment.

Patterned scaffold is important for achieving different morphology of tissues after cell cultivation. A conventional patterning device, comprising poly(methyl methacrylate) (PMMA) with a thickness of 5 mm and steel wires with a diameter of 0.7 mm, and a interval of 3 mm, is utilized to generate dot pattern of magnetic field on the surface.

A neodymium magnet (surface magnetic flux density is 0.035 T) is placed under the patterning device as shown in Fig. 8.36(a). The magnetic flux density is larger near the cylindrical steel area with maximum value of magnetic flux density at their axial directions is 48.6 mT when a neodymium magnet with the surface magnetic flux density of 0.4 T is placed under the patterning device. Magnetized MSPs aggregated in these areas and form assembled dot pattern. Contrast experiment was conducted between before and after exposing the pattern in the gradient magnetic field to evaluate the effectiveness of our proposed method in improving the pore interconnection among different pores. This pattern is chosen under the consideration that about 20 MSPs can aggregate on the steel area for MSPs motion control experiments. Figure 8.36(b) shows the optical microscope image of MSPs with controlled diameter. Commercially available magnetic powder is used in this fabrication as encapsulated particles. Compared with traditional magnetic particles of ribbon shape, the magnetic powder used is isotropic spherical structure, has a mean diameter of 1–10 μm and a maximum residual magnetic flux density of 835–865 mT after magnetization. These MSPs are with the size range of 160–200 μm and the magnetic powders inside possess a residual magnetic flux density of 36.6 mT after magnetized by a permanent magnet whose surface magnetic flux density is 0.4 T. MSPs are divided into two groups with different quantity before magnetization for the motion evaluation experiment.

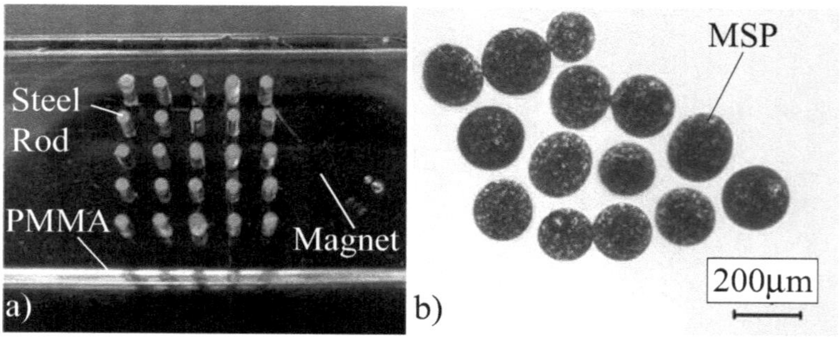

Figure 8.36 (a) Patterning device over a magnet. with a (PMMA thickness 5 mm, steel rods diameter 0.7 mm, distance between rods 3 mm). (b) Optical microscope image of MSPs.

Two groups of MSP clusters are used in the motion evaluation experiment. Group 1 is with the total MSP number of 12 and Group 2 is with the total MSP number of 37. The clusters are steered in the proposed system as shown in Fig. 8.35, and with impulse current whose peak value is 0.21 A. Figure 8.37 shows the distance of two group of MSP as a function of time, the frequency of input impulse current is 0.8 Hz with a wave duty circle of 30% to ensure enough delay time for coils. As the figure shows, linear displacement have achieved by impulse current input with average velocity of 0.25 mm/s and 0.116 mm/s, and linear regression coefficient of 0.997 and 0.998, respectively. This velocity difference between the two groups is caused by different friction force between MSP cluster and the wall of glass chamber and different magnetic force due to the unequal quantity of magnetic powders between MSPs. In a single impulse, the average moving speed of MSP cluster for group 1 and group 2 are 0.773 mm/s and 0.442 mm/s, respectively. Figure 8.38 shows the motion images of MSP cluster from Group 1 and Group 2 at different time, demonstrating good controllability to MSP/MSP cluster.

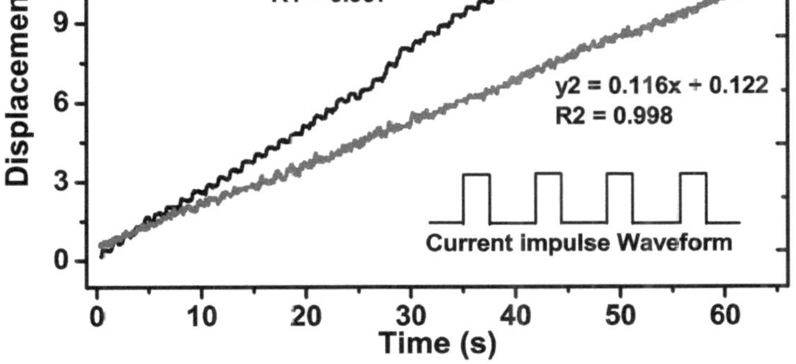

Figure 8.37 The displacement of an MSP cluster at different time. The peak value of input current impulse is 0.21 A, with a frequency of 0.8 Hz.

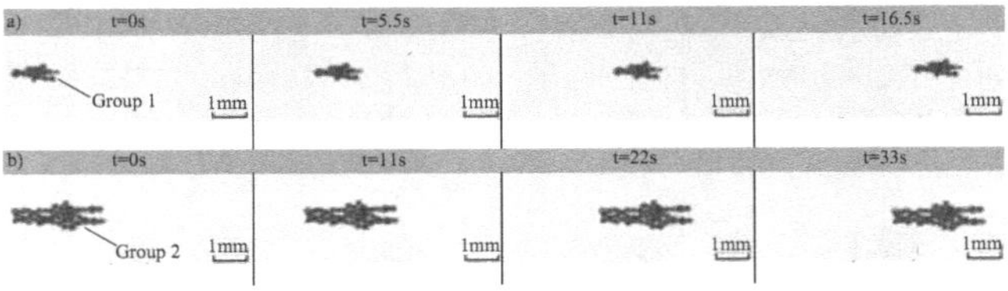

Figure 8.38 Motion images of MSP cluster at different time: (a) Displacement of Group 1, (b) Displacement of Group 2.

Magnetized MSPs aggregated in cylindrical steel areas of the patterning device to form self-assembled dot pattern due to the stronger distribution of magnetic flux density. We compare the MSPs pattern between before and after exposing the pattern in the gradient magnetic field as illustrated in Fig. 8.39, the interval between two dots is filled with MSPs at the magnetic field gradient direction, which improves the pores interconnection among different dots. After the removal of permanent magnet, MSP can still keep the pattern shape and be further used for fabrication of scaffold with the method mentioned in [51], and dot pattern on the scaffold is shown in Fig. 8.40. The scaffold shows controllable diameters and appropriate interconnection of pores, elegant pore wall morphology and high porosity in small-size scaffolds. This sheet-like scaffold can be further used for fabricating tubular scaffold by welding the boundary of inner and outer sheet of scaffold with original polymer solution (PLCL 10 w% in chloroform) as described in [51].

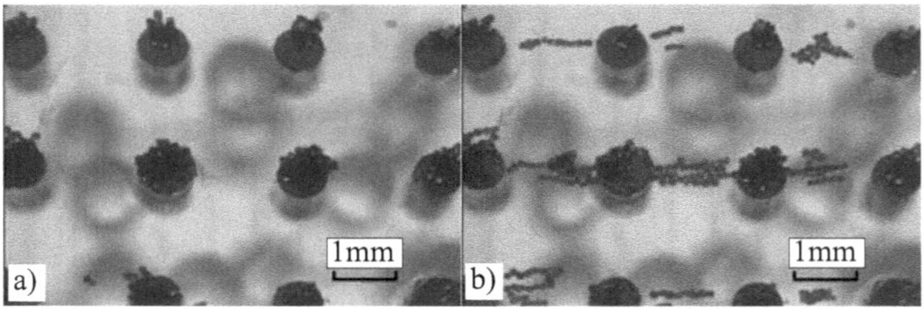

Figure 8.39 (a) Dot pattern of magnetically-guided self-assembled MSPs. (b) Self-assembled MSPs under motion evaluation system, the interval between two dots filled with MSPs at the gradient magnetic field direction.

Therefore, these results indicate that the proposed method suggests the feasibility of steering MSP/MSP cluster in hexane moving at a desired trajectory. One pairs of Helmholtz coils and Maxwell coils can generate propelling force in one dimension. By adding coils at different orientation, we are expecting to achieve 3D steady control of MSPs cluster. This method will also make sense by introducing magnetic pattern device to generate a preliminary pattern of MSPs first and then utilizing MSP motion control system for improving the MSPs distribution. Currently specific shaped dissolvable magnetic model has been fabricated by our group.

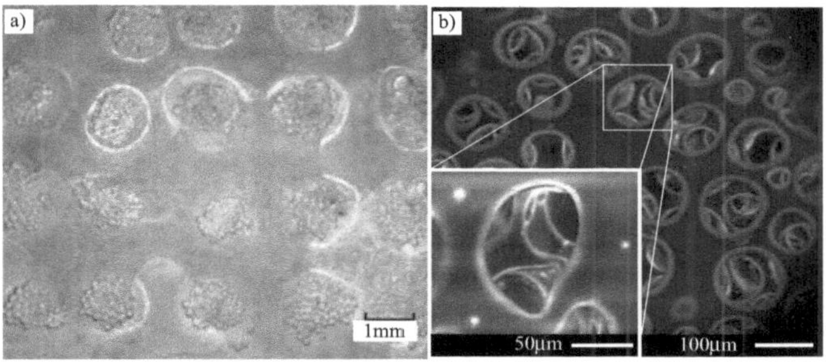

Figure 8.40 (a) Optical microscope image of PLCL scaffold with dot pattern; (b) ESEM images of the dot pattern on the PLCL scaffold, within the scaffold average pore size is 150 μm.

However, further investigations should be carried out in following:

- High gradient magnetic field is necessary to provide sufficient magnetic force for MSP's motion control, also high 3D gradient magnetic field is difficult to be generated only by combination of Maxwell coils and Helmholtz coils due to the space interference at different coil orientations, specific mechanism should be developed to cooperate with combined coils.
- External image capture system should be further developed for providing on-line position information of MSPs as feedback control.

References

1. Matsuda T. *Artif Organs,* 2004, 28: 64–71.
2. Matsuda T. *Kekkan Igaku* (Medical Review), 2004, 5: 577–586.
3. Matsuda T. *Tissue Engineering* (Nihon Igakukan), 2005, 67–75.
4. Matsuda T. *Saisei Iryou* (Medical Review), 2005, 4: 15–25.

5. Matsuda T, Ihara M, Inoguchi H, Kwon IK, Takamizawa K, Kidoaki S. *J Biomed Mater Res A*, 2005, 73A: 125–131.
6. Kidoaki S, Kwon IK, Matsuda T. *Biomaterials,* 2005, 26: 37–46.
7. Kwon IK, Kidoaki S, Matsuda T. *Biomaterials,* 2005, 26: 3929–3939.
8. Kidoaki S, Kwon K, Matsuda T. *J Biomed Mater Res B,* 2006, 76B: 219–229.
9. Inoguchi H, Kwon IK, Inoue E, Takamizawa K, Maehara Y, Matsuda T. *Biomaterials,* 2006, 27: 1470–1478.
10. Inoguchi H, Tanaka T, Maehara Y, Matsuda T. *Biomaterials,* 2007, 28: 486–495.
11. Nakayama Y, Matsuda T. *J Biomed Mater Res,* 1995, 29: 1295–1301.
12. Matsuda T, Nakayama Y. *J Biomed Mater Res,* 1996, 31: 235–242.
13. Doi K, Nakayama Y, Matsuda T. *J Biomed Mater Res,* 1996, 31: 27–33.
14. Gao J, Crapo PM, Wang YD. *Tissue Eng,* 2006, 12: 917–925.
15. Ikeda S, Arai F, Fukuda T, Negoro M, Irie K. *J Robot Mechatron,* 2005, 17; 327–334.
16. Ikeda S, Arai F, Fukuda T, Negoro M, Irie K, Takahashi I. *Lect Notes Comput Sci,* 2005, 3749: 925–932.
17. Uchida T, Ikeda S, Oura H, *et al. J Biotechnol,* 2008, 133: 213–218.
18. McDonald JC, Chabinyc ML, Metallo SJ, Anderson JR, Stroock AD, Whitesides GM. *Anal Chem,* 2002, 74: 1537–1545.
19. Whitesides GM, Ostuni E, Takayama S, Jiang XY, Ingber DE. *Annu Rev Biomed Eng,* 2001, 3: 335–373.
20. Langer R, Vacanti JP. *Science,* 1993, 260: 920–926.
21. Griffith LG, Naughton G. *Science,* 2002, 295: 1009–1014.
22. Khademhosseini A, Langer R, Borenstein J, Vacanti JP. *Proc Natl Acad Sci U S A,* 2006, 103: 2480–2487.
23. Gao J, Crapo PM, Wang YD. *Tissue Eng,* 2006, 12: 917–925.
24. Heijkants RGJC, Van Tienen VG, De Groot JH, *et al. J Mater Sci,* 2006, 41: 2423–2428.
25. Kidoaki S, Kwon IK, Matsuda T. *Biomaterials,* 2005, 26: 37–46.
26. Kwon IK, Kidoaki S, Matsuda T. *Biomaterials,* 2005, 26: 3929–3939.
27. Vozzi G, Flaim C, Ahluwalia A, Bhatia S. *Biomaterials,* 2003, 24: 2533–2540.
28. Mikos AG, Sarakinos G, Leite SM, Vacanti JP, Langer R. *Biomaterials,* 1993, 14: 323–330.

29. Bettinger CJ, Weinberg EJ, Kulig KM, *et al. Adv Mater,* 2006, 18: 165–169.

30. Fidkowski C, Kaazempur-Mofrad MR, Borenstein J, Vacanti JP, Langer R, Wang YD. *Tissue Eng,* 2006, 11: 302–309.

31. Mironov V, Boland T, Trusk T, Forgacs G, Markwald RR. *Trends Biotechnol,* 2003, 21: 157–161.

32. Xu T, Jin J, Gregory C, Hickman JJ, Boland T. *Biomaterials,* 2005, 26: 93–99.

33. Nam YS, Park TG. *Biomaterials,* 1999, 20: 1783–1790.

34. Lee M, Dunn JCY, Wu BM. *Biomaterials,* 2005, 26: 4281–4289.

35. Wu LB, Jing DY, Ding JD. *Biomaterials,* 2006, 27: 185–191.

36. Murphy WL, Dennis RG, Kileny JL, Mooney DJ. *Tissue Eng,* 2002, 8: 43–52.

37. Wei GB, Ma PX. *J Biomed Mater Res A,* 2006, 78A: 306—315.

38. Hashimoto M, Mayers B, Garstecki P, Whitesides GM. *Small,* 2006, 2: 1292–1298.

39. Xia Y, Yin Y, Lu Y, McLellan J. *Adv Funct Mater,* 2003, 13: 907–918.

40. Ito A, Ino K, Hayashida M. *Tissue Eng,* 2005, 11: 1553—1561.

41. Deng T, Whitesides GM, Radhakrishnan M, Zabow G, Prentiss M. *Appl Phys Lett,* 2001, 78: 1775–1777.

42. Uchida T, Ikeda S, Oura H, *et al. J Biotechnol,* 2008, 133: 213–218.

43. Inoguchi H, Tanaka T, Maehara Y, Matsuda T. *Biomaterials,* 2007, 28: 486–495.

44. Inoguchi H, Kwon IK, Inoue E, Takamizawa K, Maehara Y, Matsuda T *Biomaterials,* 2006, 27: 1470–1478.

45. Ino K, Ito A, Honda H. *Biotechnol Bioeng,* 2007, 97: 1309–1317.

46. Yesin KB, Vollmers K, Nelson BJ. Modeling and control of untethered biomicrorobots in a fluidic environment using electromagnetic fields. *Int J Robot Res,* 2006 (May), 25: 527–536.

47. Frutiger DR, Vollmers K, Kratochvil BE, Nelson BJ. Small, fast, and under control: wireless resonant magnetic micro-agents. *Int J Robot Res,* 2010 (April), 29(5): 613–636.

48. Pouponneau P, Leroux J-C, Martel S. Magnetic nanoparticles encapsulated into biodegradable microparticles steered with an upgraded magnetic resonance imaging system for tumor chemoembolization. *Biomaterials,* 2009, 30: 6327–6332.

49. Gao M, Hu C, Chen Z, Zhang H, Liu S. Design and fabrication of a magnetic propulsion system for self-propelled capsule endoscope. *IEEE Tran Biomed Eng*, 2010 (December), 57(12): 2891–2902.

50. Tercero CR, Ikeda S, Fukuda T, Negoro M. Image processing for sensing guide wire behaviour during endovascular surgery simulation. *International Symposium on Robotics*, 2010, 716–721.

51. Uchida T, Oura H, Ikeda S, Fukuda T, *et al.* Fabrication of biodegradable scaffolds by use of self-assembled magnetic sugar particles as a casting template in *IEEE International Conference on Robotics and Automation*, 2008 (May), 413–418.

Index

Color Insert

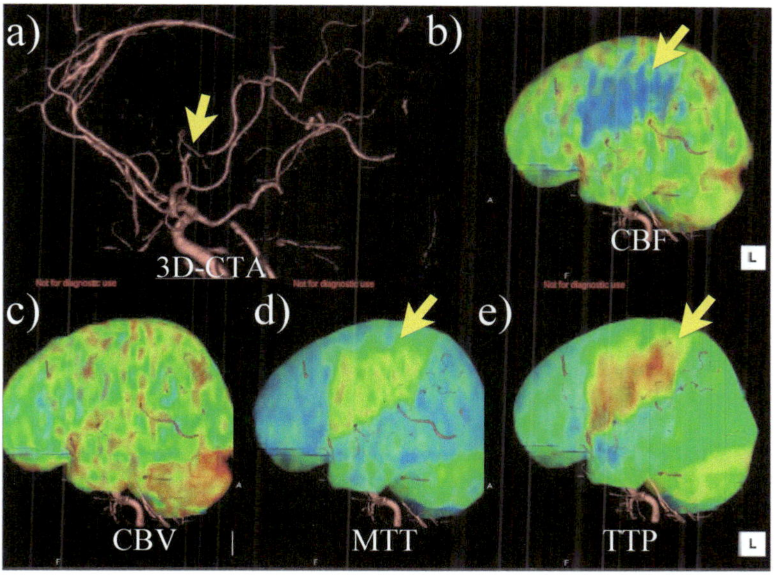

Figure 2.16 Lt.MCA superior trunk occlusion 3D-CT angiography (3D-CTA), CT perfusion (CTP). (a) 3D-CTA Lt.MCA superior trunk occlusion is shown with an arrow. (b) Cerebral blood flow (CBF) at the Lt.MCA superior trunk area (arrow) is reduced. (c) No visible change is seen for cerebral blood volume (CBV) in the region where the blood flow is obstructed. (d) Mean transit time (MTT) duration is confirmed at the Lt.MCA superior trunk area (arrow). (e) Time to peak (TTP) duration is confirmed at the Lt.MCA superior trunk area (arrow).

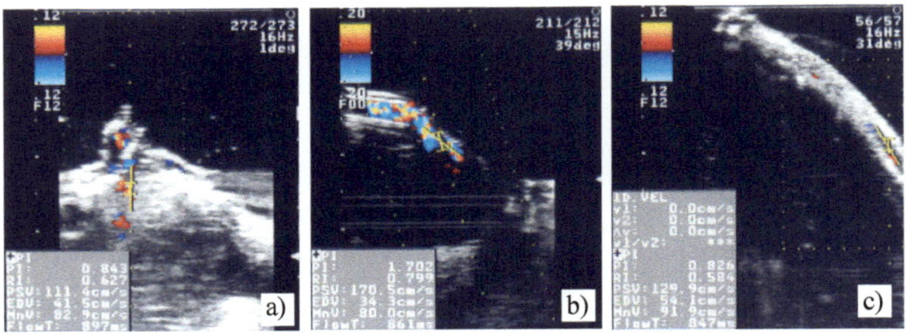

Figure 3.40 Ultrasound image obtained for the membranous model of representative case No. 2. Colored area indicates the direction of the blood streaming obtained by ultrasound Doppler method (red area indicates forward flow and blue area indicates backward flow). Ultrasound images: (a) Inside the aneurysm. (b) At the inlet of aneurysm. (c) At the outlet of aneurysm.

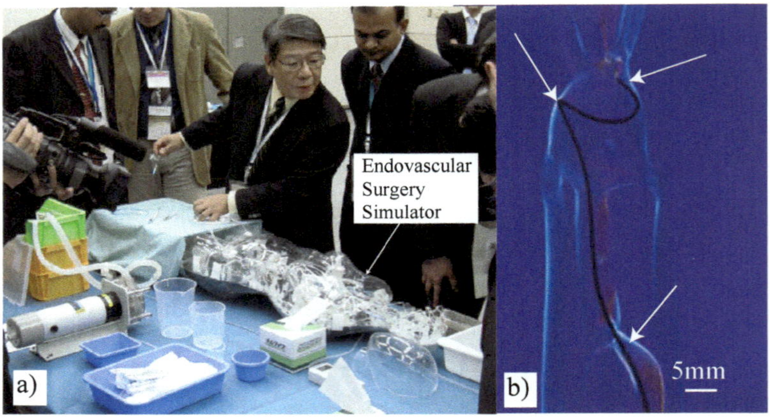

Figure 4.1 (a) Endovascular surgery simulator. (b) Birefringence consequence of stress applied to the vasculature model by a guide wire.

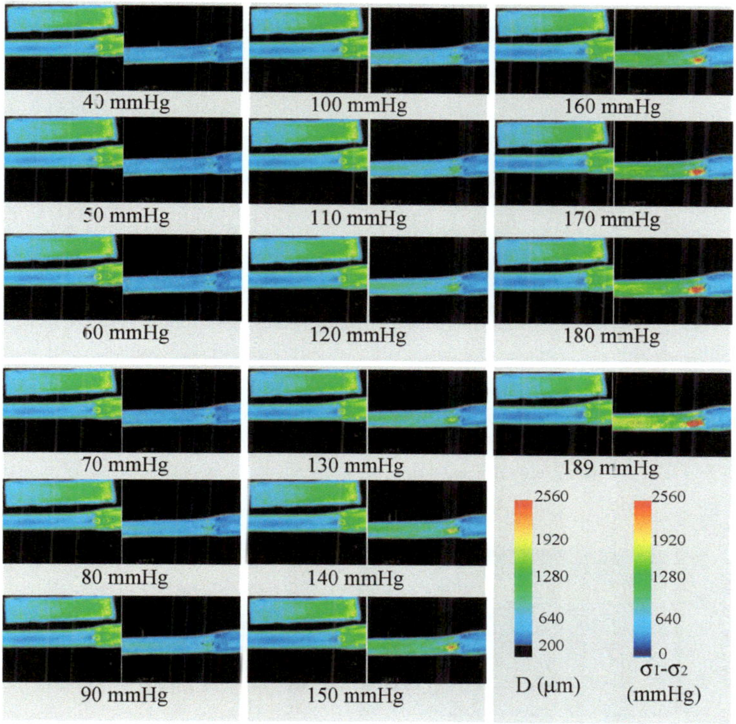

Figure 4.14 Result image of the 16 samples. Below each set of two images, the corresponding pressure is written. For each set, optical path length *D* (left) and the corresponding photoelastic stress analysis (right) are shown. A local maximum of stress appears on the right side of the phantom as pressure increases.

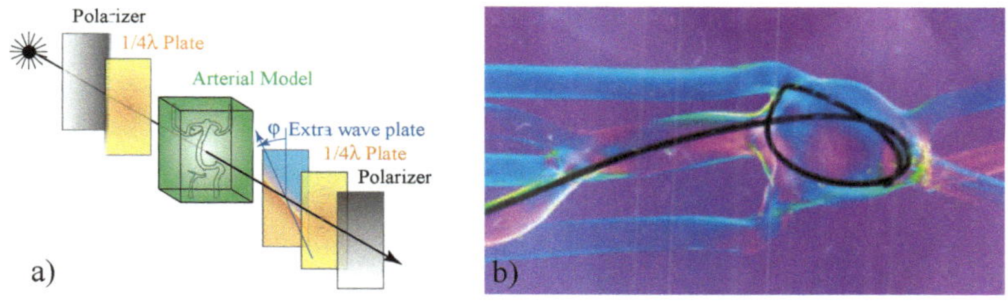

Figure 4.17 Circular polariscope for simultaneous measurement of stress magnitude and direction (left). High contrast between background and surgical instruments (right).

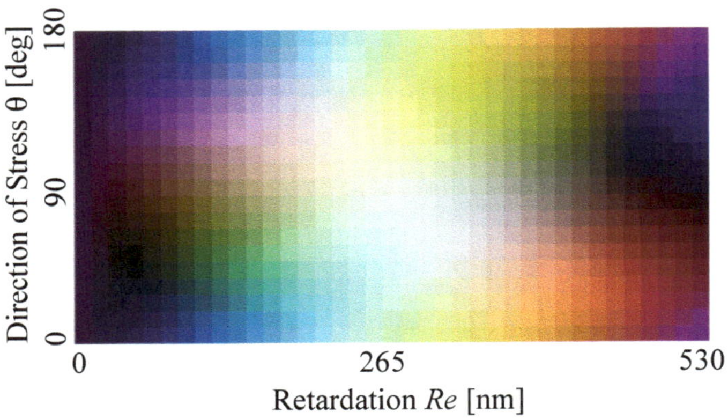

Figure 4.18 Color map calculated from equation 4.36 for RGB for retardations within the model below 530 nm and stress direction below 180°.

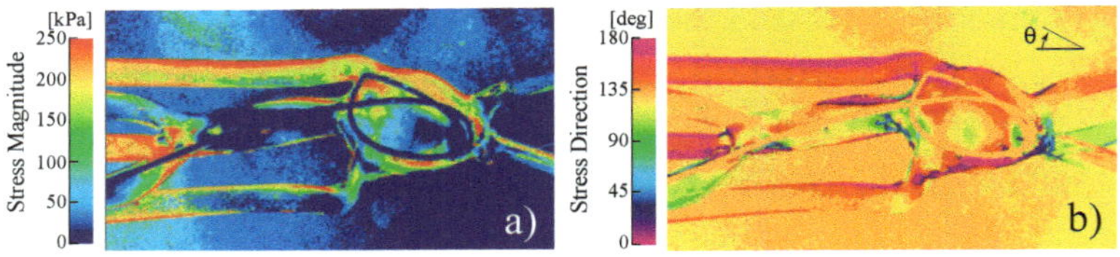

Figure 4.19 Simultaneous measurements of stress magnitude (a) and direction (b) for the single-layer model shown in Fig. 4.17.

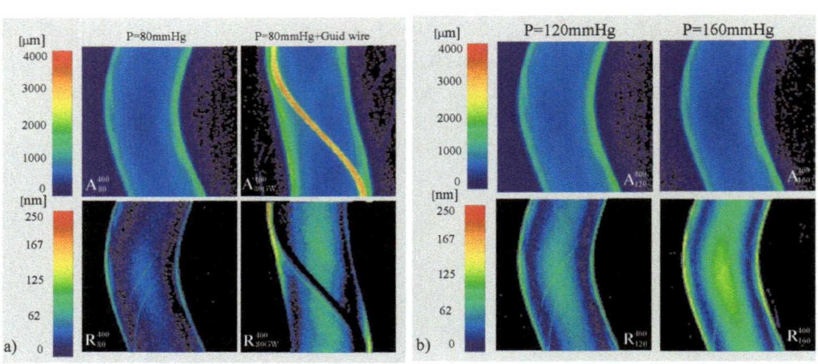

Figure 4.21 (a) Sinograms of optical path length and retardation with 80 mmHg of pressure inside the model and with guide wire inserted. (b) Sinograms of optical path length and retardation with pressures of 120 mmHg and 160 mmHg.

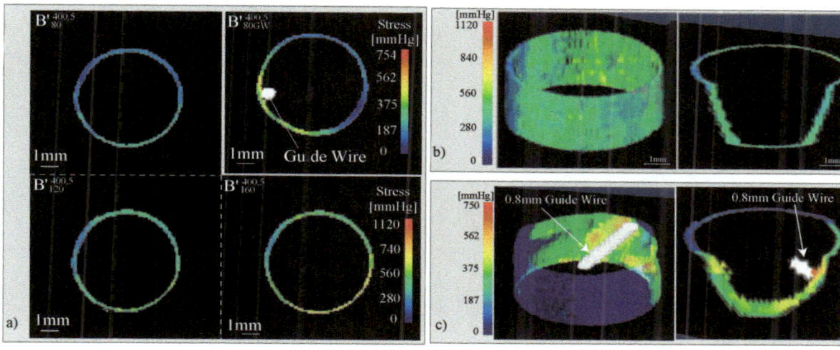

Figure 4.23 (a) Reconstruction stress distribution slices using ML-EM method in a model of vasculature slice at pressures of 80, 120 and 160 mmHg, and with guide wire inserted at 80 mmHg. (b) Three-dimensional visualization of stress and longitudinal stress variation within the model wall for a phantom with inner pressure of 120 mmHg. (c) Three-dimensional visualization of stress and longitudinal stress variation within the model wall for a phantom with a guide wire inserted and with inner pressure of 80 mmHg.

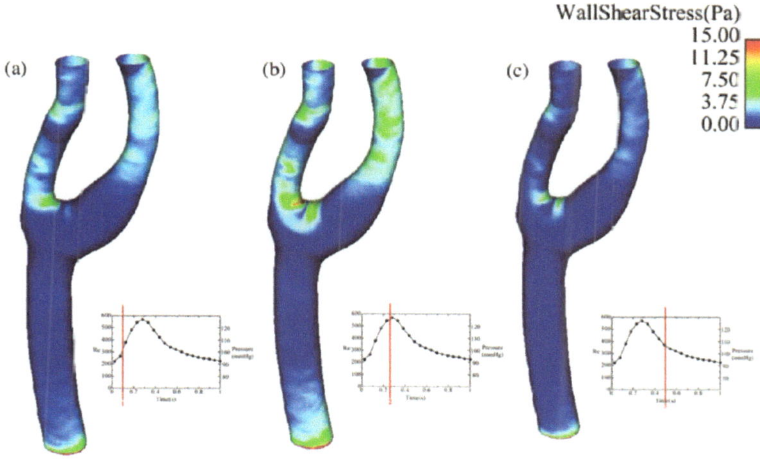

Figure 5.22 Wall shear stress.

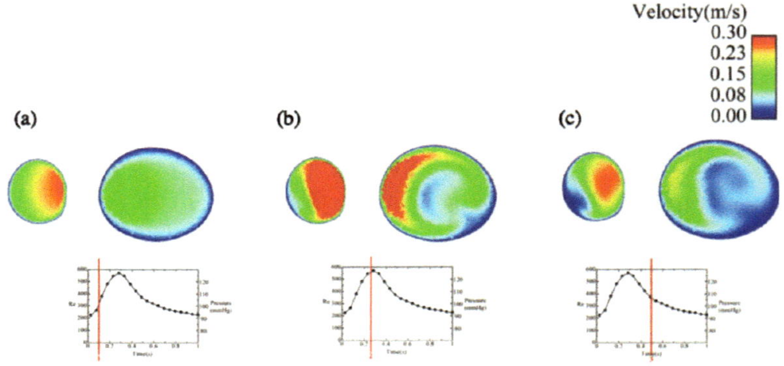

Figure 5.23 Velocity.

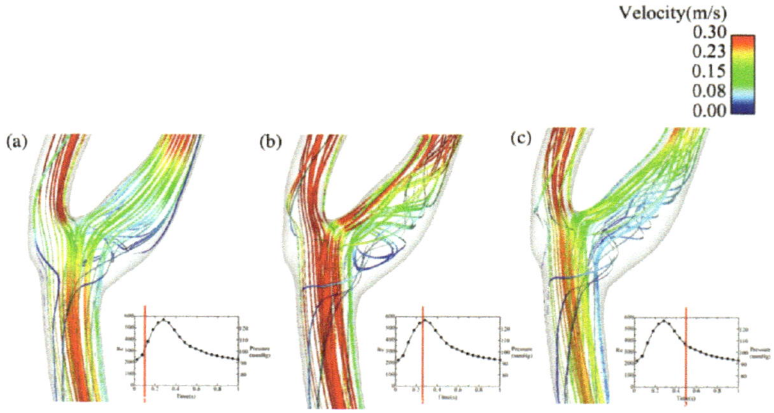

Figure 5.24 Streamline.

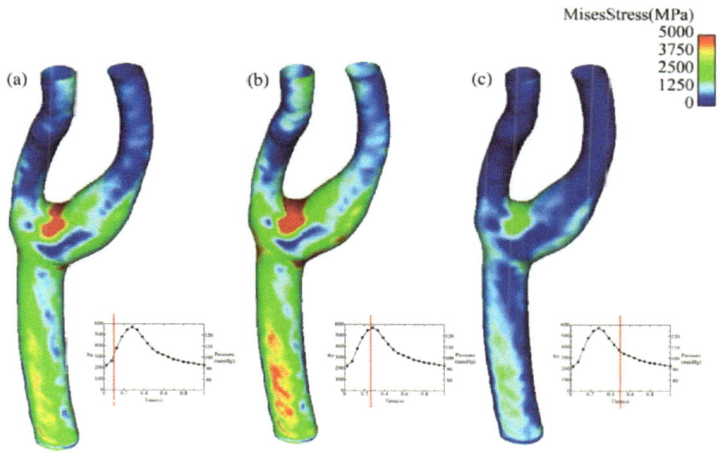

Figure 5.25 Mises stress.

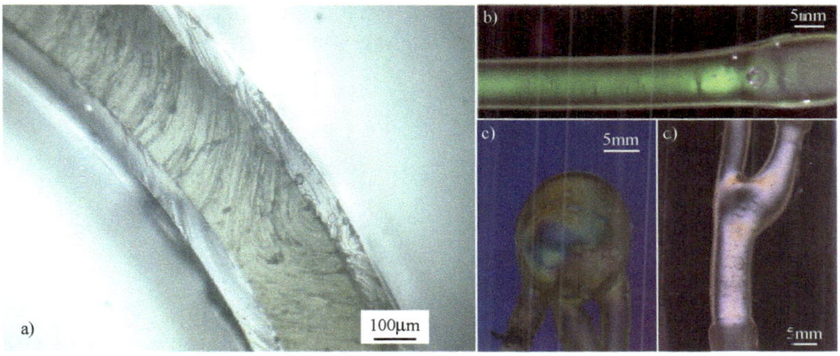

Figure 6.3 (a) Cross section of the multilayer membrane. Urethane elastomer (yellow) is shielded from flow and environment with silicone. Visualization of stress produced by inner pressure above 180 mmHg in multilayer models of vasculature, (b) straight vasculature segment within a circular polariscope and green filter, (c) giant aneurysm within elliptical polariscope, and (d) carotid artery within a circular polariscope.

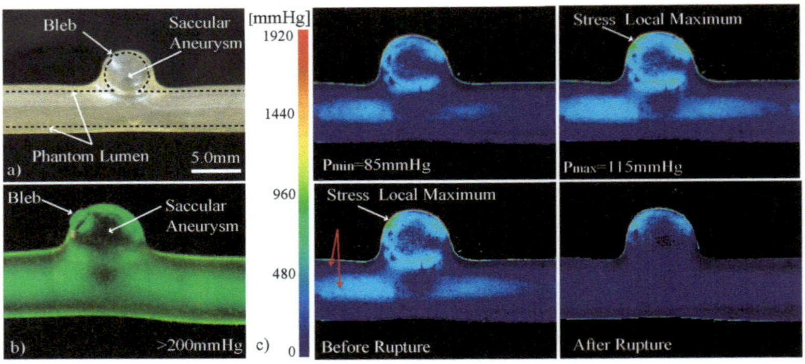

Figure 6.19 (a) Phantom of saccular aneurysm with bleb. (b) Source image example for stress measurement, green light intensity represents the stress in the phantom wall. Bleb shape is appreciated over 200 mmHg. (c) Stress analysis in Aneurysm Bleb phantom with pulsed flow circulation at: minimum pressure, maximum pressure, before rupture and after rupture. Red arrows point a measurement distortion appearing when the angle between the light beam and σ_3 increases.